Yellowstone's Wildlife in Transition

Yellowstone's Wildlife in Transition

EDITED BY

P. J. WHITE

ROBERT A. GARROTT

GLENN E. PLUMB

HARVARD UNIVERSITY PRESS   Cambridge, Massachusetts, and London, England

2013

Library of Congress Cataloging-in-Publication Data

Yellowstone's wildlife in transition / edited by P.J. White, Robert A. Garrott, Glenn E. Plumb.

pages cm

Includes bibliographical references and index.

ISBN 978-0-674-07318-0 (alk. paper)

1. Animal ecology—Yellowstone National Park. 2. Wildlife management—Yellowstone National Park. 3. Ecosystem management—Yellowstone National Park. I. White, P. J. (Patrick James) II. Garrott, Robert A. III. Plumb, Glenn E. (Glenn Edward)

QH104.5.Y44Y45 2013

591.709787'52—dc23   2012031817

Contents

Contents

INVASIVE, NON-NATIVE SPECIES

CONCLUSION

Foreword

EDWARD O. WILSON

Yellowstone! First national park in world history, birthplace of the National Park Service, and the federal institution most admired by the American people. Its very name is an iconic symbol that summons the best of the American spirit. The concept by which the great park was conceived is quintessentially American. Called civic egalitarianism, its guiding principle says that places and artifacts should not be privately owned but instead belong to the nation as a whole, to be kept open to everyone equally—street cleaner to billionaire.

Class and wealth disappear at the gates of Yellowstone. When you arrive you will be admitted, at least within reasonable hours and if the roads are not closed by winter snow. And there to welcome you will be rangers, a blessed breed of professionals whose purpose is to protect the park and offer help and information to all. At Yellowstone, they will explain that a large portion of the park is a volcanically active caldera, which if it blew in full force (not any time in the near millennia) would decimate North America. The rangers and other staff will tell you about the plants and animals, particularly the large free-ranging animals that abound, with one overriding message: this is the real thing.

There is wilderness all around the sparse trails and beyond. Yellowstone occupies a large piece of Wyoming, with extensions into Montana and Idaho. Walk a little way from the main roads, and you will experience the sight and feel of a part of America as it was during the nineteenth century and remains our priceless heritage in the twenty-first century. With farms, cultivated woodlots, and urban centers replacing the remainder of the American landscape, Yellowstone and other protected areas are now viewed in a wider perspective than originally held. They are now the logical best sites for fundamental studies in scientific natural history. They allow scientists a baseline to analyze the biology of plants and animals in as undisturbed condition as possible and to assess by comparison the impact of human activity on the remainder of the planet.

Most notable in present-day research at Yellowstone is the ecology of the megafauna, comprising grizzly and black bears, bison, moose, elk, mule and white-tail deer, mountain goats and bighorn sheep, cougars, wolves, lynx, and the large predatory and scavenging birds. As a people and nation, we consistently invest in Yellowstone science, as the wildlife populations wax and wane and as the climate changes, migrations are allowed or prevented, wolves are eliminated and restored, and invasive species of animals and plants press in. Translation of this science into twenty-first century stewardship is the intent of the present volume. The information is timeless and valuable beyond measure.

There is yet another meaning to Yellowstone and the many reserves modeled after it. Human culture, having now plunged itself into the digital Internet age, is growing superexponentially. It is, in other words, growing faster even than an exponential or "compound-interest" pace. Human populations continue to grow, and of far greater consequence, so is per capita consumption of land and resources. The space occupied by humanity on both the land and sea is undergoing radical transformation, to an end that cannot be foreseen. It is of vital importance that nature be maintained as a parallel domain, comprising the ecosystems produced by 3.5 billion years of prehuman evolution. We need to conserve this part of the planet, which created our own species, and allow it to continue evolving. It is the theater of sustainability of the rest of life on Earth and quite likely our own survival.

Yellowstone National Park is famous for its concentration of geothermal features, including geysers, fumaroles, mud pots, and hot springs. Canary Spring on the main terrace at Mammoth Hot Springs is a wonderful example of travertine-depositing hot springs. *Photo by Cindy Goeddel.*

Yellowstone's Wildlife in Transition

background and concepts

CHAPTER ONE

Ecological Process Management

P. J. WHITE
ROBERT A. GARROTT
GLENN E. PLUMB

THE NATIONAL PARK system has been promoted as one of America's best ideas. Wild places throughout the nation were set aside and protected to ensure that all citizens could enjoy the scenery, natural and historic objects, and wildlife through time. This protection was effective at providing safe havens for native species from such threats as timber logging, minerals extraction, livestock grazing, and wildlife overharvesting. Equally important, this ideal increased recognition that healthy park units enhanced the lives of people living in nearby areas by providing sustainable resources such as clean water, wildlife populations, and vegetation communities. As this understanding of the benefits of natural systems grew, however, there was also recognition that protection alone was insufficient to ensure the preservation of wildlife species into the future because parks are generally part of larger ecosystems and often insufficient in size and scope to preserve the processes necessary to sustain wildlife species over the long term. Thus, Leopold et al. (1963) suggested a new conservation ethic for national parks that involved the protection of representative samples of ecological systems, with recognition of the role of fire, other disturbances, and active management in maintaining these systems (Keiter and Boyce 1991; Boyce 1998).

The National Park Service responded to this recommendation by redesigning, or transitioning, management policies to "permit ecological processes such as fire and predators to prevail in national parks" while minimizing aggressive manipulations of wildlife populations and ecosystems by humans—which was the prevalent management approach in Yellowstone and other parks until 1960 (Keiter and Boyce 1991, 380; Sellars 1997). However, managers soon realized the futility of trying to manage ecosystems to maintain some agreed-upon stable condition given the uncertainty of future conditions and our inability to sustain any environmental state (National Research Council 2002). Also, there were problems with defining the benchmark or natural state of ecological systems because, over long periods, these complex, ever-changing systems have experienced "changing climate, major geological processes, and large-scale disturbances such as fire, floods, blow downs, fluctuations in abundances of herbivores (plant eaters) and predators, and outbreaks of diseases and insects" (Sinclair 1998; National Research Council 2002, 120).

These realizations led to another transition in park management whereby managers attempt to (1) preserve and sustain ecosystem processes (that is, rates, relationships, and variation) rather than specific species or states, while (2) monitoring system attributes that indicate potentially serious and unacceptable changes before they become severe (National Park Service 2006). There is enhanced emphasis on scientific research to improve management decision making and provide an objective basis for recommending ecosystem-based resource management policies in parks and on adjacent private and public lands (Keiter and Boyce 1991). Furthermore, there is a realization that dynamic processes such as climate and fire cannot be constrained by artificial management boundaries and that species such as grizzly bears *(Ursus arctos),* bison *(Bison bison),* bald eagles *(Haliaeetus leucocephalus),* elk *(Cervus elaphus),* trumpeter swans *(Cygnus buccinator),* and wolves *(Canis lupus)* can be managed effectively only on an ecosystem scale, which generally encompasses public and private lands outside parks (Keiter and Boyce 1991). Thus, effective management must link understanding of ecological processes, wildlife "population dynamics, and habitat relationships with social and economic concerns related to human-wildlife interactions" (Gordon et al. 2004, 1027; du Toit et al. 2004).

Yellowstone (Figure 1) is a "natural treasure of international significance," (Despain et al. 1986, 1) and what happens in the world's first na-

Greater Yellowstone Area

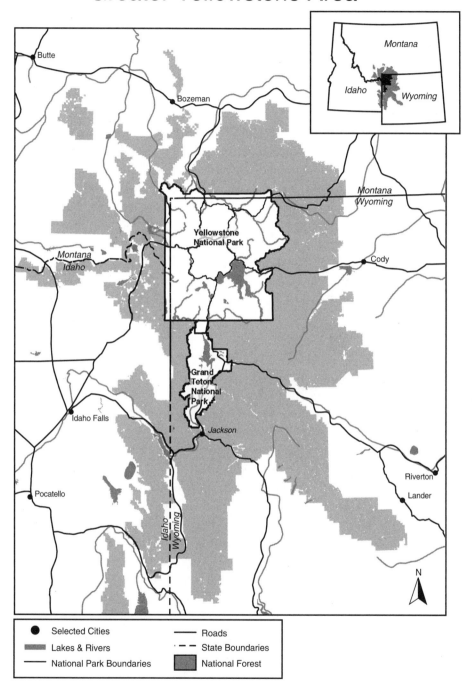

Figure 1. Map depicting Yellowstone National Park and the greater Yellowstone area.

tional park has influenced the debate on conservation approaches for protected areas throughout the world for over a century and will continue to
do so in the future. When ecological process management began being
implemented in 1969, park biologists disagreed whether an approach with
minimal human intervention would work. Thus, in 1986, biologists at
Yellowstone assessed the effectiveness of ecological process management
on wildlife and their habitats, some 20 years after this new paradigm began being implemented (Despain et al. 1986). These biologists concluded
that, in principle, the approach enjoyed great public support and was
proving constructive—conclusions that were echoed by subsequent independent assessments (Boyce 1998; Huff and Varley 1999; National
Research Council 2002). However, Despain et al. (1986) admitted that
the sustained success of this relatively hands-off approach was not clear
given the ecological uncertainties of proposed management actions, such
as wolf restoration, and the associated political and social dimensions
of such decisions. Thus, they recommended another assessment in 10 to
20 years.

Since Despain et al. (1986) published their assessment, the Yellowstone
ecosystem has been extensively modified by the fires of 1988, the recovery
of the grizzly bear and wolf populations, the expansion of bison and elk
wintering areas outside the park, the invasion of non-native diseases and
organisms, and the continued harvest of wildlife outside the park. These
and other changes, in the context of a warming climate and intensifying
development on unprotected lands, "created a degree of complexity that
makes projection of long-term conditions in and near the park difficult"
(National Research Council 2002, 128). In response, park managers initiated a comprehensive, integrated program of independent research and
monitoring that measures the consequences of changes in variables that
drive the system (Yellowstone National Park 2011). This program includes
"studies of animal and plant populations and their interactions, of predator-
prey relationships, and of changes in the behaviors of ungulates (hoofed
animals) and predators as the system adjusts to the recovery of large
predators, as well as concurrent studies of riparian and aspen *(Populus
tremuloides)* recruitment, sagebrush (*Artemisia* spp.) communities, stream
fluvial geomorphic processes in relation to riparian vegetation dynamics,
rain, snow, surface flows, and groundwater levels, and other ecosystem
components" (National Research Council 2002, 128).

This book evaluates the effectiveness of ecological process management at sustaining essential processes in Yellowstone National Park, some 25 years after the initial assessment by Despain et al. (1986). We did not attempt to present exhaustive summaries of what has been learned about ecological processes and their preservation. Rather, we provide select assessments that focus on key wildlife species and their habitats and illustrate essential processes and current management issues. We intentionally minimized tables, plots, and graphic displays to make the text more readable. These data are provided in the cited references for interested readers. The first three chapters discuss ecological process management, the history of wildlife and resource management in the Yellowstone area, and the importance of scale and perception for integrating scientific knowledge into natural resource management. Chapters 4 through 7 explore population dynamics and interactions among species, including competition, omnivory, predation, and symbiosis. Chapters 8 through 12 focus on landscape-scale processes such as natural disturbance dynamics, vegetation phenology, climate change, and migration and dispersal, as well as factors influencing grassland and riparian communities and the processes that sustain them. Chapters 13 through 15 discuss the effects of exotic, invasive organisms on native species and their aquatic and terrestrial environments, as well as the management challenges they present. Then, in the final chapter, we discuss the future of ecological process management and whether further transitions in policy may be needed.

Most land managers and range scientists now recognize that Yellowstone National Park, with its complex interactions among wildlife, plant, and microbial (meaning microorganisms; minute life forms) communities and climate drivers, cannot be compared with range lands managed for domestic livestock production because fluctuations in wildlife populations and forage conditions should be expected (National Research Council 2002). However, the debate continues about how much and what kind of human intervention is necessary and appropriate (Wagner 2006; White, Garrott, and Olliff 2009). It is our sincere hope that the collective efforts of the scientists involved with this book will invigorate strong partnerships that enable the continued cooperative collection and sharing of information, discussion of alternate management approaches, and implementation of practical measures to ensure that potentially serious and unacceptable changes to ecosystem processes do not occur. Thus, we have written this book with three

A bison crosses the Yellowstone River in the southern portion of Hayden Valley. Restoration of bison is one of the many wildlife conservation success stories in Yellowstone National Park, where the population was less than 25 individuals at its low point in the early 1900s but reached approximately 5,000 animals in 2005. *Photo by Cindy Goeddel.*

broad groups of people in mind: (1) natural resource managers in the greater Yellowstone ecosystem who will use the information to make wise decisions, implement solutions, and communicate information to stakeholders; (2) wildlife ecologists who will benefit from the integration of up-to-date information and new ideas; and (3) the park's three million visitors and additional "virtual visitors" who monitor the condition and management of resources in their park through visits, the Internet, and other outreach avenues.

CHAPTER TWO

Understanding the Past

The History of Wildlife and Resource Management
in the Greater Yellowstone Area

S. THOMAS OLLIFF

PAUL SCHULLERY

GLENN E. PLUMB

LEE H. WHITTLESEY

THE ESTABLISHMENT OF Yellowstone National Park in 1872 began a great quest to make the most of this extraordinary place. Since then, prevailing views of how a national park should be managed have undergone many changes, reflecting both the ongoing public debate about the purpose of a national park and scientific advances in our understanding of how natural ecosystems function. Changes in how the park's resources have been and currently are managed can be understood through the lens of organizational evolution. Changes in the geophysical processes, political setting, public engagement, and/or scientific knowledge have caused sweeping perturbations in the management system. Those perturbations resulted in a succession of distinctly identifiable eras in the management of the park.

We have thus divided the historical context of the management of Yellowstone National Park into five eras: the Wide-Open Era, in which park resources were treated similarly to corresponding resources elsewhere in North America (1872–1883); the Game Preservation Era, in which certain wildlife species were favored at the expense of the rest of the ecological community (conservatively, in the years 1883–1918; liberally, in 1883–

1974); the Agricultural Management Era, in which park managers embraced commercial standards for measuring the success of resource management (1918–1968); the Ecological Management Era (popularly known as natural regulation, 1968–1983); and the Native Species Restoration Era (1983–present). We discuss these eras in greater detail below, and this discussion provides context for considering the changing perceptions and use of historical study as a management tool.

These eras are characterized by dramatic shifts in how resource management was perceived and carried out by those in charge of managing the park. For example, non-native fish were once widely stocked in the park, but some of those same fish were later eradicated during a different management era (Varley and Schullery 1998). Also, native predators were once controlled and even eradicated, but later those same predators were reintroduced or restored (Schullery and Whittlesey 1999b; D. W. Smith 2005). In addition, exotic plants were once cultivated as ornamentals or crops, but those same plants were later targets for large-scale removal programs (Olliff et al. 2001). Furthermore, native ungulates were once ranched and their numbers reduced, but later these same populations were allowed to fluctuate naturally, driven by ecological processes and different management regimes across jurisdictional boundaries (D. B. Houston 1982). More recently, bison numbers have been controlled as they leave the park's boundary each winter in search of snow-free areas on which to graze (Plumb et al. 2009).

Within each of these large perturbations, smaller, more subtle changes in management have occurred, much like the changes in a forest over time after a stand-replacing fire, due to what now might be called adaptive management. One example of these changes is the stringent catch-and-release and slot fishing regulations imposed on anglers as a result of the dramatic decrease in fish populations in the 1950s and 1960s. Another example is the control program that began almost immediately after lake trout *(Salvelinus namaycush)* were discovered in Yellowstone Lake in 1994 (Varley and Schullery 1995b). Advances in trapping technology and monitoring have almost quadrupled the number of fish removed in the 16 years of the program's history (Gresswell 2009). Yet another example is the updates and changes to the grizzly bear management program since these bears were listed as threatened under the Endangered Species Act in 1975 (Gunther 1994, 2008). Park managers have completely revamped garbage collection, food storage, and camping regulations and have instituted nightly campground patrols to enforce these regulations. More recently, managers have been wrestling with how to manage roadside bears, the presence of which provides an unparalleled opportunity for visitors to observe bears in a near natural habitat but also presents visitor and bear management challenges for managers (Gunther and Wyman 2008).

Since the park's establishment in 1872, management strategies were shaped, sometimes knowingly and sometimes not, in response to several dynamic forces: (1) geophysical processes such as volcanic doming (e.g., the tilting of Yellowstone Lake), short-term weather changes (e.g., the drought of the 1930s), and long-term climate changes (e.g., the end of the Little Ice Age in the mid-1800s); (2) the public's and managers' attitudes and tolerance for change on the landscape; (3) the public's relative valuation of, and enthusiasm for, various elements of the Yellowstone setting; and (4) scientific investigation and perennial reinterpretation of the Yellowstone ecosystem. It was predominantly geophysical forces, especially climate, that forced the management directions and policies we employ today to build ecosystem resilience, and these will likely continue to do so in the foreseeable future.

The historical narratives of Yellowstone National Park are many. They have had many purposes, ranging from the earliest exercises in self-celebration by park advocates wishing to firm up their place in the park's history through progressively more sophisticated attempts to provide needful historical context on the park's founding and subsequent institutional evolution. For most of the park's first century, historians pursued traditional questions of administrative and legislative development. However, a growing number of scientists began to apply historical evidence to questions of concern to them, the foremost of which was the condition and character of the park setting prior to the arrival of Euro-Americans and their influences. The most important modern development in the study of history in the Yellowstone area is certainly the remarkably interdisciplinary broadening of the historical enterprise in the past 30 years. History is no longer defined as the written record of what white people did. Thanks to the recent contributions of a host of researchers in partner disciplines of paleontology, archeology, anthropology, and other fields, our grasp of regional history is infinitely richer than ever before, and it promises to yield important lessons and insights unimaginable to investigators even 30 years ago.

All of Yellowstone's past and present management regimes and the resources they attempt to manage have inspired passion, debate, disagreement, and occasionally consensus. Elk management, for instance, has been one of the most controversial resource issues in Yellowstone for more than a century (D. B. Houston 1982; National Park Service 1997). Management of bears, bison, elk, fish, wolves, and natural fires—and how to allow visitors to interact with these resources and events—all have stoked national debates among scientists, citizens, managers, and politicians. These issues have been the subject of uncountable scientific studies, congressional inquiries, public debate, and dramatic demonstrations, and they have resulted in numerous books as well as magazine and newspaper articles beyond counting. Yellowstone matters, among many reasons, because it is America's park: it is the forum that gathers our citizens to debate how our country cares for our wild lands and the native species that live there.

This debate will soon be more important than ever as we face the ecological and societal effects of rapid, human-forced climate change. This chapter explores two interwoven topics that are central to modern dialogues about the management of Yellowstone National Park: the history of management and the study of the history of Yellowstone's past.

ERAS OF RESOURCES MANAGEMENT

Human management of the landscape of the greater Yellowstone ecosystem is an ancient endeavor. Various groups of native people who lived in or visited the ecosystem are known or assumed to have had a variety of effects on plant and animal communities. Archeological and historical evidence indicates that native people hunted a variety of native animals and gathered numerous plant species (Haines 1977; Nabokov and Loendorf 2004; Schullery 2004; Loendorf and Stone 2006). Perhaps the most persistently discussed effect of humans on the Yellowstone landscape prior to the arrival of Euro-Americans involves fire. Native people in the American West intentionally set fires for several purposes, but our knowledge of the specifics of these activities, and how these activities may have changed over time, is regrettably slight: "direct evidence still remains too thin to make a solid case about the degree to which Yellowstone National Park proper was subject to alteration by intentional Indian fires, though some scholars have tried"

(Nabokov and Loendorf 2004, 208). Thus, in any discussion of management history in or near Yellowstone, it is essential to recognize that the prehistoric condition of the ecological setting has been an important and controversial point of debate. The extent and meaning of native peoples' activities and effects on the greater Yellowstone ecosystem are still a matter of intense interest and disagreement among researchers, managers, and advocates attempting to select or influence future management directions.

Though we are in a much better position to analyze specific directions and changes in management policy and practice since the creation of Yellowstone National Park in 1872, it must be recognized that any attempt to break a historical continuum of events into distinct segments or eras must begin with an admission of the fundamentally artificial nature of the enterprise. Such chronological organizations are often very helpful, but they always involve a certain amount of arbitrary dating of ideas, processes, and movements that are not really that tidy. History does not periodically restart itself with a clean slate; the seeds of each successive era were sewn during previous eras. However, there is still great worth in identifying general trends in the thoughts and actions that have shaped natural resource management in the greater Yellowstone ecosystem. The identification of such eras is a valuable device for clarifying past directions and considering future ones.

WIDE-OPEN ERA (1872–1883)

The act creating Yellowstone National Park gave little direction to the secretary of the interior regarding the park's biological features, beyond requiring that they be protected in their natural condition, a mandate and a term that have been debated for the last 140 years (U.S. Statutes at Large, volume 17, chapter 24, 32–33; Haines 1974; Pritchard 1999; Wagner 2006; D. N. Cole et al. 2008). Because of the widespread slaughter of thousands of animals in the new park between 1872 and 1882, in 1883 the secretary of the interior established a policy forbidding hunting in the park—in one stroke creating the world's foremost public wildlife preserve (Hampton 1971; Haines 1977; Schullery 2004). The park was created to preserve geological wonders, but from 1883 on, biological values and issues dominated management attention.

A massive pile of bison skulls at an unidentified location in 1890. During the initial years after Yellowstone National Park was established, wildlife were not protected and exploitation continued, similar to what was being experienced throughout North America. However, a prohibition against hunting in the park was soon enforced and continues to this day. *National Park Service photo taken by Hugh Lumsden.*

In the park's first 11 years of existence, a number of forces combined to set up later managers for difficult quandaries about management of this pioneering experiment in natural resource conservation. Exclusion of native people and their activities became progressively more thorough, ensuring that whatever their influences on the landscape had been at various times prior to the creation of Yellowstone National Park, those influences had diminished or disappeared (Nabokov and Loendorf 2004). The Little Ice Age, a cooler period lasting about four centuries, ended in the mid-nineteenth century, thereby ensuring that the ecological setting of the greater Yellowstone ecosystem was in for a period of adjustment and change even if Euro-Americans and their influences had not arrived in the region. Also, Euro-Americans were appearing in increasing numbers and began a wholesale overhaul of large portions of the landscape, including the sustained destruction of large mammals. It is only in the past four decades that intensive attention has been paid to the possible long-term consequences of dramatic changes that occurred in the park's first years.

GAME PRESERVATION ERA (1883–1918)

This era was characterized by protection and promotion of favored animal species. Few predators were protected, and the destruction of others was better organized (Schullery and Whittlesey 1999a, 1999b). Popular non-native fish were widely introduced (Franke 1996, 1997; Varley and Schullery 1998). Non-native plant species were increasingly introduced, both accidentally and intentionally (Despain 1990; Olliff et al. 2001; Whipple 2001). Introductions of other European and North American game animals were planned or attempted (Schullery and Whittlesey 2001). Favored scenery was likewise nurtured, and the suppression of natural fires was initiated by the U.S. Army in 1886 (Despain 1990; Franke 2000; Barker 2005).

With hindsight, we may be inclined to see these early managers as short-sighted or misguided, but their grasp of their responsibilities was often quite nuanced. It is difficult for us to fully imagine the intellectual and political realms they inhabited. Except for the first three years under civilian administration, the Game Preservation Era as we define it coincided precisely with the 32-year (1886–1918) stay of the U.S. Cavalry in Yellowstone National Park (Haines 1977; Bartlett 1985; Broadbent 1997; Barker 2005). The army provided the necessary discipline and muscle to see the struggling young park through its early years, and army officers activated much of the resource management policy that was continued by the National Park Service when that agency assumed full control of the park in 1918. Through enforcement of park regulations and some high-profile cases involving poaching, the U.S. Cavalry created the slate on which to write modern wildlife management laws and fostered the substantial recovery of Yellowstone's large ungulates, especially bison and elk, following the extensive slaughters of the 1870s.

AGRICULTURAL MANAGEMENT ERA (1918–1968)

We have named this era agricultural because many management practices tended at first to reflect prevailing professional agricultural values, especially those related to evaluating range condition and establishing so-called appropriate herd sizes of grazing animals. However, this era also witnessed the slow, comprehensive, and often bitterly resisted departure of park policies from mainstream agricultural thinking (Pengelly 1963; Tyers 1981; D. B. Houston 1982; National Park Service 1997; National Research Council 2002). Established army programs such as predator killing, ungulate feeding and population control, fire suppression, bear feeding, and fish stocking were at first embraced but eventually met with increasing scrutiny and disapproval (Schullery 2004). By 1968, the end of this era, all such programs were either eliminated or circling the drain.

Ecological thinking eased in slowly, but by the 1930s, biologist George Melendez Wright and his colleagues had laid out a series of essentially modern ecological rationales for national park management (G. M. Wright et al. 1933). The influential Leopold and Robbins reports of the early 1960s

Bison at the Buffalo Ranch in the Lamar Valley of Yellowstone National Park during 1930. From the early part of the twentieth century until 1968, the wildlife management policy in Yellowstone reflected an agricultural philosophy where wildlife restoration included intensive animal husbandry practices for the large herbivores, such as bison. National Park Service photo taken by an unknown photographer.

reinforced the earlier work of Wright (Leopold et al. 1963; W. Robbins et al. 1963). The Leopold Report was especially important, as it suggested a guiding philosophy for managing park lands: a respect for the land's character as it was before the arrival of Euro-Americans. A return to primeval wilderness, or a reasonable facsimile, was to be the guiding objective for the National Park Service (Wondrak Biel 2006). By the late 1960s, political crises in the management of several charismatic wildlife species became irresistible forces for abrupt changes in policy and led to a comprehensive and controversial reimagining of the park's potential as an institution (Craighead et al. 1982, 1995; D. B. Houston 1982; Schullery 1992; National Park Service 1997; Barmore 2003).

ECOLOGICAL MANAGEMENT ERA (1968–1983)

Though from the very first decade of the park's existence a few commentators had recognized the park's potential as a site for the study of evolutionary processes, this is the first era in which the goal of managing for naturalness, a concept implicit in the National Park Service 1916 act, and reflective of the spirit of the 1964 Wilderness Act, came to be a dominating policy presence. The concept of naturalness, like the earlier term "natural condition," is constantly still discussed and debated (Rolston 1990; Boyce 1991; Wagner 2006; D. N. Cole et al. 2008). In this era, a goal coalesced around the principle of heightening wildness, that is, of allowing ecological processes as much independence and as little obstruction by humans as possible (Despain et al. 1986).

The flagship issue of this era, in fact the issue that more than any other launched it, was the management of winter range for elk and other ungulates in the northern portion of the national park. After several decades of intractable controversy, in the 1960s park rangers killed thousands of elk to satisfy prevailing but erroneous concepts of appropriate population size and range condition. This crossed public and political thresholds of tolerance, forcing management change just as new thinking in ecology arose (D. B. Houston 1982; Barmore 2003). In 1971, ecologist Douglas Houston laid the groundwork for a new and enormously productive scientific inquiry and debate, with a hypothesis for the ecological management of northern Yellowstone elk, predicting that the population would be limited by intraspecific competition for food and associated winter mortality with no negative effects on other ecosystem elements (D. B. Houston 1971). More than 100 scientific studies later, the analysis and controversy continue (Despain 1994; National Park Service 1997; Wagner 2006).

This flagship issue was accompanied by a hefty fleet of other equally vexing and stimulating issues. Restoration of the essential functions of natural fires began in the park's centennial year (1972) and seemed a model of policy success until the fires of 1988 revealed how socially, politically, and scientifically challenging authentic natural fire processes could be (Franke 2000; Wallace 2004; Barker 2005). Brucellosis in Yellowstone bison, almost as historically venerable an issue as elk management, likewise tested attempts to break down the famous boundary mentality that so often frustrates advances in ecosystem management (Meagher 1973, 1989a; Gates et al. 2005; U.S. Government Accountability Office 2008). In addition, the separation of grizzly bears from human food sources, while revealing a sea change in public and management attitudes, has only been accomplished and sustained through continued intensive research and monitoring (Gunther et al. 2004; Schwartz, Harris, and Haroldson 2006; Gunther 2008).

During this era more than any other, National Park Service policy and management intention most clearly departed from the traditional directions of agricultural management and mainstream wildlife management. The long-established model of North American wildlife management was primarily focused on the sustained public harvest of preferred wild animals; the model was and is admirable for its numerous successes. More than that, the North American model played the essential role of keeping most elements of the Yellowstone wilderness setting in place for nearly a century, thus providing National Park Service management the opportunity to explore new realms when the agency's leadership and constituencies matured to the point where such exploration became not only desirable but urgently necessary. By the late 1960s, however, the challenges facing managers in Yellowstone National Park and elsewhere required a rethinking of the fundamental appropriateness of the North American model of wildlife management in the Yellowstone context.

The scientific legacy of the Ecological Management Era is infinitely richer than that of its predecessors. It heralded an unprecedented intensity of scientific scrutiny on the greater Yellowstone landscape, perhaps best exemplified by the three occasions on which the National Academy of Sciences was called on to analyze and arbitrate Yellowstone's scientific conversation (Cowan et al. 1974; Cheville et al. 1998; National Research Council 2002)— but also exemplified by a successful series of biennial scientific conferences and the launching, by the National Park Service in cooperation with the nonprofit Yellowstone Association, of the semitechnical quarterly educational journal *Yellowstone Science,* which has provided a forum in which Yellowstone's many natural and cultural resource topics can be responsibly addressed.

RESTORATION ERA (1983–PRESENT)

In 1976, the National Park Service made an unsuccessful attempt to restore grayling *(Thymallus arcticus)* to a small stream in the Madison River drainage, but we begin this era in 1983 because of the high-profile success of peregrine falcon *(Falco peregrines)* restoration that began that year (Baril et al. 2010). Other restoration efforts were subsequently initiated, including grayling in the mid-1990s (Kaya 2000), abandoned mine lands in 1990, wolves in 1995 (D. W. Smith 2005), bald eagles in 2005 (Baril et al. 2010), abandoned agricultural fields in the Gardiner basin in 2008 (National Park Service 2010), and westslope cutthroat trout *(Oncorhynchus clarki lewisi)* in the Gallatin River drainage in 2008 (Koel et al. 2008).

Concurrently, a series of projects aimed at protecting native species from non-native invasive species was initiated, including eradication of clandestinely introduced brook trout *(Salvelinus fontinalis)* from Arnica Lagoon and Creek in 1985 (Gresswell 1991), intensive efforts to control exotic plants in 1986 (Olliff et al. 2001), lake trout control in Yellowstone Lake in 1995 (Varley and Schullery 1995b; Koel et al. 2008; Gresswell 2009), and the implementation of the Interagency Bison Management Plan to control the spread of the exotic bacteria *Brucella abortus* in 2000 (Plumb et al. 2009). Contrary to a common perception of natural regulation as a passive, hands-off policy, these programs reveal a forceful and aggressive management effort to restore and preserve ecosystem functions.

The environmental legislation of the 1970s took hold slowly in park service culture and greatly increased the complexity of all management processes. Ironically, despite the tremendous increase in research since the 1960s, it was not until the passage of the National Parks Omnibus Management Act in 1998 that the 82-year-old agency was actually required to use science as a basis for management decisions. During this time, insights from paleoecology, long-term ecological studies, and modeling revealed that ecosystems are highly dynamic and change in complex, unpredictable ways that may be unidirectional or at least maintained at a steady state that is relatively unchanging for long periods of time. The long-held belief in the balance of nature was replaced with a new paradigm: nature in flux or switching between several steady states (Lodge and Hamlin 2006; D. N. Cole and Yung 2010a). Understanding how ecosystems are changing through time, whether those changes are transitions between alternate steady states and whether they are deterministic (nonrandom, predictable) or stochastic (random, probabalistic) is paramount in understanding how to meet management goals (D. N. Cole and Yung 2010a).

Interagency ecosystem-level management became common during this era despite strong political resistance. The Greater Yellowstone Coordinating Committee, composed of park superintendents and forest supervisors, was formed in 1964 but paid little attention to ecosystem issues until the mid-1980s, by which time Yellowstone staff were already quietly involved in dozens of cross-boundary initiatives (Congressional Research Service 1987; Barbee et al. 1991; Greater Yellowstone Coordinating Committee 1991). Today, with high-profile management programs, like those for bison, grizzly bears, and wolves, the roll call of involved agencies, tribes, and other entities is not so much a list as a directory.

THE STUDY OF YELLOWSTONE HISTORY IN THE CAUSE OF SCIENTIFIC INQUIRY

We present the foregoing narrative historical overview with an intense awareness that the human history of Yellowstone National Park and the surrounding lands has been told many times for many purposes. The first tellers of the story, in the late nineteenth century, wove a heroic tale in which civilized and foresightful Euro-American men (it was solely a male narrative) conquered a savage wilderness, gradually excluded its savage native inhabitants from the park area, and triumphed over economic and political forces to create and protect the park (Norris 1878; Chittenden 1895; Langford 1905). Although aspects of this simplistic narrative were perpetuated by subsequent generations of historical writers, some second- and third-generation historians questioned various historical details or explored nuances that had been neglected or ignored in earlier narratives—for example, revising and expanding the lists of characters who may or may not have been key to the park's establishment and protection (Matthews 1906; Dolliver 1927; Cramton 1932)

However, by the mid-twentieth century and in growing force thereafter, the savviest historical commentators—who were much more likely than their predecessors to have some training as historians—conducted deeper and more rigorous analyses, including necessary deconstructions of some portions of the tale. For example, historians questioned the creation myth of the Madison Campfire of 1870, at which altruistic regional citizens were said to have invented the idea for Yellowstone National Park. Historians also challenged the egregiously erroneous but common belief that native people were superstitiously afraid of the Yellowstone Plateau because of its devilish geothermal activity (Huth 1948; Hultkrantz 1957; Russell 1960; Haines 1974, 1977). Muddled logic, insufficient research, blatant self-promotion, and glaring racism in earlier accounts were undermined or at least identified. Such reconsiderations of the park's historical narrative have appeared at an accelerating rate since then (Hampton 1971; R. G. Wright 1992; Sellars 1997; Spence 1999; Reiger 2001; Schullery and Whittlesey 2003; Schullery 2004).

From the park's earliest years, some conservationists expressed a sentiment that the setting aside of the Yellowstone Plateau in an undeveloped state would provide future generations with a reasonably clear glimpse of an earlier North American landscape (Schullery 2004). This sentiment was later expressed more formally and became a guiding principle of the national parks. Thus, as investigators from several disciplines examined the pre-1872 historical record, it was their implied, if not explicitly stated, goal to determine just what the area had been like before Euro-American development and settlement came to the region, because that was, presumably, the state in which the area should be preserved in the future. As already mentioned, eventually this study came to include the question of the extent to which pre-1872 ecological processes had been influenced by the humans who had occupied or visited the area for 10,000 years.

The idea of Yellowstone National Park as a kind of museum of an earlier North America was accompanied by an equally important idea: there was,

or should be, a higher goal for the national parks. This goal had been identified by an occasional scientist or conservationist almost from the year of
the creation of the park. The goal was publicly expressed as early as 1874
when geologist Theodore Comstock became perhaps the first scientific
voice advocating the preservation of Yellowstone's wildness for the immense
social and scientific values to be gained from a nature reserve wherein the
processes of evolution could be observed and learned from. Each succeed-
ing generation had scientific voices expressing the same thought with increasing force and clarity (Pritchard 1999). Ecologist Charles Adams and
his colleagues in the 1920s recognized the wisdom and advantages of fos-
tering this level of wildness in Yellowstone and other parks. Historian
James Pritchard accurately characterized their contribution to national park
thinking as follows: "The scientists' proposal during the early 20th century
that Yellowstone serve as an ecological control has endured as one of the most
significant purposes for the national parks, underlying both management
and public understandings of nature in Yellowstone" (Pritchard 1999, 314).

Both of these impulses, first to preserve a segment of primitive North
American landscape for society's general edification and second to use that
landscape as a laboratory for the study of evolutionary and other unhin-
dered biological processes, would be primary drivers in the quest for a
deeper understanding of Yellowstone's long-term history. While for most of
the twentieth century professional historians confined their efforts to refin-
ing our understanding of the administrative history of Yellowstone National Park, writers in the natural sciences began addressing their own historical issues surprisingly early in the park's history. Many wildlife scientists
and naturalists have examined the Yellowstone historical record and offered
their respective but strikingly discordant conclusions about the state and
abundance of wildlife in the Yellowstone region prior to the park's creation
in 1872 (Skinner 1927, 1928; Bailey 1930; Rush 1932; Grimm 1939; Murie
1940; Koch 1941; G. Cole 1969; Lovaas 1970; Gruell 1973; Meagher 1973;
D. B. Houston 1982; Barmore 1987; Kay 1990, 1995; Craighead et al.
1995; Wagner et al. 1995; Wagner 2006). Although several of these publications are deservedly recognized as milestones in the scientific literature
regarding Yellowstone National Park, it is a revealing condition of this extended conversation that their authors invariably drew on the same small pool
of about 20 (in many cases fewer) readily accessible historical accounts of
Yellowstone-area wildlife before 1880 and yet reached remarkably dissimilar

conclusions. Historical evidence is just as difficult to analyze as any other kind and probably more difficult than many. The errors made in some of these analyses provide important lessons for scientists regarding the study of Yellowstone's history.

The relatively recent improvement and diversification of the tools available for the study of the past have resulted in a wealth of new considerations regarding Yellowstone's landscape history. In the past 30 years, not only historians and ecologists but a gratifying variety of paleontologists, anthropologists, and archeologists have addressed some of the same questions about the historic and pre–Euro-American condition of Yellowstone's wildlife, vegetation, and related landscape processes (Romme 1982; Romme and Despain 1989; Engstrom et al. 1991, 1994; Whitlock et al. 1991; Cannon 1992; Laundré 1992; Meyer et al. 1992, 1995; Schullery and Whittlesey 1992, 1999a, 1999b; Whitlock and Bartlein 1993; Barnosky 1994, 1996; Turner and Gardner 1994; Romme et al. 1995; Vale 1998; Yochim 2001; National Research Council 2002; Schullery 2004). Implicit in these studies is a modern redefinition of the meaning of the term "history," which now is seen as applying to the long haul of human (and, in some cases, even prehuman) presence on the landscape rather than just the period during which literate whites were here. This broadening of scale has been essential to appreciating the expanses of time necessary for a variety of ecological and geophysical processes to operate.

CONSERVATION IMPLICATIONS

As so often happens in Yellowstone, the demands of making the best possible management decisions test and stretch our capacity to conduct intelligent study of the issue at hand. The natural regulation hypothesis developed in the late 1960s and early 1970s has resulted in an explosion of scientific attention and conversation. Had the National Park Service attempted to handle their various wildlife crises over the past 40 years through application of the traditional agricultural and wildlife management approaches, and even had those attempts proved adequately satisfying to managers or the public, we would have wasted a unique opportunity to make the spectacular advances in our understanding of ecological processes in Yellowstone National Park that have been made.

The application of historical study to ecological issues in Yellowstone has also resulted in an almost amazing broadening of our perspective. The study of Yellowstone's environmental history began almost solely as an exercise in consulting historical documents but now includes a still-expanding array of evidence types, from historic photographs to alluvial deposits, pond sediments, packrat middens, archeological sites, and oral traditions of several native and Euro-American cultures. That interpretations of these sources of evidence continue to differ, sometimes sharply, should be no surprise. The price we will always pay for increasing complexity of inquiry is increasing opportunity for alternate opinions. However, the greater effect of increased complexity of inquiry is surely an infinitely richer grasp of the historical process than was possible even three or four decades ago.

Perhaps even more important, we have moved from a complete isolation of the two cultures of the humanities and the sciences to a complex collaboration of many disciplines, draping layer on layer of evidence while confronting the challenges of the different spatial and temporal scales that must be addressed. The partnerships are not always smooth or even productive, and practitioners of the respective disciplines still do not always grasp the subtleties and protocols of each other's approach to a given question. However, we no longer struggle to overcome quite so many of the traditional barriers that separated these realms of inquiry only a few decades ago.

Given this vastly improved—if still often problematic—grasp on the historical context of today's management issues in Yellowstone, we must continue to ask the fundamental question: So what? Why should the historical record matter so much to us? We are not concerned merely with understanding how, historically, Yellowstone National Park arrived at its current administrative, scientific, and, especially, ecological state. We are far more urgently concerned with where it must go now and how our improved understanding of its past should affect our thinking about that future. If we have indeed escaped from past generations' simplistic convictions about how Yellowstone used to be, then how are we to apply our newfound historical awareness to modern issues?

To consider that question, we propose that Yellowstone National Park may now have entered another era in its management history—Landscape Conservation. In 2009, Secretary of the Interior Ken Salazar mandated that, because of the breadth of impacts of climate change, all Interior Department bureaus must contribute to large-landscape conservation (U.S.

Department of the Interior 2009). It is almost uncanny how the eras of eco-logical management and restoration anticipated the large-landscape conser-vation approaches suggested in the recent scientific literature. The National Park Service now has co-leadership of the Great Northern Landscape Co-operative, which covers much of Idaho, Montana, Washington, and Wyo-ming, and parts of Oregon—one of 21 such science-management partner-ships established to promote large-landscape conservation.

Today, throughout the realm of natural-area management, there are calls for a fresh look at the inherent ambiguities of traditional and even more recent management strategies (D. N. Cole et al. 2008; Jackson and Hobbs 2009; National Park Service 2011). Pluralistic strategies for future management direction include various combinations of managing for natu-ralness while conserving biodiversity and resilience and mitigating ecologi-cal disturbance such as invasive species and restoring degraded systems. Yellowstone National Park will no doubt play an important role in these deliberations and in the management experiments that grow from them. In fact, Yellowstone is perhaps the only place prominent enough on the na-tional stage to hold a sustained conversation about the meaning and future of the conservation movement. National Park Service managers have long known, and often lamented, Yellowstone's unique magic in the public mind and heart. The world's affection for the park has only intensified during its 140-year history. Modern media and scientific discourse, operating at unpre-cedented and accelerating rates, are compounded by heightened levels of public attention through social networking (2 million hits annually on the Old Faithful webcam are emblematic of this change), until Yellowstone is subjected to a global scrutiny that would have seemed fantastic even 20 years ago.

However, even in this exciting and forward-looking time, there are still countless occasions when our decisions must be made with limited under-standing of historical conditions or when underlying if unarticulated con-victions about historical conditions still lurk in our minds. Given these circumstances and apparent directions, we may ask how our greatly height-ened awareness of the long-term ecological and social history of Yellow-stone might serve managers. We suggest a few tentative answers.

First, most serious participants in the dialogues regarding Yellowstone now recognize that the goal of establishing some pre-1872 state of nature in the park as an ideal toward which management should aim was simplistic, unrealistic, and, most important, unattainable. Historical study of Yellowstone National Park has taught us many things about the range of variability of the greater Yellowstone ecosystem and the potential for native and Euro-American effects on the landscape that are vital for informed management now, but it does not provide us with some cookbook-simple template we need only duplicate to get things right. It has been especially important to abandon the idea that the year 1872 somehow provided managers with a reliable baseline from which to judge subsequent changes in the landscape. Euro-American influences on the landscape, which ranged from far-reaching disease epidemics to introduced horses to various technologies to trade incentives, began arriving in the greater Yellowstone ecosystem at least a century before the park was established.

Second, early conservationists and managers almost invariably made their most foresightful and durable decisions about Yellowstone National Park when they enlarged their sense of scale. Recognition of a greater Yellowstone ecosystem and the need to manage regional landscapes across boundaries arose in the park's first 10 years and clattered around in public conversations rather obscurely from then until the idea gained scientific and public traction in the 1960s as a result of the crises in wildlife management mentioned earlier. The idea of a greater Yellowstone ecosystem provided countless members of the public with a new framework in which to imagine the park and the region and empowered two generations of researchers to consider Yellowstone National Park as part of a larger setting. Our sense of scale was permanently revamped in this process. We are now infinitely better prepared to collect and integrate data on scales that many early park managers would have found incomprehensible. Our sense of temporal scale may be most changed; in considering issues relating to climate change and ecosystem resilience, our Holocene-length perspective gives us an incalculable advantage over early managers of Yellowstone National Park, for whom park management was a year-by-year mission, and long-term climatic variation, to say nothing of wholesale global climate change, was of no interest or meaning.

Third, continued disagreements over just what the historic record of the Yellowstone area tells us about the present state of ecological and geophysical affairs in the park suggest that we are not yet done learning from history or sorting out its complexities. We generalize and theorize about historical conditions here at our considerable peril and are still frequently guilty of highly refined versions of the same kinds of sloppy thinking, of inattentive analysis of complex historical evidence, and of rushing to convenient judgments that we so lament in our predecessors on the Yellowstone historical scene. We still too often tend to oversimplify the past. A single day in 1800 or 1600 or 1200 was no less complex a period of time than a single day in 2000. The effects of a single severe winter or unusually wet spring were no less pronounced then than now; subtlety and nuance were everywhere on the landscape. Sweeping generalizations about entire decades, centuries, or millennia are no less reckless if the decade occurred in the late 1800s than if it is the decade we are currently experiencing.

CHAPTER THREE

Scale and Perception in Resource Management

Integrating Scientific Knowledge

MATTHEW S. BECKER

ROBERT A. GARROTT

P. J. WHITE

YELLOWSTONE NATIONAL PARK comprises the core of the greater Yellowstone ecosystem, a relatively large area (56,000 square kilometers) spanning multiple management regimes and constituting the world's largest intact temperate ecosystem, with a full complement of native species that were historically present (Keiter and Boyce 1991). The largest concentrations of wild ungulates and large carnivores (meat eaters) in the lower 48 states are found here, along with some of the best quality habitat for native char and salmonid fishes. The Yellowstone ecosystem is characterized by numerous high-profile and wide-ranging wildlife species and by large-scale ecological processes. These factors, combined with rapidly changing human communities (Gude et al. 2007) possessing a diversity of philosophies and goals and a strong sense of public ownership, result in tremendous amounts of controversy over the management of natural resources (A. Chase 1987; Schullery 2004; Wagner 2006; P. J. White, Garrott, and Olliff. 2009).

Such debates highlight the need for objective scientific information gleaned from research, which in turn informs and contributes to policy and management decisions. Integrated research and monitoring programs implemented in Yellowstone National Park are aimed at informing management actions, but the resolution of natural resource controversies also suffers from fundamental differences in perception of ecological dynamics or, more simply, disagreements over the nature of nature (Dizard 1996). Despite decades of scientific research indicating that ecosystems are better characterized by a multitude of complex and dynamic processes, popular public perception and media continue to perpetuate the idea of stability and equilibrium, commonly referred to as the balance of nature (Ladle and Gillson 2009; Yung et al. 2010). Similarly, while heterogeneity in time and space better describes ecological dynamics, the temporal and spatial scales at which these are measured are often inadequate, disparate, or inappropriate for effectively addressing particular management problems (Hobbs 2003). Here, using several of Yellowstone's most prominent natural resource controversies, we describe public perceptions of nature and issues of scale and discuss some ideas for altering misconceptions and integrating ecological knowledge to help resolve these issues.

THE BALANCE OF NATURE

One of the primary motivations for establishing the national parks and wilderness system in the United States was to maintain natural areas that remained relatively untouched and unchanged from what was believed to be their historical state—so-called vignettes of primitive America (Aplet and Cole 2010). Implicit in this designation was that conservation should strive to allow ecosystems to tend toward their perceived natural state of self-regulation, stability, and equilibrium (Yung et al. 2010). Concepts such as succession to a climax community in which populations of plants or animals are thought to remain stable and exist in balance with each other and their environment reflect this paradigm. Although stakeholders in natural resource controversies are often dichotomized into preservationists or utilitarians, both parties still subscribe to this equilibrium view of nature. Management actions for preservationists promote suppression or prevention of disturbance and change, while utilitarian views promote active management and harvests of resources around a perceived optimal balance reflected by concepts such as carrying capacity and maximum sustained yield (meaning the maximum number of animals the environment can sustain indefinitely and the maximum number that can be harvested indefinitely; Ladle and Gillson 2009). The concept that ecosystems were static and stable, and that management actions should tend toward restoring optimal conditions that occurred historically, was prevalent in the 1960s before ecological process management became the dominant management paradigm in Yellowstone National Park. For example, the Leopold Report (Leopold et al. 1963) recommended that national parks should maintain large examples of relatively stable climax forest communities that perpetuate themselves indefinitely.

Perceptions of a balance of nature have limited discussions of dynamic change and constrained management options. The gap in public understanding is likely due to the uncertainty and complexity characteristic of dynamic ecological processes, as well as concern that a perception of ecosystems as resilient to change and perturbation can justify harmful or intrusive management practices (Ladle and Gillson 2009). However, it is unrealistic to expect that natural resource management in Yellowstone National Park can be completely devoid of human intervention given the pervasive human influence within and around it. In addition, the perception that wildlife species in Yellowstone have historically been managed with a relative absence of intervention is erroneous. Many large mammal species have been subject to intense management through much of the park's history. For example, biologists and hunters culled and harvested tens of thousands of elk during the 1930s through 1968 because of concerns about overgrazing (D. B. Houston 1982). Likewise, bison in the Lamar and Pelican valleys of Yellowstone National Park were subject to intense animal husbandry (confinement, culling, feeding, herding) between 1902 and 1948 and reintroduced into the Hayden and Firehole valleys of Yellowstone in 1936 (Barmore 1968; Meagher 1973). Until 1934, pronghorn *(Antilocapra americana)* were subject to feeding, irrigation, and fencing efforts that reduced their distribution and apparently reinforced the tendency for some pronghorn to remain on the winter range year-round through the present (Keating 2002). Furthermore, 212 grizzly bears died as a result of human causes in the years 1967–1972, largely due to removals by state and federal agencies and private citizens defending their life and property (Craighead et al. 1988).

Several decades of productive scientific investigations under ecological process management have prompted a shift from the balance-of-nature metaphor and equilibrium paradigm to a focus on the dynamic nature of ecosystems and the processes that generate this variation (P. S. White and Bratton 1980; Botkin 1990). This shift has served to move management actions away from conserving ecosystems and species in static states and toward conserving processes such as nutrient cycling, migration, and population dynamics. This new paradigm could be called the flux of nature because it focuses on maintaining ecosystem heterogeneity in time and space (Ladle and Gillson 2009). Such a shift acknowledges that ecosystems are resilient, can exist in several different states, and are subject to change and perturbations. Furthermore, management for a historical optimum often is not possible or relevant given the extensive, rapid, and long-term human alteration of earth's ecosystems and climate (Suding et al. 2004; Aplet and Cole 2010), as well as a lack of clarity and agreement as to the specific historical period or state toward which management actions should strive. However, continued integration of this ecological knowledge into educational and interpretive messages and products will be necessary to alter public misconceptions and resolve this issue, as discussed in the remainder of this chapter.

SCALE: HETEROGENEITY IN TIME AND SPACE

Describing and drawing inferences from the dynamic processes that characterize ecosystems focuses around heterogeneity across time and space, as well as the units of measurement used to investigate them. Fundamental to evaluating this heterogeneity is the concept of scale, or the size of measurements required to study interactions (Hobbs 2003), with large-scale referring to observations over relatively long periods of time or areas of space (Allen and Hoekstra 1992). Thus, the spatial scale of a study is the total area from which observations or measurements are drawn, the number of samples, and their area, whereas the temporal scale is the duration of the study and the frequency and duration of observations or measurements (Hobbs 2003). As a result, scale is a unifying concept that considers the importance of heterogeneity as an agent driving many ecological processes and enables the comparison of data (Hobbs 2003). Valid comparisons among studies require observations across similar scales (both large and small) in space and time.

Many problems and misperceptions in ecology result from mismatches in scale (Lee 1993). For example, "depletion of ungulate populations can occur when short-term rates of harvest exceed long-term rates of renewal" (Hobbs 2003, 226). Also, densities of predators and prey often trend similarly at large scales because "predators seek prey that are spatially concentrated, and as a result, both species are often found within the same habitat types." However, the spatial distributions of predators and prey often trend in opposite directions at small scales because of predator avoidance by prey (Hobbs 2003, 227). Solving ecological problems necessitates integrating information across scales of time and space to make reliable predictions. Ecologists seek to understand how environmental heterogeneity shapes ecosystems by paying attention to scale (Hobbs 2003).

Wildlife species "traverse large areas of space in relatively brief intervals of time and, consequently, respond to landscape heterogeneity expressed across a broad range of scales" (Hobbs 2003, 223). Most wildlife issues in the greater Yellowstone ecosystem operate at a scale larger than Yellowstone National Park and involve human dimensions as well as ecological relationships (Keiter and Boyce 1991). However, ecologists generally study only fractions of this area that are managed for natural resources, which results in a mismatch between the scale of investigation and the scale of management decisions (Hobbs 2003). This mismatch is important because the findings from "ecological investigations often depend on the scale at which observations are made, a phenomenon known as scale dependence" (Hobbs 2003, 226). Variability increases with scale, and large-scale studies often encounter many sources of spatial and temporal heterogeneity not evident at smaller scales of space or time. Thus, the distributions or dynamics exhibited by populations of wildlife at large scales cannot be described as the sum of smaller-scale patterns (Hobbs 2003). Conversely, small-scale patterns can create heterogeneities at larger scales of the landscape that alter processes like nutrient cycling and community dynamics (Pastor and Cohen 1997; Frank and Groffman 1998; Knapp et al. 1999; Augustine and Frank 2001; Hobbs 2003).

Further complicating evaluations is the interaction between scale and human perception. While ecological processes can span considerable lengths of time and space, human perception is finite, typically limited to half a century or less, and focused around the scope and duration of personal

Ecological processes in the greater Yellowstone area operate across a continuum of spatial scales ranging from those that occur within the Lamar River Canyon in the foreground of this photo to processes that extend beyond the boundaries of Yellowstone National Park and onto National Forest lands, such as the Absaroka Range in the distance outside the east boundary of the park. *Photo by Cindy Goeddel.*

experience (Ornstein and Ehrlich 1989; Hobbs 2003). Such constraints limit human ingenuity and the ability of people to accept and adapt to changes beyond their experience and perception. Similarly, the application of ecological knowledge at temporal and spatial scales beyond the perceptions of researchers and managers is difficult and constrains the scope of management problems they are willing to address. Collectively these interactions discourage large-scale solutions and often lead to vitriolic opposition from the public (Hobbs 2003).

DIFFICULTIES WITH SCALE AND PERCEPTION: YELLOWSTONE EXAMPLES

The combination of public perceptions of nature as balanced and tending toward equilibrium, coupled with problems of scale dependence, tends to exacerbate natural resource controversies. Given the size of the greater Yellowstone ecosystem, the highly mobile nature of its wildlife across multiple temporal and spatial scales, and the mismatch between ecological and political boundaries, this is well illustrated in several of Yellowstone's debates over ecological processes and wildlife.

Fire Management: Historical management of wildfire in the United States perhaps best exemplifies some of the problems with scale and public perception because, for decades, the public was ingrained with the idea that wildfires were destructive and should be prevented on public lands (Jacobsen et al. 2001). Ecosystem process management in Yellowstone allowed for many fires to run their course as a natural process of disturbance, but such a policy was initially met with resistance by some agencies, managers, and sectors of the public (Lichtman 1998). This debate intensified after the massive fires in Yellowstone during the summer of 1988, but research in the wake of these fires contributed to more acceptance of the essential functions of natural fire in the ecosystem. Unfortunately, given the immediate and evident effects of wildfire on ecosystems compared with the long-term and more subtle effects of fire suppression, mismatches in scale continue to contribute to calls for fire management policies that promote suppression on some public lands (Hobbs 2003). The short-term effects of fire are obvious to land managers and the public because they "occur over scales of time and space that make them easy to perceive. Conversely, the effects of fire suppression are difficult to perceive because they accumulate slowly over time and large areas" (Hobbs 2003, 230). Fire suppression remains the prominent wilderness fire management strategy in the United States because of concerns about human safety, property damage, and timber resources (D. J. Parsons 2000), and virtually no approaches exist for dealing with effects of fire suppression, such as maturation of plant communities (Hobbs 2003). Yellowstone National Park provides an opportunity to further elucidate the essential functions of wildfire and demonstrate that management not based on immediate suppression is tenable.

Elk Management: Yellowstone National Park supports one of the highest densities of native free-ranging ungulates in North America, which have inspired one of the most productive, if sometimes bitter, dialogues on the management of a wildland ecosystem. For almost 70 years this debate focused on whether there were too many elk spending the winter on the northern grassland and shrub steppe of Yellowstone. Under ecological process management, which began in 1969, the park instituted a moratorium on elk removals and has since let a combination of weather, predators, and range conditions influence their abundance. In the absence of culling by park rangers, and with a much reduced human harvest and the securing of additional lands for wintering elk outside the park, the northern Yellowstone elk population grew rapidly, with counts increasing from approximately 4,000 in 1968 to 19,000 elk by 1988. These elk doubled the extent of their winter range north of the park from 22,179 hectares to 53,262 hectares (Lemke et al. 1998).

The policy of ecosystem process management proved to be a highly contentious approach to elk management, with criticisms primarily focused on effects of perceived overabundance on conditions of the winter range in and outside the park (National Research Council 2002). Movements by thousands of elk beyond the park boundary, which people had not seen in their lifetimes, led to claims that elk were overabundant inside the park and had degraded the range (Kay 1998; Wagner 2006). In response, the National Research Council (2002) conducted a two-year study of the winter range for ungulates in northern Yellowstone, reviewing all available science related to it and publishing its findings in a book entitled *Ecological Dynamics on Yellowstone's Northern Range*. The Council concluded that the evidence did not indicate that ungulate populations were irreversibly damaging the northern range and that migration and dispersal were natural processes rather than indications of mismanagement. Thus, they indicated that park managers could continue to use ecosystem process management, with populations of "elk, bison, and other ungulates fluctuating without direct controls inside the park, letting a combination of predators, weather, range conditions, and outside-the-park hunting, land uses, and population reduction by state agencies influence population numbers" (National Research Council 2002, 131). These findings helped to reduce misperceptions, and today most people realize that Yellowstone is not a ranch managed for animal production but that instead fluctuations in wildlife populations and forage conditions should be expected. Also, the restoration of wolves to the system changed the debate from concerns about too many elk to concerns about too few.

Bison Management: Historically, bison were a nomadic, wide-ranging species that occupied approximately 20,000 square kilometers in the headwaters of the Yellowstone and Madison rivers in what is now referred to as the northern greater Yellowstone area (Meagher 1973; Schullery et al. 1998; Gates et al. 2005; Schullery and Whittlesey 2006). By the early twentieth century, however, Yellowstone National Park provided sanctuary to the only relict, wild, and free-ranging bison remaining in the United States (Plumb and Sucec 2006). Thus, for much of the past 100 years, as Yellowstone bison recovered from near extirpation, they were constrained to two or three relatively independent breeding herds that migrated into discrete wintering areas but did not regularly and extensively venture outside the park. This has led to the popular beliefs that bison should always remain in Yellowstone and that they leave the park only because it is overgrazed and their numbers are too high. Such beliefs, in turn, have led to calls for intensive management to limit the abundance and distribution of bison inside Yellowstone, through means including fencing, fertility control, hunting, and brucellosis test-and-slaughter programs (Hagenbarth 2007; Kay 2007; Schweitzer 2007).

The belief that bison are not migratory and should remain in the park is reflected in the treatment of bison as livestock in areas adjacent to the park (Plumb et al. 2009). The Comprehensive Wildlife Conservation Strategy for Idaho mentions bison as a species of concern that is critically imperiled,

The conservation and management of Yellowstone's bison population is a good example of the challenges of scale in wildlife management. Here, a lone bull bison is seen on the vast grasslands of Lamar Valley within the core area of Yellowstone National Park that provides total protection to the animals. However, bison frequently migrate out of the park, where they sometimes come into conflict with agricultural and other pubic interests and priorities. *Photo by Cindy Goeddel.*

but the state agricultural regulations do not recognize wild bison and consider them livestock. Wyoming has designated specific areas adjacent to Grand Teton National Park and Yellowstone National Park where bison are considered wildlife—elsewhere they are considered livestock. Montana considers the Yellowstone bison population to be wildlife, with disease control management under the lead authority of the Montana Department of Livestock and hunting on lands adjacent to the park managed by the Montana Department of Fish, Wildlife, and Parks. In addition, the U.S. Department of Agriculture, through the auspices of the Animal and Plant Health Inspection Service, considers all bison removed from Yellowstone National Park, for purposes other than consignment directly to slaughter, as alternate livestock for the purposes of brucellosis management. Thus, even if the risk of brucellosis transmission could be eliminated from bison, it is unlikely these massive wild animals would be tolerated in most areas outside the park because of social and political barriers, such as human safety concerns (motorists), conflicts with private landowners (property damage), depredation of agricultural crops, competition with livestock grazing, lack of local public support, and lack of funds for state management (Boyd 2003). Since the evolution of an expanded bison conservation area outside of Yellowstone is the prerogative of the states comprising the greater Yellowstone area, the prevailing social carrying capacity (meaning the number of bison that will be tolerated by local communities), is perhaps most limiting (Plumb et al. 2009).

Wolf Management: Wolf restoration to the greater Yellowstone ecosystem has been described as one of the greatest conservation successes in the last century (D. W. Smith et al. 2003), but it contributed to a 60 to 90 percent decrease in the numbers of elk in three populations in central and northern Yellowstone and adjacent areas of Montana (Hamlin et al. 2009). This decreased elk abundance has evoked considerable controversy given the economic significance of elk hunting in the greater Yellowstone area, with some constituencies suggesting that the decreases are indicative that wolves will decimate all elk populations in the region if numbers are not intensively regulated through harvest (Dickson 2002; M. L. Miller 2005). However, it is possible that recent findings regarding wolf–elk dynamics are incomplete given that the majority of wolf–elk studies have been confined in spatial scale to high-elevation, deep-snowpack environments where wolf effects on elk demography may likely be greatest and where transitional dynamics are likely still occurring. If the historic abundance, distribution, and predator avoidance mechanisms of elk differed from what we observe today, with many elk historically residing in, or undertaking winter migrations to, lower-elevation valley bottoms now converted to human use, then wolves may be having an elevated effect on contemporary elk populations in national parks and wilderness areas relative to what might occur in lower-elevation valleys.

Elk are coursing animals whose morphology is structured for open habitats in which they can outrun predators (Geist 2002). Prior to European settlement and the large-scale decimation of elk herds, historical accounts suggest that elk and wolves were abundant in plains habitats in North America (Hornaday 1913; Seton 1927; O. J. Murie 1951; Burroughs 1995).

Also, many elk undertook large-scale migrations to lower-elevation valley bottoms during winter (Skinner 1925; O. J. Murie 1951), which may have served, in part, to decrease their vulnerability to wolf predation. Thus, it is possible that the abundance of low-elevation areas in the greater Yellowstone area with little snow and high plant productivity historically provided wintering elk with good nutrition and improved escape potential from wolves, lessening the impacts of landscape and climate on wolf predation (Garrott et al. 2005; Dunkley 2011). In contrast, most contemporary elk herds in western North America, including those in the greater Yellowstone area where wolves were restored, are associated with protected public lands in mountainous regions (Rodman et al. 1996; Hansen et al. 2000; J. M. Scott et al. 2001). J. Berger (2004) estimated that 58 percent of the migration routes to and from lower-elevation winter ranges have been lost in this area, and increased wolf predation in areas with deeper snowpack is well demonstrated (Mech 1970; R. O. Peterson 1977; M. E. Nelson and Mech 1986; Huggard 1993; Mech et al. 2001; Becker, Garrott, White, Gower, et al. 2009, Becker, Garrott, White, Jaffe et al. 2009). Consequently, although elk in Yellowstone National Park exhibit numerous antipredator strategies in the presence of restored wolves, these elk primarily reside in areas where they are most vulnerable to predation as a result of severe winter conditions.

There is a widespread public perception that wolf reintroduction has restored balance to Yellowstone by returning a top predator (Clifford 2009). Although wolves have indeed exerted strong top-down ecosystem influences that were absent for approximately 70 years, other factors have likely substantially altered wolf–elk dynamics from historical times, including (1) altered dynamics in landscape-level movements of large mammals as a result of human activities; (2) large-scale conversion of land for human use; (3) loss of traditional migratory routes; and (4) disproportionate protection of high-elevation, deep-snowpack areas in the national park and wilderness system.

Garrott et al. (2005) suggested that elk reductions in Yellowstone National Park were not indicative of what will occur with wolf range expansion throughout the greater Yellowstone area. Rather, they suggested that wolf–elk dynamics within and outside reserves would reflect the complexities of land management, landscape, and climate variables that collectively can prohibit wolves from reaching the densities observed in protected areas.

Hamlin et al. (2009) used a landscape approach to compare the population dynamics of seven elk populations in the greater Yellowstone ecosystem. They documented low survival and recruitment for elk that spent winter in mountainous, protected areas with high densities of predators, and high survival and recruitment for elk that spent winter in valley-bottom mosaics of private and public lands with low densities of predators and more favorable landscape and climate conditions (Hamlin et al. 2009). Thus, the histrionic claim that wolves will decimate all elk populations in the region has little support to date.

Disease Management: It is often difficult and, at times, ineffective to apply plans with the dual objectives of reducing disease infection while conserving robust populations of wildlife because risk management and conservation actions generally occur at different scales. The Yellowstone bison population has been infected with brucellosis since at least 1917, likely from cattle (Meagher and Meyer 1994). Brucellosis is a bacterial disease caused by *Brucella abortus* that may induce abortions or the birth of nonviable calves in livestock and wildlife (Rhyan et al. 2009). Infected livestock lead to economic loss, as infected cattle must be slaughtered, testing requirements increase, and marketability can be reduced. Brucellosis has been declared eradicated from cattle herds in the United States, but bison and elk persist as the last known reservoirs of infection in the greater Yellowstone area (Cheville et al. 1998). Approximately 40 to 60 percent of Yellowstone bison have been exposed to *Brucella abortus* (Treanor et al. 2007), and some of these animals migrate to winter ranges in Montana, where there is a risk of brucellosis transmission to cattle that graze on public and private lands (Plumb et al. 2009). Thus, the federal government and the state of Montana agreed through a court-mediated settlement to an Interagency Bison Management Plan in 2000 that established guidelines for cooperatively managing the risk of brucellosis transmission from bison to cattle, while allowing some migration out of Yellowstone National Park to winter ranges in Montana (U.S. Department of the Interior, National Park Service and U.S. Department of Agriculture, Forest Service, Animal and Plant Health Inspection Service 2000). The agencies agreed to (1) enforce spatial and temporal separation between bison and cattle; (2) use hazing (i.e., humans on horseback, all-terrain vehicles, or in helicopters) to prevent bison egress from the park; (3) if hazing is unsuccessful, capture all bison attempting to leave the park and test them for brucellosis exposure; and (4) send test-positive bison to slaughter.

Intensive management near the park boundary has maintained separation between bison and cattle, with no transmission of brucellosis. However, management culls of bison to maintain separation once hazing has failed have differentially affected breeding herds, altered gender structure, created reduced female cohorts, and dampened productivity of bison (P. J. White, Wallen, et al. 2011). These findings demonstrate the difficulties of balancing competing objectives that are based on limited understanding of bison ecology and disease dynamics and assessed at different spatial and temporal scales. Managers act to prevent disease transmission annually in localized areas, but demographic and genetic effects to bison may not be detectable for decades, and as a result, unintended consequences may occur. Thus, managers should continually review and integrate conservation into management policies to better protect migratory ungulates and facilitate their restoration. Also, best management practices for preventing disease transmission should be conservative regarding effects to wildlife to avoid undermining long-term conservation efforts where impacts are more subtle and occur over a longer time period.

CONSERVATION IMPLICATIONS

Given the dynamic nature of Yellowstone's ecological processes, the multiple scales of space and time traversed by many of its most high-profile wildlife species, and the increasing human pressures and interests within and around the greater Yellowstone area, adaptive management has frequently been promoted as one of the cornerstones of effective ecosystem management in Yellowstone and elsewhere and as an effective means of resolving management controversies. Ecosystem management necessitates adaptive management as its method of implementation because complex ecological components and processes are constantly changing, and as a result, there is substantial uncertainty regarding how systems will respond to disturbances and various human interventions.

Adaptive management is a continuous decision-making process whereby "(1) system of interest is modeled based on the best available scientific information to screen credible hypotheses, (2) potential actions are incorporated into a research and management plan designed to evaluate them, and (3) the potential actions are implemented and their impacts and effectiveness are evaluated at scales relevant to decisions they are intended to support" (Holling 1978; Walters 1986; Walters and Holling 1990; Hobbs 2003, 232; Williams et al. 2007). Actions are refined with information to enhance progress toward objectives and minimize adverse environmental consequences. Such an approach provides ecologists with the ability to integrate knowledge and test hypotheses across scales in a way that is useful to managers and interested constituents (Hobbs 2003).

However, although the underlying premise of adaptive management is appealing, successful applications of it in ecosystem management are rare in Yellowstone and elsewhere. This is likely because such an approach requires clearly articulated goals and objectives agreed on by stakeholders in order for management actions to be effectively directed. Given the diversity and stark differences in values and philosophies among various constituencies in Yellowstone, such failures are not surprising, and consensus on management goals and objectives are extremely difficult to achieve. For example, elk hunters and ranchers may have fundamentally different values, preferences, and perspectives on managing predators than wolf enthusiasts, and consequently, these different groups will be unlikely to reach agreements on goals underpinning adaptive management. Thus, arguments over science in Yellowstone are often replaced by untimely arguments over values and preferences (Lackey 2007). Also successful management of wildlife species around historically derived baselines is usually controversial and often unsuccessful given that such baselines are usually extremely vague, stakeholders rarely benefit from abundant and accurate historical information, and target numbers, time periods, and conditions are poorly defined or agreed on. In addition, the actual conditions that supported historical wildlife numbers rarely are known, still present, or possible to restore. Thus, although adaptive management can address issues of scale, human values, preferences, and subsequent perceptions contributing to management controversies often obstruct effective implementation of this approach (Lackey 2007; P. J. White, Garrott, and Olliff 2009).

While values and preferences of different stakeholders in resource management controversies are extremely difficult to influence and change, and contribute heavily to failures of adaptive management, it is more feasible to alter faulty perceptions regarding public views of nature. Undoubtedly, one of the most effective measures for integrating scientific knowledge into management controversy resolution is to improve the public understanding of ecology, moving away from the balance-of-nature paradigm and toward a better appreciation of the importance of maintaining the diverse and dynamic array of ecological processes underpinning the greater Yellowstone ecosystem. Without such a public education effort, discussions over resource management will be inherently in gridlock, given that many proposed management actions will be opposed or perceived poorly by the public despite their ecological value, as these management actions may neglect public perceptions of balance and stability in favor of ecological process maintenance. Although such educational efforts may not fundamentally alter people's values and preferences for Yellowstone's wildlife, ideally this integration of scientific knowledge would shift the focus of controversies to relevant issues of scale and ecological process management and away from disagreements over the nature of nature in Yellowstone.

population dynamics and
interactions among species

CHAPTER FOUR

Population Dynamics

Influence of Resources and Other Factors on Animal Density

P. J. WHITE

KERRY A. GUNTHER

POPULATIONS, OR AGGREGATIONS of individuals of the same species that interact, are often the basic unit of concern in wildlife management. A population can be managed to make it increase or decrease or provide a sustainable harvest, or it can be left unmanipulated and simply monitored over time (Caughley and Sinclair 1994). Each of these options has been pursued for various wildlife populations in and near Yellowstone National Park, with management objectives for a given population often changing over time. Animals from many populations migrate outside the park for a portion of the year and into areas managed by other state and federal agencies, where they are exposed to different management actions, such as hunting, or different land-use practices, such as agriculture and housing development, than those realized within the park.

The number of animals in a population at any time is equal to the number of animals in the population at some previous time period, plus the number of animals added through births and immigration, minus the number of animals removed through deaths and emigration (dispersal).

Any changes in the rate of increase or decrease of population size are indicative of some underlying change in births, deaths, emigration, immigration, or some other characteristic of the population that can interact with these factors, such as age distribution (Caughley and Sinclair 1994). The manner in which a population increases or decreases is determined by its relationship with one or more resources that animals need to survive and reproduce. If resources are unlimited, then animal populations tend to grow at their biological maximum as determined by their life history characteristics and components of the environment that are not resources, such as temperature (Caughley and Sinclair 1994). However, if resources such as food or space are limited, then population growth will eventually be constrained in some manner based on animal abundance, the renewal rate of resources, or numerous other factors (Caughley and Sinclair 1994).

Increasing density, or the number of animals inhabiting a given area, can slow the growth rate of wildlife populations by decreasing the resources available for each individual and, in turn, decreasing nutrition, body condition (fat, protein), reproduction, and survival (Sinclair 1975; Caughley 1976; Eberhardt 1977, 2002). Other factors not related to density, such as drought or deep snowpack, can worsen these effects by further decreasing the availability of forage and increasing energetic costs (Clutton-Brock et al. 1985; Sæther 1997; Gaillard et al. 2000). Thus, understanding the importance of these feedbacks on population growth is essential for developing management strategies to conserve wildlife.

Yellowstone National Park and surrounding public and private lands of the greater Yellowstone ecosystem comprise the largest, intact ecosystem in the lower 48 states. All native animals are present, and natural disturbances, native species, and ecological processes interact with relatively little human intervention. Thus, the system serves as a critical benchmark, or point of reference, for making comparisons with more disturbed systems and monitoring long-term regional changes in climate, wildlife diversity, and other natural resources (Boyce 1998). Herein we examine the population dynamics of several wildlife species in Yellowstone to illustrate resource relationships that lead to changes in their population sizes. We also discuss the importance of animal density and other factors, such as climate, conflicts with people, disease, harvest, and predation on their population dynamics.

ELK

The northern Yellowstone elk population spends winter on grasslands and shrub steppes along the northern boundary of Yellowstone and nearby areas of southwestern Montana (D. B. Houston 1982; Lemke et al. 1998). These elk migrate seasonally, moving from the low-elevation winter range to higher-elevation summer ranges throughout the park (Skinner 1925; Craighead et al. 1972; P. J. White et al. 2010). About 10,000 elk were counted in the early 1930s, and managers believed that overgrazing was having a deleterious effect on vegetation in the park (D. B. Houston 1982). Thus, from 1930 through 1968, managers relocated about 13,500 elk, shot or trapped about 13,000, and permitted hunters to harvest another 45,000 elk north of the park in Montana (D. B. Houston 1982). This culling decreased elk counts to fewer than 4,000 by 1968 (D. B. Houston 1982).

Culling inside the park ended in 1968, and elk numbers were allowed to fluctuate in response to weather, predators, resource availability, and outside-the-park hunting (G. F. Cole 1971). Hunter harvest was reduced to fewer than 210 elk per year, and the population grew rapidly, with abundance increasing about 20 percent each year during the years 1969 to 1975 (Eberhardt et al. 2007). Elk counts increased to approximately 12,000 by the mid-1970s, which resulted in the state of Montana resuming liberal harvests of elk that migrated outside the park. These harvests focused on female elk after 1985 because harvesting females is more effective at reducing population growth (Lemke et al. 1998; P. J. White and Garrott 2005a, 2005b).

Prior to 1968, northern Yellowstone elk spent winter primarily within Yellowstone National Park, likely due to the previous decades of aggressive hunter harvest in Montana and a lack of tolerance for elk on what was likely traditional winter range outside the park. During the mid-1970s through the 1980s, however, distributions of wintering elk again spread down the Yellowstone River Valley north of the park, with the extent of winter range outside the park more than doubling (Lemke et al. 1998). This restored migratory pattern probably occurred because of high elk densities within the park, acquisition and management of former agricultural land as elk winter range, and changes in the timing of hunts in Montana to enable easier movements by elk out of the park (Lemke et al. 1998). This range expansion provided additional resources for continued population growth, and elk counts increased to approximately 19,000 by the winter of 1988 (Eberhardt et al. 2007). However, between 3,000 and 5,800 elk starved during the winter of 1989 following an extended drought and major summer fires that burned approximately 34 percent of their winter range (Singer et al. 1989). Also, about 2,900 elk were harvested outside the park during the winter of 1989 and another 9,400 elk were harvested from 1990 through 1994 (Coughenour and Singer 1996; Taper and Gogan 2002). Thereafter, counts again increased to about 19,000 elk by 1994 (Eberhardt et al. 2007).

The relatively slower growth rate for northern Yellowstone elk during the years 1976 to 1994 (compared with 1965 to 1975) was interpreted as the population approaching its estimated carrying capacity of 20,000 to 25,000 elk. Carrying capacity is a term that describes the limits of available food, space, and other resources on the number of animals that can be supported by the environment. Fewer resources are available for each elk when the population is near carrying capacity, which contributed to decreases in pregnancy rates, calf survival, recruitment, and adult survival as elk counts increased toward 19,000 (D. B. Houston 1982; Coughenour and Singer 1996; Singer et al. 1997; Taper and Gogan 2002; Barmore 2003).

Environmental conditions such as spring precipitation, forage production, and snow depth also strongly influenced the population dynamics of Yellowstone elk by affecting the availability of forage and, in turn, survival and recruitment (Coughenour and Singer 1996). Increased energy expenditure and reduced forage intake during winters with deep snows can reduce the condition of pregnant females, resulting in calves with low birth weights and poor survivorship (J. G. Cook 2002; J. G. Cook et al. 2004; R. C. Cook et al. 2004). Calves are at a competitive disadvantage if they begin winter with lower fat reserves, and as a result, they are more susceptible to mortality (J. G. Cook 2002). Furthermore, colder winters, which lead to cycles of icing and thawing, can increase the hardness of the snowpack and, in turn, reduce forage availability and increase energy expenditure by ungulates (Wang et al. 2006). In addition, deep snowpack influences elk vulnerability to hunting by increasing the number of animals migrating outside the park and hunter success (P. J. White and Garrott 2005a).

In 1995 and 1996, 66 wolves were released into Yellowstone National Park and the central Idaho wilderness. This restored population rapidly increased in abundance and distribution throughout the greater Yellowstone ecosystem (D. W. Smith 2005). Elk are the most abundant ungulate in Yellowstone and comprised 90 percent of wolf kills (Mech et al. 2001; D. W. Smith 2005). Yellowstone already supported several large predators, including grizzly bears, black bears *(Ursus americanus),* cougars (also known as mountain lions or pumas; *Felis concolor*), and coyotes *(Canis latrans).* Bears were known to be significant predators of newborn elk (Singer et al. 1997), and previous studies of systems with one or two large predators suggested that wolf predation could be more effective in locales where other large carnivores were present (Crête and Manseau 1996). Thus, the debate regarding northern Yellowstone elk changed from concerns about too many elk to speculation about there being too few elk in the future to support hunting opportunities following the reintroduction of wolf predation (National Research Council 2002).

Counts of northern Yellowstone elk were approximately 17,000 when wolves were reintroduced in 1995 but decreased to approximately 12,000 in 1998 following the starvation and harvest of more than 3,300 elk outside the park during the severe winter of 1997. Counts varied between 11,500 and 14,500 elk from 1999 to 2001 but then decreased to about 6,600 elk in 2006 and fewer than 5,000 in 2011. This decrease in abundance was attributed to several factors, including the continuation of liberal harvests after wolf restoration, predation of elk by wolves and bears, a substantial winter-kill of elk during 1997, and drought-related effects on maternal condition and the survival of calves. Elk harvests were not reduced following wolf restoration to compensate for predation, and the number of permits for antler-less elk (females, calves) during a late season hunt was maintained between 2,660 in 1995 and 2,882 in 2000. This consistent and efficient harvest regime significantly decreased the survival rates of prime-age females that produce calves and, in turn, the growth rate of the population (Vucetich et al. 2005; P. J. White and Garrott 2005a; Evans et al. 2006; Eberhardt et al. 2007). In essence, hunters and wolves were in competition for decreasing numbers of elk, thereby resulting in an accelerated rate of decrease in elk abundance. However, permits were gradually reduced to 1,400 by 2004 and to 140 in 2006 through 2009 and eliminated entirely by 2010 because of decreased elk abundance and low recruitment (P. J. White et al. 2012).

Wolf predation had a lower effect on elk population growth rates than hunter harvests from 1996 to 2002 because wolf kills were concentrated on calves and older individuals that produce few calves (G. J. Wright et al.

2006; Eberhardt et al. 2007). However, wolf predation has exceeded the number of harvested elk since 2003, and the survival of adult female elk decreased from 99 to 85 percent as wolf numbers increased (P. J. White and Garrott 2005a; Hamlin et al. 2009). Recruitment also decreased from an average of 32 calves per 100 adult females (range = 6 to 48) before wolves (1968 to 1994) to 21 calves per 100 females (range = 11 to 33) after wolf restoration (1995 to 2009). However, wolf predation may compensate somewhat for elk starvation during winter, given that wolves are highly selective for younger and older elk (G. J. Wright et al. 2006; Eberhardt et al. 2007). Also, Barber-Meyer et al. (2008) reported that grizzly and black bears accounted for 60 percent of the deaths of radio-tagged elk calves during an intensive research project conducted from 2003 through 2005, while wolves accounted for only an additional 15 percent of deaths.

There is no consensus regarding the future population dynamics of elk and wolves, including to what extent losses through wolf predation will be offset by reductions in elk mortality from other causes. N. Varley and Boyce (2006) predicted that wolf predation would have a stabilizing influence on elk abundance by eliminating massive starvation during severe droughts or winters with low forage availability. However, Hamlin et al. (2009) suggested that high numbers of predators relative to elk had reduced calf survival below levels that would have been experienced in the absence of predators. Thus, elk numbers may remain low in coming years until predator numbers decrease substantially—even if human harvests remain low (P. J. White and Garrott 2005a).

BISON

Bison were nearly extirpated by the early twentieth century as a result of market hunting, with Yellowstone providing sanctuary to fewer than two dozen animals in 1901 (Meagher 1973). The population was gradually restored through supplementation, protection, husbandry, and the relocation of bison (Cahalane 1944; Meagher 1973). Supplemental feeding of hay to the northern herd during winter contributed to bison numbers increasing from 21 in 1902 to 1,200 in 1929. Park managers then began periodic culling of the northern herd that decreased abundance to fewer than 200 bison (Meagher 1973; Fuller et al. 2007). The central herd was not fed supplemental food and numbered fewer than 100 bison through the mid-1930s. To stimulate population growth, park managers augmented the central herd with 71 bison from the northern herd in 1936 (Cahalane 1944). Thereafter, the number of bison in the central herd increased to 1,300 by 1954 (Fuller et al. 2007). Park managers then began periodic culling of the central herd during 1954 to 1968 that decreased abundance to fewer than 400 bison (Meagher 1973).

Prior to 1968, bison spent winter within Yellowstone National Park, likely due to the previous decades of aggressive culling and a lack of tolerance for bison on traditional winter range outside the park (Plumb et al. 2009). Park managers ceased the culling, augmentation, and feeding of bison inside Yellowstone in 1969 (G. F. Cole 1971), and bison numbers increased to 600 in the northern herd and 1,700 in the central herd by the mid-1980s. As numbers increased, seasonal migrations became more expansive, with some bison in both herds moving from higher-elevation summer ranges to lower elevations in and outside the park during autumn through winter (Meagher 1989b, 1993; Bjornlie and Garrott 2001; Bruggeman et al. 2009). These migratory movements and range expansion provided additional forage for bison and contributed to sustained population growth in both herds (Taper et al. 2000; Gates et al. 2005; Fuller et al. 2007).

The Yellowstone bison population is infected with brucellosis, which may induce abortions or the birth of nonviable calves and can be transmitted between wildlife and livestock (Rhyan et al. 2009). Thus, between 1984 and 2000, the state of Montana culled more than 3,000 bison that left the park to prevent the possible transmission of brucellosis from bison to cattle. Also, there was evidence that the population growth rate slowed from 1970 to 2000 as numbers of bison increased to more than 3,000 animals in the central breeding herd (Fuller et al. 2007). Adult female survival decreased when the central herd exceeded 2,000 to 2,500 bison, and this decrease was exacerbated during winters with severe snowpack because more bison moved outside the park and were culled (Geremia et al. 2009). Also, bison moved from the central herd to the northern herd when resources became limited in central Yellowstone as a result of increasing bison numbers and severe snowpack (Fuller et al. 2007; Bruggeman et al. 2009).

The federal government and the state of Montana agreed to a bison management plan in 2000 that established guidelines for cooperatively managing the risk of brucellosis transmission from bison to cattle. The plan emphasized conserving the bison population and allowing some bison to occupy winter ranges on public lands in Montana (U.S. Department of the Interior, National Park Service and U.S. Department of Agriculture, Forest Service, Animal and Plant Health Inspection Service 2000). However, more than 3,600 bison were removed from the population from 2001 to 2010, with more than 1,000 and 1,700 bison being culled from the population during the winters of 2006 and 2008, respectively. These culls unintentionally removed more calf and female bison from the central herd, which, if continued over time, could result in alterations of the sex and age structure of the population and consequent changes in demographic processes that could persist for decades (P. J. White, Wallen, et al. 2011). Thus, management adjustments were developed to reduce large removals of bison and their potential long-term effects by implementing smaller, more frequent culls, including increased public and tribal hunting opportunities in Montana, selectively shipping likely infectious bison to slaughter, and relocations of brucellosis-free bison after quarantine (U.S. Department of the Interior et al. 2008).

PRONGHORN

Yellowstone pronghorn inhabit mountain slopes and valley bottoms in northern Yellowstone and adjacent areas of Montana. Thousands of pronghorn once migrated 80 to 130 kilometers along the Yellowstone River from higher-elevation summer ranges in Yellowstone to lower-elevation winter ranges in Montana (Skinner 1922). However, human settlement and market hunting reduced pronghorn abundance to approximately 200 animals by the late 1880s (Keating 2002). The U.S. Army arrived at the park in 1886 and began protecting pronghorn and initiated predator control, which led to increased (500 to 1,000) pronghorn numbers by 1900. Also, in 1902 the army began cultivating and feeding pronghorn alfalfa inside the northern boundary of the park, and abundance increased to about 2,000 pronghorn by 1908 (Keating 2002). However, a series of severe winters, starvation, and dispersal led to a decrease in abundance to 200 to 300 pronghorn by 1917 (Keating 2002).

Feeding, irrigation, fencing, and predator control were continued by the army and subsequently the National Park Service until 1934, and counts of pronghorn increased from approximately 250 to more than 600 (Keating 2002). However, these actions further reduced the distribution of pronghorn and apparently reinforced the tendency for some pronghorn to remain on their winter range year-round. Thus, the pronghorn migration from the park to the north was effectively eliminated sometime before 1920 (Skinner 1922), which also severed meaningful connectivity and genetic exchange with neighboring pronghorn populations (M. D. Scott and Geisser 1996). The purchase and annexation of boundary lands near Gardiner, Montana, for pronghorn habitat in 1932 increased the winter range for pronghorn inside the park, and counts increased to a high of 811 pronghorn between 1936 and 1946 after husbandry ceased and pronghorn were allowed to expand their distribution (Keating 2002). However, between 1947 and 1968 at least eight culls totaling 1,144 pronghorn were conducted by park staff to reduce abundance because of concerns about overgrazing, and counts decreased to approximately 200 animals (Keating 2002).

A moratorium on culling pronghorn was instituted in 1969, but counts did not increase and varied between 100 and 190 until 1981 (P. J. White, Bruggeman, and Garrott 2007). D. B. Houston (1982) speculated that culling had reduced pronghorn abundance from levels limited by resources such as food to lower levels maintained by coyote predation, vehicle strikes, and dispersal. There was also a severe drought from 1974 to 1980 and an increase in counts of elk from 4,000 to 16,000 (D. B. Houston 1982), but evidence suggests high predation rather than increasing competition for a dwindling food supply as the primary factor limiting population growth (P. J. White, Bruggeman, and Garrott 2007).

Pronghorn numbers increased rapidly to more than 594 from 1982 to 1991, possibly due to favorable weather and forage conditions, increased predator harvest outside the park, and increased irrigated cropland outside the park (Singer and Norland 1994; Keating 2002). There was evidence of a slowing in population growth as pronghorn counts exceeded 500, followed by a precipitous decrease in counts from 536 to 235 between 1992 and 1995 (P. J. White, Bruggeman, and Garrott 2007). There was some indication that this crash in pronghorn numbers was related to diminished food resources on the winter range (Boccadori et al. 2008). Intense browsing associated with the rapid growth in elk numbers contributed to a decrease in sagebrush on the pronghorn winter range, which was the major component of pronghorn diets and has high protein content during winter (Singer and Norland 1994; Singer and Renkin 1995; Wambolt and Sherwood 1999). The portion of sagebrush in winter diets of pronghorn decreased from approximately 70 percent from 1985 to 1988 (Singer and Norland 1994) to less than 10 percent from 2000 to 2001 (Boccadori et al. 2008). Pronghorn numbers have remained at a relatively low level (200 to 300) following this crash in abundance, suggesting that the winter range may not support populations of more than 500 pronghorn for sustained periods, as occurred historically (P. J. White, Bruggeman, and Garrott 2007; Boccadori et al. 2008).

The dynamics of the pronghorn population support the idea that irruption, or periods of sustained population growth resulting in high numbers followed by substantial mortality and rapid decreases in numbers, is a natural pattern of growth when forage and weather conditions become favorable after range expansion or release from harvesting (Caughley 1979; Forsyth and Caley 2006; P. J. White, Bruggeman, and Garrott 2007). Yellowstone pronghorn have high fecundity, with most does older than two years twinning each year (O'Gara 1968). The irruption between 1982 and 1991 was spurred by high recruitment (80 fawns per 100 does) during 1982 and 1986, whereas recruitment decreased precipitously to between 8 and 22 fawns per 100 does during the crash from 1992 to 1995.

BIGHORN SHEEP

Bighorn sheep *(Ovis canadensis)* in the northern portion of Yellowstone and nearby areas of Montana are organized into 10 to 13 bands with movements and gene flow among them (D. B. Houston 1982; Keating 1982; Legg 1996; Ostovar 1998). Most of these sheep are migratory and spend winter in lower-elevation areas before moving to higher-elevation summer ranges in May through October. Approximately 500 sheep were counted during 1978 and 1981, but an outbreak of infectious keratoconjunctivitis or "pinkeye" caused by the protozoan *Chlamydia* resulted in the deaths of at least 60 percent of these sheep during the winter of 1982 (Meagher et al. 1992).

Bighorn sheep counts did not increase during the next 13 years, even though there was no sign of pinkeye in sheep (P. J. White et al. 2008). Many bighorn sheep populations that spend winter at high elevations are small and slow growing (Buechner 1960; D. R. Stevens and Goodson 1993; Wishart et al. 1998). Adult survival is not affected substantially by changes in population density and, as a result, year-to-year variations in lamb and yearling survival profoundly affect population dynamics (Jorgenson et al. 1997). Bighorn sheep in northern Yellowstone do not display vigorous recruitment. Since 1992, recruitment has ranged from 6 to 32 lambs per 100 ewes (average = 23), with poor recruitment in many years due to recurring lamb pneumonia and starvation in this deep-snow environment (P. J. White et al. 2008). Low recruitment limits the ability of the population to recover quickly from decreases in abundance caused by disease outbreaks or starvation during severe winters.

Also, competition with abundant elk (14 to 16 per square kilometer) may have hampered the recovery of the bighorn sheep population after the pinkeye epidemic. Bighorn sheep and elk have similar diets and habitat use in northern Yellowstone, which possibly resulted in lower forage intake rates for bighorn sheep (Singer and Norland 1994). Also, large numbers of elk spending winter in northern Yellowstone provided carrion that supported abundant coyotes, golden eagles *(Aquila chrysaetos)*, and cougars that also preyed on bighorn sheep (Ostovar and Irby 1998). The bighorn sheep population increased slowly during the decade following the 1995 to 1997 restoration of wolves to Yellowstone. Counts and recruitment for bighorn sheep decreased due to a severe winter in 1997 that led to substantial starvation and poor maternal condition and high lamb mortality the following year (Ostovar and Irby 1998). During the years 1998 to 2005, however, recruitment increased to an average of 30 lambs per 100 ewes, and there was a slow to moderate increase (2 to 11 percent) in bighorn counts (P. J. White et al. 2008). Survival of adult bighorn sheep was high (89 to 97 percent) in the presence of wolves as bighorn sheep comprised less than 1 percent of wolf kills (D. W. Smith 2005). Bighorn sheep may escape attention by wolves because they are much less abundant than elk and inhabit steep, rugged terrain where they are difficult to capture.

During the decade after wolf restoration, elk numbers decreased 60 percent (P. J. White and Garrott 2005a, 2005b; Eberhardt et al. 2007), while counts of bighorn sheep increased slowly after the severe winter of 1997.

Approximately 350 bighorn sheep were counted in the northern Yellowstone area from 2008 through 2011. It is enticing to suggest that increasing bighorn sheep numbers are due to decreased competition for resources with elk. However, continued monitoring across a greater range of bighorn sheep densities will be necessary to evaluate this relationship (Ostovar and Irby 1998; P. J. White et al. 2008).

GRIZZLY BEARS

By the 1950s, the greater Yellowstone ecosystem was one of the last areas inhabited by grizzly bears in the lower 48 states (Craighead and Craighead 1967). From 1959 to 1967, approximately 234 to 360 bears occupied 5 million acres centered on Yellowstone National Park (Cowan et al. 1974; Craighead et al. 1974; G. F. Cole 1976). Human-caused grizzly bear mortality averaged 15 bears per year, which appeared sustainable because abundance was relatively stable (Cowan et al. 1974; Craighead et al. 1988). Most of the grizzly bears killed by humans during this period had become conditioned to human foods and garbage in developments and campgrounds (Schullery 1992).

To break this association, four garbage dumps that had aggregations of grizzly bears in Yellowstone and one in Montana were closed between 1968 and 1971 (Haroldson et al. 2008; Meagher 2008). Also, regulations prohibiting the feeding of bears were strictly enforced and bear-proof garbage cans and dumpsters were placed throughout the park (G. F. Cole 1976). These actions eliminated most human food and garbage from bear diets and, over time, substantially reduced bear-inflicted human injuries, property damage, and management removals of problem bears (Meagher and Phillips 1983; Gunther 1994; Haroldson et al. 2008). However, in the short term, the closing of the garbage dumps increased bear deaths because they had to travel over a much larger area to forage, which made them more susceptible to human sources of mortality. From 1967 to 1972, a minimum of 212 grizzly bears were known to have died because of human causes, largely due to management removals of problem bears, private citizens defending life and property, and hunting outside of the park (Craighead et al. 1988). Also, the elimination of garbage as a stable, dependable food source that buffered bears against occasional shortages of natural foods resulted in decreased litter sizes and cub survival (Craighead et al. 1995; Haroldson et al. 2008).

A grizzly bear sow leans back against a fallen tree and nurses her cubs while she keeps a vigilant watch for any threats. The population dynamics of Yellowstone's iconic wildlife varies among species, depending on their life history characteristics. Reproductive rates of grizzly bears are relatively modest compared with other wildlife species because grizzly bears are slow to become sexually mature and invest a great deal of time and energy in raising their young. A mother will generally keep her cubs with her for two winters following their birth. *Photo by Cindy Goeddel.*

Mortality remained high from 1972 to 1974, with at least 59 bears being killed by people, resulting in bear abundance in the greater Yellowstone ecosystem decreasing to a range between 136 and 218 animals (Cowan et al. 1974; Craighead et al. 1974, 1988, 1995). As a result, grizzly bears in the greater Yellowstone ecosystem were designated a federally threatened species in 1975 because of the high level of human-caused bear mortality combined with low reproduction, loss of habitat, and significant habitat alteration (U.S. Fish and Wildlife Service 1975). From 1975 to 1979, the last two remaining garbage dumps in the Yellowstone area that had aggregations of bears were closed, 45 more grizzly bears died from human causes, and the population decreased to 99 bears.

By the early 1980s, management to reduce bear–human conflicts began to pay dividends, and the population started to recover, with approximately 120 bears in 1985, 204 bears in 1990, 259 bears in 1992, and 344 bears in 1998 (Haroldson et al. 1998; Haroldson 1999). By 2004, there were about 431 grizzly bears, and they had expanded their range by over 48 percent to more than 8.5 million acres (Haroldson and Frey 2005; Schwartz et al. 2002; Schwartz, Haroldson, et al. 2006). Female survival is highest within Yellowstone National Park and on surrounding Forest Service lands with secure habitat (Haroldson et al. 2006). Conversely, bear survival decreases as road density and the number of developed sites increase. Grizzly bear numbers inside Yellowstone increased, and as they approached the capacity of the environment to support them, there was a decrease in litter size, higher starvation, and more predation of cubs (Schwartz, Haroldson, and Cherry 2006; Schwartz, Haroldson, and White 2006; Haroldson et al. 2008). In 2005, the U.S. Fish and Wildlife Service designated the grizzly bear population in the greater Yellowstone ecosystem as a distinct population segment on the basis that it is geographically isolated from all other grizzly bear populations, it has different genetic characteristics, and its bears have a tendency to feed on ungulates as a primary source of nutrition (Mattson 1997; Paetkeau et al. 1998; Jacoby et al. 1999; C. R. Miller and Waits 2003). The Fish and Wildlife Service also published a rule to remove these bears from the protection of the Endangered Species Act in 2007 (U.S. Fish and Wildlife Service 2007a), but this action was challenged in court. In September 2009, the U.S. District Court of Montana vacated the rule removing the Yellowstone grizzly bear population segment from the list of threatened species. The government challenged this decision before the 9th U.S. Circuit Court of Appeals, but the decision was upheld, in part, during November 2011. The total grizzly bear population in the greater Yellowstone ecosystem was estimated at 602 in 2010 (Haroldson 2011). Grizzly bears currently occupy over 14 million acres in the greater Yellowstone ecosystem.

There is some concern about the future for grizzly bears in the greater Yellowstone ecosystem because of recent decreases in some of their foods that could potentially affect reproduction and survival. Cutthroat trout *(Oncorhynchus clarkii)* were once a high-quality and highly preferred spring food for grizzly bears with home ranges adjacent to Yellowstone Lake (Haroldson et al. 2005). Mortality caused by non-native lake trout and whirling disease *(Myxoblus cerebralis)*, combined with reductions in recruitment caused by drought, have reduced the cutthroat trout population to less than 10 percent of historical numbers (Koel et al. 2005). From 2007 through 2009, no cutthroat trout were detected in bear scat collected around the lake (Fortin 2011), indicating they are no longer an important diet item. Also, whitebark pine *(Pinus albicaulis)* seeds are a high-quality, highly preferred autumn food for grizzly bears in the greater Yellowstone ecosystem. However, whitebark pine stands have experienced significant mortality in recent years, and 69 percent of tagged trees used to index cone production in 2002 were dead by 2009 (Haroldson and Podruzny 2011). That said, grizzly bears are generalist omnivores (i.e., they eat both plants and animals) that exhibit significant flexibility in their diet, and their numbers and distribution have continued to increase despite these changes in food availability (Schwartz et al. 2002; Schwartz, Haroldson, et al. 2006; Harris et al. 2006).

Though reproduction in grizzly bears is still linked to natural factors, such as food abundance, survival is now more driven by human factors, such as numbers of developments, people, and roads, and the amount of protected habitat (Schwartz, Haroldson, and West 2010; Schwartz, Haroldson, and White 2010). Yellowstone National Park and surrounding Forest Service lands have successfully conserved bears that are now venturing outside these protected areas and expanding the distribution of the species (Schwartz, Harris, and Haroldson 2006). In the greater Yellowstone ecosystem, about 85 percent of known bear deaths are due to humans (Haroldson et al. 2006). Thus, human values more than environmental factors will likely determine where and how many grizzly bears persist in the greater

Yellowstone ecosystem (Gunther 2008; Schwartz, Haroldson, and White 2010). In turn, grizzly bears will likely always require intensive management of human–bear interactions to conserve the species in today's society (J. M. Scott et al. 2005).

TRUMPETER SWANS

Trumpeter swans were nearly extirpated in North America by 1900 due to a combination of overharvest and habitat destruction (Banko 1960). However, a small group of swans survived by remaining year-round in the vast wilderness of the greater Yellowstone ecosystem. These swans helped facilitate the restoration of the species in the region, along with harvest regulations and habitat protection (U.S. Fish and Wildlife Service 1998). Trumpeter swans in Yellowstone National Park are part of the Rocky Mountain population, which includes genetically similar subpopulations that breed elsewhere in the greater Yellowstone ecosystem or in western Canada from the southeastern Yukon territories to eastern Alberta (Oyler-McCance et al. 2007). Both subpopulations use common wintering sites in the greater Yellowstone ecosystem. Yellowstone National Park supports resident, nonmigratory trumpeter swans through the year, as well as regional migrants from the greater Yellowstone ecosystem and longer-distance migrants from Canada during winter.

The number of trumpeter swans that resided and nested in Yellowstone National Park increased to between 40 and 50 adults during the 1940s, 50 and 60 adults by the 1950s, and a high of 69 adults in 1961. Cygnet production was typically 10 to 20 cygnets per year during the 1950s before decreasing to fewer than 10 cygnets per year during the 1960s and fewer than five cygnets per year thereafter. As a result, numbers of resident swans in the park gradually but consistently decreased to fewer than 10 swans by 2007 (Proffitt, McEneaney, et al. 2009; Baril et al. 2010). This decrease was more rapid following the termination of supplemental feeding and draining of winter ponds at the Red Rock Lakes National Wildlife Refuge in the Centennial Valley of Montana—located about 120 kilometers west of Yellowstone (Proffitt, McEneaney, et al. 2009). Supplemental feeding occurred during winter from 1936 to 1992, and during the late 1980s and early 1990s trumpeter swans at the refuge were relocated to other regions, wintering ponds were drained, and swans at the refuge and elsewhere were hazed to reduce winter concentrations (U.S. Fish and Wildlife Service 2003). The population history of trumpeter swans in Yellowstone and adjacent areas suggests that swan dispersal from the larger Rocky Mountain subpopulation in the greater Yellowstone ecosystem may be an important factor for maintaining resident swans in Yellowstone by filling vacant territories or providing mates for single adult birds (Banko 1960; Gale et al. 1987; McEneaney 2006).

The decrease in resident trumpeter swans in Yellowstone did not appear to be due to competition with migrants to the park in the winter that could potentially reduce food resources for resident swans during breeding (U.S. Fish and Wildlife Service 1998). The abundance of migrant swans that spent winter in the park increased from 1967 to 1992 but leveled off following the termination of supplemental feeding at Red Rock Lakes National Wildlife Refuge and hazing operations outside the park (Proffitt, McEneaney, et al. 2009). Despite this stabilization in numbers of migrant swans during winter, the resident flock continued to decrease in abundance because of low nesting and fledging success. Trumpeter swan productivity in Yellowstone was consistently low from 1987 to 2007, with an average of only three cygnets per year surviving until their first autumn (Proffitt, McEneaney, et al. 2010). Average clutch size in the park (four) was lower than clutch sizes in other parts of the greater Yellowstone ecosystem and Canadian breeding territories (Gale et al. 1987; Proffitt, McEneaney, et al. 2010). Also, only about one cygnet per pair was fledged in Yellowstone from 1977 to 2007 (Shea 1979; Proffitt, McEneaney, et al. 2010), which is low compared with other subpopulations in the Rocky Mountain region (Gale et al. 1987).

The reproductive success of resident trumpeter swans in Yellowstone is strongly influenced by annual variations in environmental conditions and naturally occurring events, such as severe winters and droughts. Fifty-three percent of egg failures between 1987 and 2007 were due to nest flooding (Proffitt, McEneaney, et al. 2010). Thus, clutch sizes and fledging success were higher in years with lower April precipitation, and the abundance of resident swans in the park increased following drier springs that reduced flooding and increased the availability of pre-nesting food resources (Proffitt, McEneaney, et al. 2009, 2010). Also, a lower portion of adult swans attempted to nest during colder springs in May and June. Conversely, cooler summers prevented wetlands used for foraging and nesting from drying until later in the season and increased aquatic plant production (Proffitt, McEneaney, et al. 2009, 2010). Molting adult trumpeter swans and cygnets are flightless during a good portion of the summer and rely on ponds for protection from predators.

In addition, long-term reductions in wetlands due to a warmer, drier climate over the past 40 years (Wilmers and Getz 2005; Pederson et al. 2010) reduced the amount of nesting habitat for trumpeter swans and led to decreased reproductive success. Nesting habitats for swans in the park are separated and differ in physical characteristics such as size, depth, and elevation. Since 1931, resident swans have nested in at least 94 wetlands in 18 different areas. However, less than 20 percent of the wetlands used by swans for nesting contribute to more than 60 percent of all fledged young (Proffitt, McEneaney, et al. 2010). In addition, clutch sizes and fledging success were higher at nesting sites within larger wetland areas that may have more abundant food resources, which could increase the ability of females to accrue nutrients prior to laying eggs and, in turn, increase clutch size (Proffitt, McEneaney, et al. 2010).

Furthermore, increased predation due to the recovery of predator populations in Yellowstone may have resulted in decreases in egg or cygnet survival and contributed to the long-term population decrease of resident trumpeter swans (McEneaney 2006). The abundance of predators, such as grizzly bears, black bears, coyotes, ravens *(Corvus corax),* eagles, and wolves, has increased substantially since the 1970s (Stahler et al. 2002; D. W. Smith 2005; Schwartz, Haroldson, et al. 2006). Fifty-four percent of trumpeter swan nests hatched at least one cygnet during 1977 and 1978, with only one of 26 nests failing because of predation by a grizzly bear (Shea 1979). However, nesting success from 1987 to 2007 was only 32 percent, with 41 percent of all egg failures and 18 detected incidents of pre-fledging cygnet mortality attributed to predation (Proffitt, McEneaney, et al. 2010).

In summary, evidence suggests that Yellowstone provides marginal conditions for nesting, and the resident flock is sustained by swans dispersing from more productive areas within the greater Yellowstone ecosystem. This effect has been compounded over the last several decades by decreases in wetlands as a result of drought and the recovery of predator populations. Thus, barring aggressive interventions such as predator-proof fencing of wetlands or manipulations of water levels that are not aligned with cur-

A nesting trumpeter swan patiently sits while her mate keeps watch in the water nearby. Trumpeter swan reproduction in Yellowstone is a major conservation concern because the number of nesting swans has plummeted from a peak of approximately 60 adults in the early 1960s to fewer than 10 individuals in recent years. The causes of the recent declines are uncertain but likely involve a number of mechanisms, including changes in swan management beyond the boundaries of Yellowstone National Park. *Photo by Cindy Goeddel.*

rent National Park Service guidelines to minimize human intervention, trumpeter swan presence in the park may be limited to occasional residents and wintering aggregations of migrants from outside the park (Proffitt, McEneaney, et al. 2009, 2010).

MOUNTAIN GOATS

Since 1990, mountain goats *(Oreamnos americanus)* from Montana have successfully colonized Yellowstone National Park. Colonists in the northeast portion of the park are likely descendents of mountain goats that were introduced in the Absaroka Range between 1956 and 1958 and then colonized areas south and east toward the park (Lemke 2004). Colonists in the northwest portion of the park are likely descendents of mountain goats that were introduced at the north end of the Madison Range on several occasions between 1947 and 1959, from the Absaroka Range, or both sources (Lemke 2004).

There is no evidence that mountain goats are native to Yellowstone National Park (Laundré 1990; Walpole 1997; Schullery and Whittlesey 2001). Historical accounts of mountain goat presence in the region prior to the known introductions described above consist of one possible sighting by unreliable observers and a few casual mentions of mountain goat presence by people of limited or unknown familiarity with the ecosystem (Schullery and Whittlesey 2001). Native mountain goat range most closely approached Yellowstone National Park to the west near the Montana–Idaho border in the Centennial Mountain Range, but no modern authority claims that mountain goats were resident in the area now encompassed by Yellowstone National Park during recent centuries (Schullery and Whittlesey 2001).

Since the late 1980s, mountain goat numbers have increased substantially in the Absaroka and Gallatin Mountains north of Yellowstone National Park. In turn, aerial counts of mountain goats inside or within 1 kilometer of the park increased from 24 to 178 animals from 1997 to 2009. The highest mountain goat densities in the park occur along the Gallatin Crest Divide, on the northwest boundary, and Cutoff Mountain, Barronette Peak, and the Thunderer, near the northeast boundary (Lemke 2004). Mountain goats have established breeding populations in the northeast and northwest portions of the park, with relatively high recruitment (23 to 36 kids per 100 adults). This is typical of colonizing populations with relatively younger age structures and high population growth rates (Eberhardt 1985).

The management objectives for mountain goats in Montana and Wyoming are to sustain populations in suitable habitats while providing a conservative harvest (Lemke 2004). In response to increasing mountain goat abundance adjacent to Yellowstone, Montana Fish, Wildlife, and Parks established two new hunting districts north of the park (Lemke 2004), and Wyoming Game and Fish enlarged a hunt area to harvest mountain goats in drainages bordering the park boundary. These harvests may retard immigration of mountain goats into the park, but current hunting intensity in these boundary areas is unlikely to reduce mountain goat abundance. Thus, resident mountain goats have become established in the park, and mountain goats from populations in areas of Montana and Wyoming outside the park will likely continue to disperse into the park for the foreseeable future. Lemke (2004) speculated that habitats inside and within 1 kilometer of Yellowstone National Park could support 200 to 300 mountain goats.

The recent colonization and rapid increase in abundance and distribution of mountain goats in and near Yellowstone raises important policy and management questions for the park. Current National Park Service (2006) policy directs that all exotic, non-native species that are not maintained to meet an identified park purpose will be managed—up to and including eradication—if control is prudent and feasible and the exotic species interferes with natural processes and the perpetuation of natural features, native species, or natural habitats. Thus, natural resource managers are currently working to evaluate the species' effects on resources and develop management plans based on these findings (Garrott et al. 2010). Studies of alpine vegetation in the northeast portion of the park (Aho and Weaver 2003) suggest that ridgetop vegetation cover is lower, and barren areas along alpine ridges are more prevalent, in areas with high mountain goat use. In addition, Laundré (1990) predicted that resource competition from mountain goats would negatively affect bighorn sheep. N. Varley (1996) noted that mountain goat and bighorn sheep diets overlapped substantially in the northeast portion of the park, where mountain goat numbers are high.

CONSERVATION IMPLICATIONS

Yellowstone's natural state is a dynamic one, and over long periods the park has experienced changing climate, major geological processes, and "large-scale disturbances, including fire, floods, blowdowns, fluctuations in abundances of ungulates and predators, and outbreaks of diseases and insects that affect plants and animals" (National Research Council 2002, 120). Animal abundance is influenced by the amount of food available for each individual, variations in weather, and competitive or symbiotic relationships with other species. In addition, the park is part of the larger greater Yellowstone ecosystem where human activities are increasing (National Research Council 2002). Because many animals migrate in and out of the park, their range often spans areas where they are exposed to different management actions, such as hunting, and different land-use practices, such as agriculture and housing development, than inside the park.

Ecological process management in Yellowstone has been effective at sustaining dynamic processes and variation in wildlife populations. Thus, park managers "could continue to let the populations of animals fluctuate without any direct controls, letting a combination of weather, predators, range conditions, and outside-the-park hunting, land uses, and management actions by state agencies influence population numbers" (National Research Council 2002, 131). Some native species, such as trumpeter swans and pronghorn, could be lost (local extinction) or persist at minimal numbers with this hands-off approach. Thus, park managers will need to decide whether to intervene to protect and enhance the population growth of certain species (National Research Council 2002). Also, there is persistent controversy and conflict in boundary areas regarding species such as bison, grizzly bears, and wolves, and their population dynamics have been strongly influenced by human-induced mortalities. Furthermore, the effects of warming climate have already been detected, and "it is not too soon for managers to start thinking about how to deal with potential changes" (National Research Council 2002, 137; Wilmers and Getz 2005).

CHAPTER FIVE

Predation

Wolf Restoration and the Transition of Yellowstone Elk

P. J. WHITE

ROBERT A. GARROTT

PREDATION IS AN interaction in which one organism (the predator) attacks and feeds on another (the prey). During the 1930s, biologists were convinced that predators kept ungulate populations at low numbers (Leopold 1933). However, as predators became scarce due to control by humans, there was increasing concern about the role of predators in biological communities (Leopold 1949). Calls to protect predators emerged after Errington (1945, 1946) proposed that predators mostly removed prey animals that would have died anyway due to starvation or other factors, such as severe weather events. Wildlife populations produce a surplus of young that cannot be supported by the resources (food, habitat, water) available to them and are doomed to die (Darwin 1859). If predation removes these doomed individuals, then it simply compensates, or is a substitute, for the other forms of mortality that would have killed prey in the absence of predators (known as compensatory mortality). Forage productivity and availability also affect the degree to which predation reduces prey survival. Animals living in crowded conditions (high densities) are generally in poorer physical condition as a result of competition for limited food and are more susceptible to predation compared with animals in low-density populations that are in better condition (Kie et al. 2003). Thus, more deaths are compensatory as the number of animals living in an area increases to a level (known as carrying capacity) where there are not enough resources in the environment to sustain them all (Kie et al. 2003).

This perception that predators did not have significant influence on prey abundance became dogma during the 1940s through the 1960s. However, Keith (1974) proposed that Errington (1945, 1946) had oversimplified predator–prey dynamics by drawing on studies of only a few species such as muskrats *(Ondatra zibethicus)* and bobwhite quail *(Colinus virginianus)*. Animals in populations that are at low density and well below the food-limited capacity of environments to sustain them are generally in "good physical condition and have high rates of productivity and survival because most females conceive and have adequate body reserves to successfully provision young" (Kie et al. 2003, 313, 321). If predators kill numbers of animals that would not have died from starvation, then these deaths are additive to other forms of mortality and will reduce prey abundance below the level that would have been realized without predators (Keith 1974). Thus, whether mortality is compensatory or additive likely changes with the number of animals living in an area and whether their abundance is near the capacity of resources in the environment to sustain them all (Kie et al. 2003). Indeed, studies of large predators and their ungulate prey during the 1970s and 1980s indicated that predators frequently influenced the abundance of their prey (Gasaway et al. 1983, 1992; Bergerud and Elliot 1986; Ballard et al. 1987, 1991; Messier 1994). Thus, predators are now considered important elements of biological systems whose influence cannot be ignored (Crête and Manseau 1996; Sæther 1997; Gaillard et al. 1998, 2000; Hayes et al. 2003).

Interest in predator–prey interactions has focused on (1) predator responses to changes in prey abundance, (2) the effects of predators on prey abundance and distribution, and (3) the effects of these responses on ecosystems. When prey are abundant compared with the number of predators in the system, each predator should kill and consume a higher number of prey than when prey are scarce, and the predator population should have high survival and reproduction (Solomon 1949; Holling 1959a, 1959b). Conversely, when predator kill rates are low because prey is scarce, then predator nutrition and body condition should be poor, which leads to lower survival, reproduction, and abundance.

Distinguishing the effects of different predators on populations of prey is challenging in systems with multiple predator and prey species. Competition for prey can reduce the efficiency by which predator species access and use shared resources (Begon et al. 2005). These effects can alter the abundance, behavior, and distribution of some predator species, which, in turn, can influence their predation rates on particular prey (Polis et al. 1989, 1997). Also, similarity among different species of predators in their selection of prey can result in one predator species replacing another to some extent in its effect on prey survival, which is known as functional redundancy (Chalcraft and Resetarits 2003). For example, Griffin et al. (2011) reported that cougar predation decreased as wolves and grizzly bears were restored to systems in the northwestern United States.

Furthermore, prey species could alter their behavior, distribution, and resource selection in response to predation risk (Gower, Garrott, and White 2009; Gower, Garrott, White, Cherry, and Yoccoz 2009; Gower, Garrott, White, Watson, et al. 2009). These anti-predator responses could affect the foraging behavior of prey and lead to inadequate nutrient intake (nutrition), which in turn could decrease body condition, pregnancy, and survival. Thus, many ecologists are interested in the consequences of anti-predator strategies used by prey in response to predation risk (Preisser and Bolnick 2008).

The high mountains and plateaus of Yellowstone National Park provide summer range for an estimated 20,000–25,000 elk from eight populations, most of which spend winter at lower elevations outside the park. Only part of the northern Yellowstone elk population and the entire Madison headwaters elk population spend winter in the park. The northern population is the largest in Yellowstone and spends winter on a series of interconnected, broad grassland-sagebrush valleys, known as the northern range (1,500 square kilometers total), that extend along the northern boundary of the park and into Montana (D. B. Houston 1982; Lemke et al. 1998). Northern Yellowstone elk migrate seasonally, moving to higher-elevation ranges during summer (Craighead et al. 1972; P. J. White et al. 2010). Hunting in the park is prohibited, but elk that spend winter on the portion of the northern range outside the park are exposed to archery and rifle hunts during autumn and early winter (P. J. White and Garrott 2005a). In contrast, the Madison headwaters population is not migratory and occupies 280 square kilometers in the upper Madison River drainage in the west-central portion of the park throughout the year (Garrott and White 2009). Therefore, hunting does not influence the dynamics of this population. The Madison headwaters area is characterized by much deeper snowpack than the northern range and also contains the major geothermal basins in the park. The heat associated with these areas provides considerable relief from deep snowpacks for many wintering ungulates.

Numbers of northern Yellowstone and Madison headwaters elk were likely near the food-limited carrying capacity of the environment to sustain them prior to wolf restoration (Coughenour and Singer 1996; Taper and Gogan 2002; Garrott et al. 2003; Garrott, White, and Rotella 2009a). As prey density approaches carrying capacity, animals increasingly compete for limited food, which leads to a decrease in nutrition and body condition (J. G. Cook et al. 2004; R. C. Cook et al. 2004). Such populations are also more influenced by weather extremes, such as severe winters and summer droughts, that further reduce forage availability. Breeding females are abundant, but recruitment and survival tend to decrease because of the females' poor condition (Kie et al. 2003). Thus, the survival of adult female elk in northern Yellowstone decreased as density approached carrying capacity due to starvation of older animals and hunter harvest of elk that migrated out of the park. Also, decreases in pregnancy rates, calf survival, and recruitment were detected as elk counts increased and more young born at high-population densities were predisposed to mortality (D. B. Houston 1982; Coughenour and Singer 1996; Singer et al. 1997; Taper and Gogan 2002; Kie et al. 2003). Survival and pregnancy rates of adult females in the Madison headwaters remained high in the absence of hunting, but recruitment was highly variable from year to year due to annual variation in depth and duration of snowpack, with the hardest winters resulting in the virtual elimination of that year's calves and many older elk as a result of overwinter starvation (Garrott et al. 2003; Garrott, White, and Rotella 2009a). Thus, much of the predation on young and older elk in these populations was likely compensatory because the animals would have died anyway (Kie et al. 2003).

Between 1995 and 1997, wolves were restored to Yellowstone and rapidly increased in abundance and distribution. Wolves using the winter range for northern Yellowstone elk increased to a high of 10 packs totaling

106 animals by winter 2004 (D. W. Smith 2005). At that time, the northern portion of the park supported one of the highest densities of native, free-ranging ungulate assemblages in North America. Elk were numerically dominant at 10 to 16 per square kilometer (Eberhardt et al. 2007). Other sympatric ungulates included bighorn sheep, bison, moose *(Alces alces)*, mule deer *(Odocoileus hemionus)*, pronghorn, and white-tailed deer *(Odocoileus virginianus)*, all numbering less than about 3,000 in combination (P. J. White and Garrott 2005a). These ungulates, in turn, already supported a diverse predator–scavenger complex, including black bears, bobcats *(Lynx rufus)*, cougars, coyotes, grizzly bears, red foxes *(Vulpes vulpes)*, and wolverines *(Gulo gulo)*.

In 1997, a resident wolf pack became established in the Madison headwaters area and grew to 45 wolves in four packs by 2004 (D. Smith et al. 2009). During that time, the Madison headwaters supported approximately 500 to 800 elk (Eberhardt et al. 1998; Garrott et al. 2003). Other ungulates included a few mule deer and moose and approximately 300 to 1,200 bison, as increasing numbers of bison migrated into the area for winter as snowpack accumulated on their higher-elevation summer ranges (Bjornlie and Garrott 2001). Grizzly bears were present from early spring through autumn, coyotes were common throughout the year, and cougars and black bears were rare.

There is compelling evidence that wolf predation limits or reduces ungulate numbers in North America but considerable debate over whether wolves can keep ungulate numbers at suppressed densities (Gasaway et al. 1983, 1992; Ballard et al. 1987, 1991; Dale et al. 1994; Hayes et al. 2003). That is, are there mechanisms that reduce predation as prey density decreases, especially at low-prey densities (Messier 1994; Murdoch 1994; Sinclair 2003)? Much of the research on this question has occurred in moose–wolf systems of northern North America. Messier (1994) and Hayes and Harestad (2000) concluded that wolf predation alone could keep moose at low densities, whereas others concluded that only when both wolf and bear predation were combined could predators keep moose at low densities (Gasaway et al. 1992; Ballard and Van Ballenberghe 1997; Crête 1998, 1999; Boertje et al. 2009). There is even less consensus regarding whether wolves can keep elk at low densities, with some studies finding strong effects of wolves (Jędrzejewski et al. 2002; Hebblewhite et al. 2005) while others reported weak or uncertain effects (Vucetich et al. 2005; Hebblewhite and

Smith 2010). There has been more than 15 years of research on this question in Yellowstone by federal, state, and academic ecologists since wolf restoration. Herein, we examine the responses of wolves restored to a high-density ungulate prey base in Yellowstone that had not been exposed to wolves for 70 years, synthesizing and interpreting intensive research conducted by a diverse group of ecologists on the effects of wolf predation on their primary prey, elk.

PREY SELECTION BY WOLVES

Wolves in northern Yellowstone preyed primarily on elk, which comprised more than 90 percent of kills during winter and approximately 85 percent of kills during spring and summer (Mech et al. 2001; D. Smith et al. 2004, 2011). Overall, calves were the preferred prey and comprised 22 percent of kills during late winter and spring, 62 percent during summer, and 49 percent during early winter (Metz et al. 2012). However, wolf prey selection shifted somewhat after 2000 to include more adult bull elk, which comprised about 40 percent of kills during late winter and spring but only 18 percent of kills during summer and early winter (Metz et al. 2012). Adult female elk comprised about 30 percent of wolf kills during winter and spring but only 16 percent of kills during summer (Metz et al. 2012). Yearling elk comprised 3 percent of kills by wolves, while bison comprised 1 to 4 percent of kills. There was an increase in kills of deer from less than 2 percent during winter to 10 to 15 percent during spring and summer, when deer migrated from their lower-elevation winter ranges in Montana and Wyoming to higher-elevation summer ranges in Yellowstone, where more wolves had territories (Metz et al. 2012). Wolves acquired about 7 percent of their food annually from scavenging (Metz 2010).

Prey selection by wolves in the Madison headwaters during winter was influenced by the abundance of predators and prey types and the duration of snowpack (Becker, Garrott, White, Gower, et al. 2009). Wolves strongly preferred elk calves, because of their high vulnerability relative to other prey types, but killed more adult elk, calf bison, and adult bison as wolf numbers increased. Predation of adult elk and bison also increased with the duration and accumulation of snow, likely due to its debilitating influence on ungulate condition, which in turn increased their vulnerability

The reintroduction of wolves into Yellowstone National Park beginning in 1995 reestablished an important predator–prey process that had been absent from the ecological community for approximately seven decades after wolves were extirpated from the park, when wildlife managers were focused on the restoration of large ungulate populations that had been decimated by commercial exploitation. In the winter of 2008, a black Druid Pack female and a gray male interloper tested this bull elk, who successfully stood his ground and kept facing the wolves. *Photo by Cindy Goeddel.*

to predators. Wolves also killed more bison as their numbers increased relative to elk, but there was not a corresponding decrease in wolf preference for elk (Becker, Garrott, White, Gower, et al. 2009).

In summary, wolves in Yellowstone commonly selected for old age, juvenile, and injured or weakened prey, such as adult male elk after the rut (D. W. Smith 2005). Thus, prey selection was strongly influenced by prey vulnerability, which is determined by the behavioral and physical attributes of prey, such as armament (antlers, horns), body condition, grouping tendencies, habitat selection, size, speed, and vigilance (Becker, Garrott, White, Gower, et al. 2009). Weather events such as drought and snowpack that influenced anti-predator responses and nutrition also affected prey vulnerability (Becker, Garrott, White, Gower, et al. 2009). However, wolves also killed a significant number of prime-age females (3 to 15 years old) each winter (P. J. White and Garrott 2005a; Becker, Garrott, White, Gower, et al. 2009).

WOLF KILL RATES

The number of ungulate prey (90 to 96 percent elk) killed per wolf per day in northern Yellowstone was lowest in early winter (0.041), increased gradually through June when more neonates were eaten (0.070), and then decreased (Metz et al. 2012). In contrast, the biomass (kilograms) of prey killed per wolf per day was highest in March through May (8.5), decreased through July (4.1), and then increased through winter (Metz et al. 2012). Thus, rates of biomass acquisition by wolves during summer were only about 50 percent of those during mid- to late winter (Metz et al. 2012). The number and biomass of ungulate prey consumed by wolves during late winter (March) was relatively constant across a wide range of elk densities (D. Smith et al. 2011), suggesting that elk were sufficiently abundant in northern Yellowstone to support relatively high acquisition and kill rates by wolves.

In contrast, the number of elk consumed by wolves in the Madison headwaters decreased significantly as elk abundance decreased to low densities (Becker, Garrott, White, Jaffe, et al. 2009), similar to findings from other multiple-prey systems where elk were the primary prey (Hebblewhite et al. 2003). Kill rates of bison in the Madison headwaters area increased with deepening snowpack as more nutritionally deprived bison migrated into the area from higher elevations. Also, kill rates of bison were higher when calves were more numerous. Bison kills did not reduce kill rates of elk but instead increased total kill rates (Becker, Garrott, White, Jaffe, et al. 2009).

The highest kill rates (0.058 to 0.061 kills per wolf per day) for primarily elk (90 to 100 percent of kills) during winter in Yellowstone were remarkably similar to those calculated by Dale et al. (1994) for caribou *(Rangifer tarandus)* and Eberhardt (1997) for moose when the various prey species were standardized for differences in body mass (Becker, Garrott, White, Jaffe, et al. 2009). This finding suggests wolf kill rates are relatively constant across a variety of wolf–ungulate systems and a wide range of ungulate densities (Eberhardt 1997; Ballard and Van Ballenberghe 1998; Eberhardt et al. 2003).

POPULATION GROWTH OF WOLVES

The wolf population on the winter range for northern Yellowstone elk increased rapidly from 1996 to 2004, with a 22 percent annual growth rate (Figure 2). However, wolf numbers decreased substantially during the winters of 2006 and 2009 due to outbreaks of canine distemper virus during 2005 and 2008 that were associated with high pup mortality (Almberg et

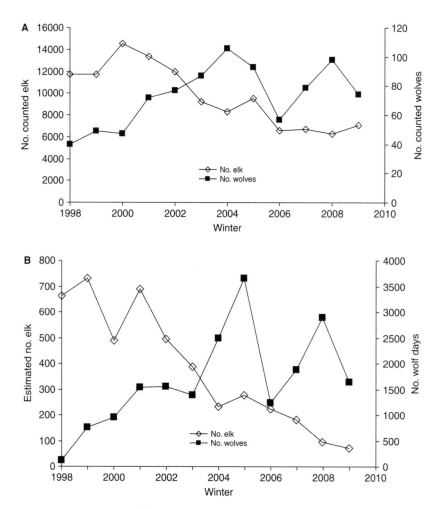

Figure 2. The number of elk and wolves during winter from 1998 through 2009 in the northern portion of Yellowstone National Park and adjacent areas of Montana (A) and in the Madison headwaters of Yellowstone National Park (B).

al. 2009, 2010). Also, a substantial decrease from 98 to 74 wolves occurred during 2009 as a result of intraspecific strife, food stress, and mange (D. Smith et al. 2011). This trend continued in subsequent years, which suggests that the wolf population may be beginning to decline in response to decreased elk availability.

Rapid population growth for restored wolves also occurred in the Madison headwaters, with a 30 percent annual growth rate from 1997 to 2005 (D. Smith et al. 2009). However, wolf use of the area decreased by 50 to 60 percent through 2009 as elk abundance decreased to low densities (Becker, Garrott, White, Jaffe, et al. 2009). No wolf pack resided entirely in the Madison headwaters, suggesting that the low number of elk prevented wolves from being able to survive and reproduce entirely within this area. Intraspecific strife was the leading cause of death for wolves (D. Smith et al. 2009).

EFFECTS OF WOLVES ON ELK ABUNDANCE

Wolf predation was one mortality factor that reduced elk numbers after wolf restoration in 1995, along with other predators (bears), human harvest, and severe weather events (Vucetich et al. 2005; Eberhardt et al. 2007; Barber-Meyer et al. 2008; Garrott, White, and Rotella 2009b). Wolf abundance was likely too low during the initial years following restoration to have a substantial effect on elk recruitment and survival, especially when the elk population was near the food-limited capacity of their environment (P. J. White and Garrott 2005a; Garrott, White, and Rotella 2009b). However, decreasing elk abundance, along with increasing wolf abundance, gradually resulted in wolves killing a higher proportion of each elk population each year, which contributed to rapid decreases in elk numbers in northern Yellowstone and the Madison headwaters. The population growth of wolves in both areas was remarkably similar and continued well after the elk numbers began to decrease—similar to findings for moose and wolves in Alaska (Gasaway et al. 1983) and on Isle Royale, Michigan (Peterson and Page 1983).

Wolves killed approximately 5 percent of adult female elk in the northern Yellowstone population and more than 20 percent of the Madison headwaters elk population from 2004 through 2009, with predation becoming more additive as elk density decreased well below carrying capacity

(Garrott, White, and Rotella 2009b; Hamlin et al. 2009). Also, recruitment was insufficient to replace the losses of adult female elk from all
sources (Garrott, White, and Rotella 2009b; Hamlin et al. 2009). Together,
additive mortality and low recruitment can reduce prey populations to low
densities that can be sustained by predation (Ballard et al. 2001). However,
the Yellowstone system is still transitioning, and a decrease in numbers of
wolves or other predators could allow increased survival, recruitment, and
abundance of elk under favorable weather conditions.

EFFECTS OF MULTIPLE PREDATORS ON ELK

Completely separating the effects of wolf predation on elk from the effects
of other predators, including humans, is difficult due to competition, carrion subsidies, and functional redundancy among different predator species
for available prey. High hunter harvests of northern Yellowstone elk continued in Montana for a decade after wolf restoration (Hamlin et al. 2009).
This harvest significantly decreased the survival rates of prime-age female
elk that produce the most calves and, in turn, the abundance of elk (P. J.
White and Garrott 2005a). Hunters and wolves were essentially in competition for decreasing numbers of elk, which resulted in a higher proportion
of the elk population removed each year and, in turn, an accelerated rate of
decrease in elk abundance (P. J. White and Garrott 2005a).

Mortality tends to become increasingly additive as animal density decreases away from the carrying capacity of the environment and animals
are in better condition and able to better withstand weather extremes (Kie
et al. 2003). Thus, as the density of northern Yellowstone elk decreased by
60 percent during the years 1995 through 2006, predation and harvest
mortality likely became more additive regardless of the predator that did
the killing because many adult elk would have survived and many young
elk would have been recruited if not killed (Kie et al. 2003; P. J. White
and Garrott 2005a). High harvest of breeding females and future offspring, combined with intense predation on elk calves by bears and other
predators (see below), substantially lowered recruitment and rapidly decreased elk density well below carrying capacity. Montana Fish, Wildlife,
and Parks reduced the number of permits for hunting antler-less northern
Yellowstone elk by 95 percent, from 1,102 in 2005 to 104 in 2006 through

2009 and zero in 2010, owing to decreases in elk abundance and recruitment. However, kills by wolves and other predators appear to have replaced deaths due to hunting somewhat, and a substantial loss of young elk to predation each year has continued (Barber-Meyer et al. 2008; Hamlin et al. 2009; Metz et al. 2012). Thus, additional years of data collection will be necessary to determine whether survival and recruitment will increase as densities of wolves and other predators decrease.

Grizzly bears were a significant predator of elk calves prior to wolf restoration (Singer et al. 1997), and there was an increase in the numbers of grizzly bears observed in northern Yellowstone during the elk calving season from 1999 through 2008—though numbers of females with cubs were relatively constant (Schwartz, Haroldson, and Cherry 2006; M. Haroldson, U.S. Geological Survey, unpublished data). The Madison headwaters had even higher numbers of grizzly bears per elk during spring (Hamlin et al. 2009). Although no studies of newborn elk survival were conducted in the Madison headwaters system, intensive studies on the northern range revealed that grizzly bears and black bears accounted for about 60 percent of newborn elk deaths from 2003 through 2005, while wolves and coyotes each accounted for an additional 10 to 15 percent (Barber-Meyer et al. 2008). Cougars also prey on elk, with elk comprising approximately 80 percent of the winter diets of cougars in northern Yellowstone (Ruth and Buotte 2007). Estimated numbers of adult and subadult cougars using portions of the northern range were relatively low from 1987 to 2003 and did not differ before (range = 14 to 23) and after wolf restoration (17 to 21; Murphy 1998; Ruth 2004). Thus, cougar predation on elk was relatively low, removing about 2 percent of the elk population annually (Murphy 1998). Also, few cougars were detected in the Madison headwaters.

Hamlin et al. (2009) found that elk populations in the greater Yellowstone ecosystem decreased in areas where high numbers of wolves and grizzly bears occurred in relation to numbers of elk. Conversely, elk populations remained stable or increased where low numbers of wolves and grizzly bears coexisted with elk and moderate levels of hunter harvest occurred. Consistently low recruitment contributed to decreasing population trends in areas with high numbers of predators relative to prey. Survival studies of newborn elk calves indicated that predation by bears accounted for most mortality during the first 30 to 45 days (Singer et al. 1997; B. L. Smith et al.

2006; Creel et al. 2007; Zager et al. 2007; Barber-Meyer et al. 2008). Predation by wolves and cougars occurred through the year and increased after calves became more active and visible (Barber-Meyer et al. 2008). Relative numbers of predators to prey appeared to be more important than the total numbers of predators in affecting elk calf survival. Thus, the lowest survival of elk calves occurred in the smaller elk populations, such as in the Madison headwaters and Gallatin Canyon, with relatively high numbers of both wolves and grizzly bears (Hamlin et al. 2009). These findings are similar to those of many other studies of wolf–bear–ungulate systems in the north temperate region (Messier and Crête 1985; Bergerud and Elliot 1986; Ballard et al. 1987; Gasaway et al. 1992; Dale et al. 1994; Crête 1998, 1999; Kunkel and Pletscher 1999).

EFFECTS OF WEATHER ON ELK

The population dynamics of Yellowstone elk are linked in a complex fashion to variations in weather that alter forage availability and quality, which in turn affect their nutrition, body condition, vulnerability to predators, reproduction, and survival (D. B. Houston 1982; Coughenour and Singer 1996; Singer et al. 1997; Taper and Gogan 2002). Survival for adult female elk and their calves the following spring is higher if females leave the summer range with at least 15 percent body fat and retain at least 5 percent body fat through late winter (J. G. Cook et al. 2004; R. C. Cook et al. 2004). Thus, rates of predation vary with factors that influence prey food, nutrition, condition, and vulnerability (Becker, Garrott, White, Gower, et al. 2009; Becker, Garrott, White, Jaffe, et al. 2009). Also, weather events such as severe winters can substantially lower the density of prey populations, after which the combined kills of multiple predators could outpace recruitment and prolong low prey densities (Bergerud et al. 1983; Messier 1994). Kill rates by wolves in Yellowstone increase in deeper snows, even at relatively low elk densities (Jaffe 2001; Mech et al. 2001; Becker, Garrott, White, Jaffe, et al. 2009). Moreover, snowpack strongly influences elk vulnerability to human hunting because more elk migrate to lower-elevation ranges outside the park as snowpack increases and are exposed to hunter harvest (P. J. White and Garrott 2005a).

Prior to wolf restoration, the mortality of juvenile elk in Yellowstone varied widely from year to year in response to interactions among population density, forage production, and snowpack that limited forage availability and increased energy expenditures (Coughenour and Singer 1996; Taper and Gogan 2002; Garrott et al. 2003; Garrott, White, and Rotella 2009a). During years with poor forage production or increased snowpack, more elk calves die and calves born the following spring are often at a lower birth weight and less likely to survive, especially when elk density is high and less forage is available for each mother (Singer et al. 1997; Garrott et al. 2003). Conversely, the survival of newborn elk may be increased by warmer spring temperatures that shorten winter and initiate the growth of new vegetation earlier, which in turn speeds the recovery of maternal condition lost over winter and contributes to better milk production and growth rates of calves (Lubow and Smith 2004; B. L. Smith et al. 2006; Griffin et al. 2011). Wilmers and Getz (2005) found that winters in Yellowstone are getting shorter, with earlier snowmelt that allows soils to warm faster and advance new plant growth that is much more nutritious than dormant vegetation.

Severe, sustained drought conditions existed in Yellowstone following wolf restoration (1999 through 2006), and the severe winter of 1997 resulted in a large human harvest of 3,320 elk and starvation of more than 530 elk on Yellowstone's northern range. Until 2011, other winters were among the mildest on record, with low snowpack that allowed elk greater availability to grassland forage. Northern Yellowstone elk captured during 2000–2002 were in relatively good condition for middle to late winter, with 85 percent having more than 5 percent body fat that indicated good-to-excellent survival probability through the rest of winter (R. C. Cook et al. 2004). However, non-lactating females were twice as fat as lactating females, which suggested that nutrition on the summer ranges was insufficient to support both lactation and fat accretion (R. C. Cook et al. 2004). Warming climate and drought has resulted in earlier peaks in the growing season followed by earlier curing of vegetation on higher-elevation grasslands, which results in less productive grasslands and perhaps poorer summer–autumn nutrition for migrant elk. Nutritional limitation in summer could result in reduced calf growth and make elk more vulnerable to predation (Merrill and Boyce 1991; J. G. Cook et al. 2004; K. M. Stewart et al. 2005).

Data from the Madison headwaters also indicate that environmental effects can substantially temper the respective strengths of food limitation and predation. Wolf predation on elk in the Madison headwaters was primarily compensatory, or substituted for starvation, during winters with deep snowpack because many elk were debilitated by severe nutritional deprivation, and starvation was pervasive under these conditions prior to wolf reestablishment (Garrott, White, and Rotella 2009b). Thus, predation reduced the survival probability of elk only by a modest 1 to 3 percent for each age class. Conversely, wolf predation during winters with mild snowpack conditions was strongly additive because few elk died under these conditions in the absence of wolves (Garrott, White, and Rotella 2009b). Thus, predation reduced survival probability by 2 to 4 percent for young females and a substantial 8 to 10 percent for adult females. This pattern was even more pronounced for overwinter calf survival. Pre-wolf data revealed a strong correlation between snowpack severity and calf overwinter survival, with relatively high survival during the lightest snowpack years and the starvation of virtually all calves during the most severe snowpack year (Garrott et al. 2003; Garrott, White, and Rotella 2009a). However, additive wolf predation under mild snowpack conditions reduced calf survival to between one-half and one-quarter of the rates observed during pre-wolf years, while wolf predation on calves during severe snowpack conditions was almost entirely compensatory because few calves would have survived in the absence of wolves (Garrott, White, and Rotella 2009b).

BEFORE AND AFTER WOLF COMPARISONS

The effects of wolves on elk population dynamics can vary considerably within the same region because of differences in land use, vegetation communities, densities of large predators and their management, local environmental conditions, elk migratory patterns, and human harvests (Garrott et al. 2005). Thus, Hamlin et al. (2009) used a landscape approach to compare the population dynamics of seven elk populations in the greater Yellowstone ecosystem, including northern Yellowstone and the Madison headwaters, before and after wolf restoration. These populations were within 115 kilometers of each other and affected by similar weather patterns, such as drought and winter severity. However, predator densities varied considerably among elk populations, ranging from some of the highest densities reported in North America to areas where wolves were transient. This before-after-control impact (BACI) comparative approach provides relatively strong conclusions regarding the effects of predation on elk survival and population growth rates (Hebblewhite et al. 2005; Hebblewhite and Smith 2010).

Hamlin et al. (2009) demonstrated that the survival of adult female elk decreased at high numbers of wolves relative to elk and that a substantial portion of the wolf predation on adult female elk contributed to increased mortality or, in other words, additive mortality. Survival of adult female elk was low in the northern Yellowstone, Madison headwaters, and Gallatin Canyon populations of Yellowstone, with ratios of more than 20 wolves per 1,000 elk, despite a reduction in hunting mortality in the Gallatin Canyon and northern Yellowstone populations to less than 3 percent of preseason numbers after 2003 and 2004, respectively, and no hunting of the Madison headwaters population. In contrast, the survival of adult female elk in the Blacktail-Robb Ledford, lower Madison, Wall Creek, and Yellowstone Valley populations, with ratios of no more than 3 wolves per 1,000 elk, was not significantly lower during the same time period (Hamlin et al. 2009).

Hamlin et al. (2009) also found some evidence that recruitment of calves was consistently low for populations with high numbers of wolves relative to elk, especially in areas with relatively high grizzly bear numbers. This consistently low recruitment dominated trends in the northern Yellowstone, Madison headwaters, and Gallatin Canyon elk populations because it was substantially below replacement levels for concurrent adult female mortality. Recruitment decreased somewhat during the post-wolf period on the other areas, but elk populations were stable to increasing in the Blacktail-Robb Ledford, lower Madison, Wall Creek, and Yellowstone Valley areas, where ratios of wolves and/or grizzly bears per 1,000 elk did not consistently exceed about 4 to 5 per 1,000 elk. Thus, elk populations appeared to be able to sustain adequate recruitment with low levels of wolf and bear predation and moderate hunter harvest, but the combined effects of wolf and grizzly bear predation when both predators are abundant relative to the number of elk in the population appears to depress elk abundance.

Wolves primarily prey on ungulates that are considerably larger and possess formidable defenses in their antlers, hooves, and behavioral responses to predator attacks. Thus, many wolf predation attempts are unsuccessful. During the harsh winter of 2010, elk were often seen high on windswept hills. Although thin and weakened by the winter, this female elk successfully outmaneuvered the wolves from the Lamar Canyon pack by facing them and even charging at them several times. *Photo by Cindy Goeddel.*

The findings of Hamlin et al. (2009) are similar to those of many other studies of wolf–bear–ungulate systems in North America (Van Ballenberghe 1987; Ballard 1991; Gasaway et al. 1992; Crête 1998, 1999; Kunkel and Pletscher 1999; Hebblewhite et al. 2002, 2005). A few ecologists have provided contrasting conclusions, suggesting that hunting and climate alone explain the decrease in northern Yellowstone elk (Vucetich et al. 2005; Hebblewhite and Smith 2010). However, their conclusions have been challenged by other scientists suggesting several fundamental flaws in the analyses. Furthermore, climate conditions in the years following these studies were favorable, and hunter harvest was reduced by more than 95 percent— yet the northern Yellowstone elk population remains at low numbers, providing additional evidence that the recovery of wolves and grizzly bears is likely strongly limiting elk abundance.

INDIRECT EFFECTS OF PREDATION RISK ON ELK

Elk could potentially alter their activity budgets, distribution, grouping tendencies, movements, resource selection, and vigilance in response to predation risk. Thus, anti-predator responses to wolf presence could affect elk energy expenditures and foraging behavior and lead to inadequate nutrition, which in turn could decrease elk body condition, pregnancy, and survival (R. C. Cook et al. 2004; Christianson and Creel 2010).

Resource Selection: Northern Yellowstone elk did not substantially alter their habitat selection after wolf restoration and continued to use the same areas they had historically (Mao et al. 2005; Kauffman et al. 2007; P. J. White et al. 2010). In summer, however, when wolf activity was still centered around dens and rendezvous sites on the elk winter range, elk avoided wolves by selecting higher elevations, less open habitat, more burned forest, and, in areas of high wolf density, steeper slopes (Mao et al. 2005). In winter, elk avoided deeper snow areas and selected relatively open habitat types, even though these areas were strongly selected for by wolves (Kauffman et al. 2007). These habitats likely permitted easier locomotion and increased the probability of escape if elk were attacked.

Similarly, elk in the Madison headwaters were less likely to occupy deeper snow areas. However, there was no evidence of strong shifts in habitat use or foraging in response to predation risk that altered their diets or nutrition (Gower, Garrott, and White 2009; Gower, Garrott, White, Watson, et al. 2009). Elk continued to select geothermal areas with higher food quantity and quality during winter, even though wolves also selected for these areas (Bergman et al. 2006; P. J. White, Garrott, Cherry, et al. 2009; P. J. White, Garrott, Borkowski, Berardinelli, et al. 2009). Indices of energy intake for elk in the Madison headwaters were not lower after wolves colonized the system, and there was no relationship between nutrition and wolf use of the area (P. J. White, Garrott, Borkowski, Hamlin, and Berardinelli 2009).

These results contrast with those from the Gallatin Canyon along the northwestern boundary of Yellowstone, where elk shifted habitats from grasslands to forests to reduce their level of predation risk and sustained a decrease in diet quality as a consequence (Creel et al. 2005; Christianson and Creel 2010). During winter, elk experience lower energy intake and diminishing fat reserves and, as a result, need to access as high a quality of forage as available (J. G. Cook et al. 2004; R. C. Cook et al. 2004; P. J. White, Garrott, Borkowski, Berardinelli, et al. 2009; P. J. White, Garrott, Borkowski, Hamlin, and Berardinelli 2009). Elk in northern Yellowstone and the Madison headwaters area continued to forage in areas where they could obtain necessary food resources after wolves became established in the system. Thus, there was no indication of any considerable change in foraging time or overwinter nutrition (R. C. Cook et al. 2004; Gower, Garrott, and White 2009; P. J. White, Garrott, Borkowski, Hamlin, and Berardinelli 2009; P. J. White, Garrott, et al. 2011).

Group Size: Mao et al. (2005) found that group sizes of northern Yellowstone elk were larger in areas of higher wolf density, even after accounting for differences in vegetation cover (forests, grasslands). Increased group sizes may have enhanced the probability of elk detecting predators or escaping attack by being one of many prey with less chance of selection by a predator (Pulliam 1973; Pulliam and Caraco 1984). Elk in the Madison headwaters also aggregated into somewhat larger groups in response to wolf predation risk (Gower, Garrott, White, Cherry, and Yoccoz 2009). Elk altered their grouping behavior among years, within winters, and daily depending on wolf presence and whether a kill had occurred in a given drainage (Gower, Garrott, White, Cherry, and Yoccoz 2009). Slightly larger group sizes and more dynamic grouping behavior may be an effective strategy when other defensive tactics, such as fleeing, do not work well in deep snow or thick vegetation that hinder efficient escape (Gower, Garrott, White, Cherry, and Yoccoz 2009).

These findings are similar to studies of ungulates in other areas with high predator density (Fryxell 1991; Heard 1992; Jędrzejewski et al. 1992; Molvar and Bowyer 1994). However, other studies in the greater Yellowstone ecosystem have documented either no detectable change in elk group sizes or elk aggregating into smaller groups in the presence of wolf predation risk. J. A. Gude et al. (2006) reported that wolves had no detectable effect on the size of elk groups in the lower Madison Valley located 40 kilometers west of the Madison headwaters area. Also, elk in the Gallatin Canyon decreased group sizes and shifted habitats from grasslands to forests to reduce their level of predation risk (Creel et al. 2005). These differences in behavioral responses are likely due to substantial differences in snowpack severity, habitat types, and other factors that influence predation risk and prey vulnerability. Thus, landscape differences among areas may strongly influence the behavioral responses of elk to wolf presence and the degree to which these behaviors are manifested (Gower, Garrott, White, Cherry, and Yoccoz 2009).

Vigilance: Northern Yellowstone elk expressed a high level of vigilance (increased visual awareness) when wolves were present (Laundré et al. 2001; Childress and Lung 2003; Wolff and Van Horn 2003; Lung and Childress 2006). Similar findings were reported for elk from the Gallatin Canyon (Winnie and Creel 2007; Liley and Creel 2008). Thus, these studies concluded that the risk of predation reduced time for foraging by elk, which could decrease their nutrition, body condition, pregnancy, and survival. In contrast, the likelihood of a foraging bout by elk in the Madison headwaters was slightly higher in the presence of wolves, possibly because wolves hunted primarily during twilight and nighttime and were inactive during the day when behavioral studies were conducted (Gower, Garrott, and White 2009). Thus, elk could sacrifice foraging for predator vigilance or avoidance during the high-risk nighttime period but compensate by increasing foraging during the lower-risk daytime hours. Behavioral scans by elk during feeding were short in duration and often occurred while elk were chewing food prior to swallowing. Thus, there was little reduction in foraging efficiency (Gower, Garrott, and White 2009).

The contrasting results between northern Yellowstone and the Madison headwaters likely reflect emphasis on differing behavioral activities, the subjective nature in classifying behaviors such as vigilance, the potential for several behaviors to occur simultaneously, and monitoring of elk that are in different physiological states (Gower, Garrott, and White 2009). Regardless, heightened behavioral alertness was an obvious response by elk when the threat of attack by a wolf was looming. Thus, loss of foraging efficiency would be expected during or immediately following a direct encounter with a wolf (Gower, Garrott, and White 2009). This interruption of feeding may last only a short time during winter because elk experience continual nutritional deprivation and need to forage efficiently (J. G. Cook 2002).

Movements: Northern Yellowstone elk did not substantially alter their migration patterns or summer use areas after wolf restoration, and there was no evidence of an increase in long-distance dispersal movements (P. J. White et al. 2010). Elk mortality from wolves was low during migration and summers because elk migrated away from the most concentrated area of wolf activity and selected habitats at higher elevations that allowed them to obtain high-quality green forage (Mao et al. 2005). Kauffman et al. (2007) found that during winter, landscape factors strongly influenced elk distribution and wolf kill sites, creating distinct hunting grounds for wolves and places of relative safety for elk. Thus, elk attempted to lower their risk of predation by movement and habitat selection across the landscape.

Most nonmigratory elk in the Madison headwaters maintained their home range within the same river drainage following wolf colonization (Gower, Garrott, White, Watson, et al. 2009). However, elk movements were more frequent and variable across the landscape as wolves increasingly encountered and attacked them. Thus, there were modest increases in elk home-range sizes and some changes in the locations of home ranges after wolf colonization (Gower, Garrott, White, Watson, et al. 2009). Also, some elk left the Madison headwaters area or became migratory following wolf colonization (Gower, Garrott, White, Watson, et al. 2009). These dispersal and migratory movements were not observed prior to wolf colonization and may have been an attempt to avoid predation risk by spatially segregating themselves from high wolf-use areas (Bergerud et al. 1984; Fryxell et al. 1988).

These findings suggest that elk were making movement decisions based on the competing risks of predation and starvation. During winter, elk in Yellowstone are constrained by deep snowpack that limits their ability to lessen predation risk by moving widely across the landscape (Gower, Garrott, White, Watson, et al. 2009). However, elk moved among different areas within short periods of time as wolves coursed the landscape looking for prey or after a nearby predation event (Bergman et al. 2006). Thus, elk movements likely reflected landscape characteristics, such as the distribution of foraging areas, and their assessment of attributes of these areas that made elk safer or more vulnerable to attacks by wolves (Kauffman et al. 2007; Gower, Garrott, White, Watson, et al. 2009).

Distribution: The number of northern Yellowstone elk on upper-elevation portions of their winter range, where elk were exposed to high predation risk from wolves during winter and bears during spring, decreased by 80 percent from 1995 to 2010. Conversely, the portion of the population that spent winter at lower elevations outside the park increased (P. J. White et al. 2012). This trend may be exacerbated in future years by continued attrition of adult elk and calves from higher-elevation portions of the winter range, the elimination of the antler-less harvest outside the park (Hamlin et al. 2009), and two to three times higher recruitment in the lower-elevation portion of the winter range than in higher-elevation areas (Barber-Meyer et al. 2008).

A similar pattern of changes in the distribution of elk occurred in the Madison headwaters after wolf recovery (P. J. White, Garrott, Cherry, et al. 2009). Prior to wolf colonization in 1997 and 1998, elk were distributed throughout the Firehole (40 percent), Gibbon (37 percent), and Madison (23 percent) river drainages. However, elk were eliminated from the Gibbon drainage by 2004 and almost eliminated from the Firehole drainage by 2010, with approximately 95 percent of the population concentrated in the Madison drainage. This shift in distribution was primarily due to wolf predation removing animals rather than elk redistributing among drainages. Attributes of the Gibbon and Firehole drainages made elk occupying these landscapes areas particularly vulnerable to predation by wolves, while the attributes of the Madison drainage lessened elk vulnerability to wolf predation (P. J. White, Garrott, Cherry, et al. 2009; Dunkley 2011).

Distribution shifts by Yellowstone elk following wolf restoration were driven primarily by the attrition of elk from higher-elevation areas on their winter ranges as a result of increased vulnerability to wolf predation in deeper snowpacks and decreased recruitment to replace these animals. Wolves frequently traveled through areas of Yellowstone's winter ranges occupied by relatively high densities of elk without making kills and appeared to frequently concentrate their hunting in lower elk-density areas (Bergman et al. 2006; Kauffman et al. 2007). Thus, differences in predation risk among areas were likely due not to differences in detection or encounter probabilities but rather to differences in the vulnerability of elk once attacked (P. J.

White, Garrott, Cherry, et al. 2009; Dunkley 2011). Even though elk were predictable for wolves to locate, elk appeared to occupy and perhaps persist in areas with landscape attributes that gave them a relatively high probability of escape (Mech et al. 2001; P. J. White, Garrott, Cherry, et al. 2009; Dunkley 2011).

Body Condition and Pregnancy: Elk in Yellowstone have made a number of behavioral adjustments to the presence of wolves, but these changes generally have not been drastic because elk need to access the highest-quality forage available during the severe winter months when they experience lower energy intake and diminishing fat reserves (Kauffman et al. 2007; P. J. White, Garrott, Borkowski, Berardinelli, et al. 2009; P. J. White, Garrott, Borkowski, Hamlin, and Berardinelli 2009). Comparisons of pre- and post-wolf restoration data indicate that habitat use, time spent foraging, and nutrition for elk were quite similar. These findings suggest that elk did not implement anti-predator behaviors that reduced their forage intake and prevented them from meeting critical nutritional requirements (Mao et al. 2005; Kauffman et al. 2007; Gower, Garrott, and White 2009). Despite these findings, it has been suggested that anti-predator responses by elk to wolf presence has reduced their nutrition, body condition, and reproduction. On the basis of concentrations of progesterone in feces, Creel et al. (2007) reported that pregnancy rates of four elk populations in the greater Yellowstone ecosystem, including northern Yellowstone elk, decreased in the presence of wolves because of changes in foraging patterns that decreased nutrition (Christianson and Creel 2010). Decreased forage intake and nutrition reduces body condition and could result in elk failing to conceive during the autumn rut or losing the fetus during winter.

There are several lines of evidence that challenge the idea that Yellowstone elk suffered from substantially poorer nutrition and body condition as a result of the presence of wolves. Annual mean weights of northern Yellowstone elk calves harvested after wolf restoration were at or above those recorded during the decade prior to wolf restoration, suggesting no wolf-induced decreases in nutrition (Hamlin et al. 2009). In addition, P. J. White, Garrott, et al. (2011) compared the body condition of northern Yellowstone elk before (1962 to 1968) and after (2000 to 2006) wolf restoration. There were no decreases in the body fat of elk during late winter between the pre-wolf and post-wolf sampling periods (P. J. White, Garrott, et al. 2011).

Likewise, indicators of energy intake for elk in the Madison headwaters area were not lower after wolves colonized the system, and there was no decrease in nutrition when wolves were using an area (P. J. White, Garrott, Borkowski, Hamlin, and Berardinelli 2009).

In addition, there is little evidence supporting a widespread and substantial decrease in pregnancy rates of elk in the greater Yellowstone ecosystem as a result of the presence of wolves. Independent studies of the same elk populations during the same time periods reported by Creel et al. (2007) found that post-wolf pregnancy rates derived from pregnancy-specific protein B assays of blood serum, which are the definitive standard for nonlethal pregnancy assessment in elk, were equal to or higher than pre-wolf pregnancy rates (Hamlin et al. 2009). Also, the portion of adult female elk in northern Yellowstone that were pregnant in the presence of wolves was similar to or higher than prior to wolf restoration (P. J. White, Garrott, et al. 2011). These assertions are disputed by Creel et al. (2011), and continued assessments along multiple lines of evidence will be necessary during the coming decades to validate and resolve the potential for wolves to indirectly affect the pregnancy rates of elk.

CONSERVATION IMPLICATIONS

The effects of large predators were not fully realized in Yellowstone when the policy of ecological process management was initiated in 1969 (G. F. Cole 1971). Wolves were extirpated by 1930, and bear numbers were relatively low during the 1970s as a result of removals following the closing of garbage dumps in and near the park (Haroldson et al. 2008). Thus, elk populations were at higher densities than may have been expected with a full complement of large predators.

There was certainly an expectation that restoring wolves to Yellowstone would reduce elk abundance. Wolf kill rates and population growth in Yellowstone's prey-rich environment were close to the maximum rates predicted prior to restoration, and as expected, elk were their primary prey (P. J. White and Garrott 2005b). However, counts of northern Yellowstone and Madison headwaters elk decreased more (60 to 90 percent) than predicted (5 to 30 percent; P. J. White and Garrott 2005b). Contrary to expectations, harvests of northern Yellowstone elk in Montana were not reduced appreciably for a decade after wolf restoration, which significantly decreased the survival rates of prime-age female elk that produce the most calves and, in turn, contributed to reduced elk abundance (P. J. White and Garrott 2005a). As numbers of northern Yellowstone elk decreased, wolf predation and hunter harvest mortality likely became more additive regardless of the predator that did the killing because many of these elk otherwise would have survived. Also, predation of newborn elk by bears was high following wolf restoration, which contributed to consistently low recruitment that was insufficient to replace losses due to all mortality sources (P. J. White and Garrott 2005a). Furthermore, wolf restoration coincided with a period of warming climate and extreme drought conditions during 1999 through 2006 that reduced summer–autumn nutrition and may have contributed to somewhat lower recruitment (R. C. Cook et al. 2004). These findings highlight the importance of considering the combined effects of abundant and diverse predators, human harvest, and climate.

The restoration of wolves to Yellowstone and the associated predator–prey research conducted in response to this ecological experiment allowed managers to gain a better understanding of the many interacting natural processes that influence elk abundance. This improved ecological knowledge suggests that the successful restoration of the full complement of large predators to Yellowstone has eliminated the perceived need for human intervention through culling to reduce elk densities and prevent overgrazing and, ultimately, mass starvation. This basic transition in wildlife management in Yellowstone is in keeping with the general principles adopted by the National Park Service for managing biological resources under ecological process management, including (1) preserving the dynamics, distributions, habitats, and behaviors of native populations, communities, and ecosystems; (2) restoring native plant and animal populations; and (3) minimizing human intervention to native populations and the processes that sustain them (National Park Service 2006).

CHAPTER SIX

Competition and Symbiosis

The Indirect Effects of Predation

ROBERT A. GARROTT

DANIEL R. STAHLER

P. J. WHITE

THE DIRECT EFFECTS of predation comprise a "fraction of the potential interactions between species" in ecosystems, and numerous factors can indirectly affect aspects of communities other than just prey species (Estes et al. 2004; Hebblewhite and Smith 2010, 75). These effects include competition with other predators that consume the same food resources, changes in competitive interactions among prey species, and interactions with species of decomposers, parasites, and scavengers. Competition occurs when two animals of the same or different species interact and the capability of an individual to survive and reproduce is lowered by the other (Begon et al. 2005). Thus, competition could decrease the abundance of one or both species or alter their distributions, habitat use patterns, or other behaviors. Symbiosis describes the close and often long-term interactions between different species that can be categorized as mutualistic when both species benefit, commensal when one benefits and the other is not harmed or helped, or parasitic when one benefits and the other is harmed (Begon et al. 2005).

The greater Yellowstone ecosystem supports a diverse and abundant group of large predators and ungulates, with six large predators (black bears, coyotes, grizzly bears, humans, cougars, wolves) and eight large ungulates (bighorn sheep, bison, elk, moose, mountain goats, mule deer, pronghorn, and white-tailed deer). The concentration of ungulates on lower-elevation winter ranges in northern and west-central Yellowstone increases spatial overlap and the potential for competition between cougars, coyotes, wolves, and bears in the spring after hibernation. Conversely, the migration of ungulates to summer ranges and a substantial increase in ungulate numbers in the park, including alternate and smaller prey, may alleviate dietary and spatial overlap between predator species (Murphy 1998; Metz 2010; P. J. White et al. 2010). Previous studies of systems with one or two predator and prey species suggested that diverse predator associations can have substantial effects on ungulates by decreasing their abundance and altering their distribution, movements, and feeding patterns (Crête and Manseau 1996). Such effects have significant implications for managers because they could alter competitive and symbiotic relationships among predators and, also, among ungulates (P. J. White and Garrott 2005b).

SCAVENGING

Carrion, or the tissues from dead animals, is an important food source for carnivores and omnivores in the greater Yellowstone ecosystem. Though the presence of carrion is unpredictable and temporary on the landscape, it represents a critical food source for species ranging from grizzly bears, to beetles, to bacteria in the soil. With wolves absent from Yellowstone for most of the twentieth century, carrion abundance was highly seasonal, with abundance strongly linked to winter severity that influenced the number of ungulates that succumbed to the effects of deep snow and cold temperatures (Gese et al. 1996). However, the restoration of wolves resulted in carrion from wolf kills being more widely and consistently available for scavengers (animals that feed on dead or injured animals) throughout the year (Stahler et al. 2002; Wilmers, Crabtree, et al. 2003; Hebblewhite and Smith 2010).

Kleptoparasitism, or food stealing, is the most common competitive interaction between Yellowstone's predators and scavengers. Here carrion is taken by scavengers that would otherwise be eaten by the predator(s) that killed the prey. Commensalism can also characterize scavenging when some species benefit by feeding on the remains of carcasses after predators have eaten their fill and moved on. This is particularly the case for smaller species and decomposers (organisms that break down dead or decaying matter) whose use of carrion, bone, and other nutrients has little to no competitive effect on predators. Even mutualistic interactions can occur at carcasses, particularly for scavengers like ravens and coyotes who benefit from each others' vigilance for predators that could attack them while they are feeding.

Although all the species in Yellowstone's diverse carnivore assemblage contribute to carrion that is eaten by scavengers, our best knowledge comes

Winter ungulate carrion, whether a result of starvation or predation, represents a critical food source for many species ranging from grizzly bears to beetles to bacteria in the soil. Ravens and magpies are some of the most common vertebrate scavengers that are found on the carcasses of ungulates killed by wolves but are easily displaced by returning predators, as is illustrated by this photo of the Lamar Canyon wolves returning to the carcass of a bull elk they had killed near the confluence of the Lamar River and Soda Butte Creek. *Photo by Cindy Goeddel.*

from wolf-killed prey. Twelve different vertebrate scavengers are observed commonly feeding on wolf kills, with bald eagles, coyotes, golden eagles, magpies *(Pica pica)*, and ravens visiting most carcasses (Stahler et al. 2002; Wilmers, Crabtree, et al. 2003). The most regular and abundant scavengers at wolf kills are ravens and magpies. With strong flight capabilities and keen eyesight, ravens forage over several wolf packs' territories, sometimes following individual packs throughout the day (Stahler et al. 2002). Magpies range over smaller distances when foraging but exhibit a similar foraging strategy by following wolves to rapidly locate carrion (Stahler 2000). The grouping behavior and calls of these birds in the air and on the ground near carcasses attracts other scavenger species and increases competition over the carrion (Stahler et al. 2002).

Coyotes are the quintessential scavenger in the Yellowstone ecosystem and are nearly as common at carcasses as ravens and magpies because they are particularly tuned in to information on carrion location inadvertently advertised by birds. However, coyotes face greater risk at carcasses than other species because they are frequently injured or killed by wolves (Crabtree and Sheldon 1999; Ballard et al. 2003; Merkle et al. 2009). Despite this risk, the commonness of coyotes at wolf kills suggests the nutritional benefits outweigh the risks.

Bald and golden eagles also regularly scavenge at predator kills, especially in winter. In Yellowstone, golden eagles are more prevalent at carcasses associated with forested and steep terrain, whereas bald eagles are more common in open habitat. However, both species can be simultaneously present at wolf kills. Eagles, being relatively large birds, apparently pose somewhat of a competitive threat to coyotes and wolves, which occasionally chase and sometimes kill eagles at carcasses.

Black and grizzly bears are the largest members of Yellowstone's scavenger associations. Bears use their finely tuned sense of smell to discover carcasses on the landscape, and carrion is an important component of their diet in Yellowstone—particularly during autumn and spring before and after hibernation (Haroldson et al. 2008). Historically, bears primarily used carrion from carcasses of ungulates succumbing to starvation in late winter (Mattson 1997). Now, with wolves back on the landscape, carrion has become more available to some bears during other critical times—namely, late summer and autumn. This may be especially important when other food sources, such as whitebark pine nuts, fail or become erratic (Hebblewhite and Smith 2010). Adult male grizzlies are almost always successful at displacing wolves from their kills (Ballard et al. 2003).

Less common but potential beneficiaries of wolf-provided carrion are some of Yellowstone's smaller scavenging carnivores: red fox, pine marten *(Martes martes)*, bobcat, lynx *(Lynx canadensis)*, wolverine, badger *(Taxidea taxus)*, long-tailed weasel *(Mustela frenata)*, short-tailed weasel *(Mustela erminea)*, and least weasel *(Mustela nivalis)*. Of these, foxes are the only species seen in association with wolf kills on a regular basis. While cougars can also scavenge on carrion, they are less likely to attend wolf-killed prey sites because of the accompanying risks of injury or death. Cougar-killed prey, however, are commonly visited by wolves, with cougars often being driven away from carcasses by superior numbers of wolves (Ruth and Murphy 2009a).

Ungulate carcasses are used by an even greater number of invertebrate species. Sikes (1994) found approximately 440 species of beetles in the vicinity of elk carcasses in northern Yellowstone, with 57 of these strongly associated with carrion. Also, thousands of species of decomposers, such as bacteria, fungi, insects, invertebrates, and mites, help break down carcasses and contribute to nutrient surges (nitrogen, phosphorus, potassium) at kill sites (Hebblewhite and Smith 2010).

Before wolf restoration, bears emerging from hibernation often fed on carcasses of ungulates that died during winter and still contained substantial biomass because there were fewer scavengers. Following wolf restoration, there was still substantial carrion available in spring as a result of relatively high wolf kill rates during winter and inefficient consumption of carcasses by wolves that left substantial biomass for scavengers. As elk numbers decreased, however, wolves began to consume carcasses more efficiently, and as a result, less carrion may be available to bears emerging from hibernation in some years. Keith (1983) reported that the percentage of carcasses consumed by wolves increased as the abundance of prey decreased. Thus, if wolves continue to reduce elk numbers and consume more of each carcass as numbers of prey decrease, somewhat less total carrion might be available for scavengers over time. For example, the portion of adult bison carcasses consumed by wolves in the Madison headwaters increased from 46 percent (95 percent confidence intervals = 10 to 83) from 1998 through 2003, to 85 percent (73 to 96) from 2004 through 2006, to 93 percent

(83 to 100) from 2007 through 2010. Likewise, the consumption of adult elk carcasses increased from 79 percent (73 to 85) from 1998 through 2003, to 86 percent (81 to 91) from 2004 through 2006, to 88 percent (83 to 93) from 2007 through 2010.

EXPLOITATIVE AND INTERFERENCE COMPETITION

The use of common resources, such as food and space, by predators living in the same area can lead to competition, which in turn can contribute to complex and variable influences on prey species and communities (Polis et al. 1989). Exploitative competition occurs when one species indirectly reduces the efficiency by which another species accesses and uses a shared resource, such as the depletion of foods used in common (Ruth and Murphy 2009). Interference competition occurs when individuals of one species are directly excluded from a resource by another species (Begon et al. 2005). Examples of interference competition are killing (Polis et al. 1989; Palomares and Caro 1999; Donadio and Buskirk 2006), kleptoparasitism or food stealing (Brockman and Barnard 1979; D. B. Houston 1979; Cooper 1991; Ballard et al. 2003), and avoidance (Karanth and Sunquist 1995; Creel et al. 2001). The factors that most strongly influence competition are the diversity and abundance of carnivore and prey species in an area (Karanth and Sunquist 1995; Creel et al. 2001). Exploitative and interference competition can suppress the abundance of some predators or alter their behavior and use of the landscape, which could affect the abundance and distribution of their prey (Polis and Holt 1992; Polis et al. 1997). Numerous factors influence the outcomes of interactions between predators, including similarities of diet and habitat use, abundance and availability of resources, prey abundance, vulnerability to predation, and other factors that could increase competitive ability, such as body size and social behavior (Palomares and Caro 1999; Creel 2001; Ruth and Murphy 2009a).

Cougars and wolves in Yellowstone rely on migratory deer and elk that aggregate on low-elevation winter ranges when snow accumulates at higher elevations (Murphy 1998). Thus, substantial dietary overlap exists between cougars and wolves, with elk constituting 80 percent of annual cougar diets and 90 percent of wolf diets (D. W. Smith 2005; Ruth and Buotte 2007). However, there appears to be little exploitative competition for food between cougars and wolves because prey selection and kill rates of cougars were similar before and after the restoration of wolves (Ruth and Murphy 2009a, 2009b). Also, cougars follow their migratory prey to higher elevations during summer, while "wolves tend to restrict movements to denning and rendez-vous areas in lower-elevation valleys (Ruth 2004; Mao et al. 2005). Thus, the use of different areas and the greater availability of smaller, alternate prey may relax competition and enhance coexistence in summer" (Ruth and Murphy 2009a, 170; 2009b).

Likewise, bears and wolves may compete for resources such as elk calves. Grizzly bears are a significant predator of elk calves in Yellowstone (Singer et al. 1997). Grizzly bears and black bears accounted for about 60 percent of newborn elk deaths in northern Yellowstone in 2003 through 2005, while wolves accounted for an additional 15 percent of newborn deaths (Barber-Meyer et al. 2008). However, there appears to be little exploitative competition for elk calves because bears are specialist predators during early calving, and their predation peaks within the first 15 days of a newborn elk's life when female elk hide their calves. In contrast, most mortality by wolves and other predators follows the hiding period, when female elk and their calves form larger nursery groups.

Predator kill sites and ungulate carcasses in general often serve as focus points of carnivore interference competition in Yellowstone. On a broad scale, the level of carcass dispersion can influence the degree of competi-tion. For example, more highly dispersed carcasses can lead to more intense competition at each carcass than if numerous carcasses are aggregated in an area at the same time (Wilmers, Stahler, et al. 2003). Interspecific competi-tion has potentially increased in Yellowstone because the restoration of wolves has led to carcasses being more dispersed across the landscape and through the year (Wilmers, Crabtree, et al. 2003). Instead of having a boom in carcasses at the end of severe winters because of ungulate starva-tion, with competition likely reduced due to the abundance of carrion, carcasses occurring at dispersed sites through the year may lead to increased competitive interactions at each carcass.

Competition can also influence predation rates because the loss of car-rion to scavengers may have energetic costs and, possibly, result in addi-tional predation (Creel et al. 2001; Kaczensky et al. 2005). Conversely, fre-quent scavenging of kills may potentially reduce predation rates because wolves in northern Yellowstone had lower predation rates during seasons when they scavenged carcasses that were more readily available during severe winters (D. W. Smith et al. 2008). Also, wolves visited or scavenged about 23 percent of cougar kills and took over about 6 percent of kills in Yellowstone (Ruth and Buotte 2007). "Social nature of wolves enhances their competitive ability, and they tend to be dominant during direct interactions with cougars" (Ruth and Murphy 2009, 166). Wolves have killed numerous cougars, and the range used by cougars in and near Yellowstone has decreased since wolf restoration (Ruth 2004). Also, grizzly bears and wolves often compete for carcasses, with bears prevailing at about 85 percent of carcass disputes and often stealing carcasses from wolves (Hebblewhite and Smith 2010). However, wolves have killed grizzly bear cubs (Gunther and Smith 2004). Meat from ungulate carcasses increases in importance to grizzly bears during years with poor whitebark pine nut crops and during the spring when they emerge from hibernation (Green et al. 1997; Felicetti et al. 2003). As a result, increased wolf and grizzly bear interactions at carcasses have been documented in years with poor whitebark pine nut production (Hebblewhite and Smith 2010).

Though competition is predicted to be strongest between species most similar in body size, ecology, and behavior (Palomares and Caro 1999), significant effects of competition can also result from interactions between species that are quite different—interactions between ravens and wolves are an excellent example. Though wolves spend some effort at resource defense by flushing these bird scavengers off carcasses, rarely do they successfully catch and kill them, and the displacement is only momentary. As such, ravens are highly effective at obtaining carrion at kill sites in the direct presence of wolves and larger scavengers. In fact, as a result of ravens' behavior of removing and hiding meat pieces for future use, or caching, significant losses of carrion to wolves can occur (Stahler et al. 2002; Wilmers, Crabtree, et al. 2003; Kaczensky et al. 2005). Consequently, because they remove carrion that would otherwise benefit wolves, scavengers like ravens are thought to play a role in influencing wolf kill rates, especially for smaller wolf packs that lose more of their kills to scavengers (Wilmers, Crabtree, et al. 2003; Vucetich et al. 2004; Kaczensky et al. 2005).

Competition among wolves and coyotes is common as a result of their ecological and behavioral similarities. With wolves being larger and dominant, the abundance and distribution of coyotes often decreases in areas with numerous wolves (Paquet 1991; R. O. Peterson 1995; K. M. Berger and Gese 2007). Merkle et al. (2009) found that 79 percent of 337 observed wolf–coyote interactions in Yellowstone from 1995 to 2007 occurred at wolf-killed ungulate carcasses and were usually initiated by wolves. These findings indicate the importance of carcasses as a carrion source to coyotes, but there is a substantial mortality risk, as evidenced by the fact that at least 25 coyotes were killed during these encounters (Merkle et al. 2009). Such interactions contributed to a 39 percent decrease in coyote abundance in northern Yellowstone following wolf restoration (Crabtree and Sheldon 1999; K. M. Berger and Gese 2007). It is interesting to note that the frequency of these aggressive interactions decreased with increasing wolf numbers (Merkle et al. 2009). Though wolves dominate most encounters with coyotes in Yellowstone, coyotes sometimes prevail when they outnumber wolves (Atwood and Gese 2008; Merkle et al. 2009).

P. J. White and Garrott (2005b) speculated that this interference competition between wolves and coyotes could possibly contribute to higher recruitment of pronghorn in Yellowstone if coyote numbers were substantially reduced in pronghorn fawning areas. Coyotes are a significant predator of pronghorn, contributing to substantial fawn mortality and decreased recruitment (Barnowe-Meyer et al. 2009). Indeed, the restoration of wolves apparently increased pronghorn fawn survival in some areas by reducing the numbers and distribution of coyotes (Barnowe-Meyer et al. 2010). However, coyotes appear to be adapting somewhat to wolf presence through changes in landscape use and by living in smaller groups. Thus, the number of coyote packs in the Lamar Valley of northern Yellowstone has recently increased to pre-wolf levels (Hebblewhite and Smith 2010), and it is uncertain whether predation on pronghorn fawns will be significantly reduced in the long term.

APPARENT COMPETITION

In systems with multiple prey species for a predator, the rate of predation on a given prey species can be strongly influenced by the presence of other prey species that allow the predator to survive and reproduce at higher numbers than could be sustained by one prey species alone—a phenomenon known as apparent competition (Holt and Lawton 1994). For example, Wittmer et al. (2005) concluded that a range-wide decrease in the abundance of woodland caribou *(Rangifer tarandus caribou)* throughout the Canadian Rocky Mountain region was due to apparent competition with moose, a preferred prey of wolves. Likewise, wolf predation in Banff National Park in Canada led to decreased numbers of caribou and moose because wolves were sustained by high numbers of elk (Hebblewhite et al. 2007). Wolves were able to keep killing caribou even as caribou numbers plummeted because wolves were sustained by abundant elk, leading to wolves killing a larger portion of the caribou population as caribou numbers decreased.

In Yellowstone National Park, elk are the preferred prey of wolves and comprise 90 percent of kills despite the presence of numerous other ungulates (Mech et al. 2001; D. Smith et al. 2004, 2011; Becker, Garrott, White, Gower, et al. 2009). The number of elk in Yellowstone has decreased by more than 60 percent since wolf restoration, but it is uncertain whether predation and other factors will interact to cause further decreases (P. J. White and Garrott, 2005a; Garrott, White, and Rotella 2009b). One important unknown is the role of bison as an alternate prey for wolves (Garrott, White, Becker, and Gower 2009). There were about 4,200 bison in Yellowstone during the summer of 2012, but they are considerably less vulnerable to wolf predation than elk because of their formidable size and group defense (MacNulty et al. 2007; Becker, Garrott, White, Gower, et al. 2009). However, some wolf packs in central Yellowstone are effective at killing bison, especially during late winter when chronic undernutrition makes bison more vulnerable. Thus, Garrott, White, Becker, and Gower (2009) predicted that bison could subsidize the wolf population, thereby lessening the decrease in wolf numbers as the abundance of their preferred prey (elk) decreased. Indeed, wolves in the Madison headwaters killed more bison as bison numbers increased relative to elk, but there was not a corresponding decrease in wolf preference for elk (Becker, Garrott, White, Gower, et al. 2009; Becker, Garrott, White, Jaffe, et al. 2009). Thus, wolves killed a larger portion of the elk population as elk abundance decreased, with the prospect that wolf predation could eventually extirpate the Madison headwaters elk population. However, wolf use of the area decreased by 50 to 60 percent through 2009, and the portion of the elk population removed by wolves in 2009 was lower than predicted. These findings suggest that decreasing predation by wolves in response to scarce elk or other factors, such as elk learning to inhabit areas that make them less vulnerable to predation, could allow elk to persist at low numbers and with a smaller distribution than prior to wolf restoration (Sinclair et al. 1998; DeCesare et al. 2009; Garrott, White, Becker, and Gower 2009).

The restoration of wolves may also have altered competitive interactions between elk and other ungulates in Yellowstone. For example, P. J. White and Garrott (2005b) speculated that wolf recovery would contribute to increased bison abundance by decreasing elk numbers. Coughenour (2005) suggested there was competition between bison and elk for food resources during the 1980s and 1990s due to high overlap in diets and habitat use (Singer and Norland 1994). Indeed, bison numbers in northern Yellowstone tripled after wolf restoration (866 bison in 1996; 2,669 bison in 2012), while elk numbers decreased 70 percent. Wolf predation focused on elk in winter and elk and deer in summer, while bison comprised less than 5 percent of wolf diets (D. W. Smith 2005; Metz et al. 2012). Though these trends are not conclusive regarding the strength of apparent competition between elk and bison, continued predation by wolves and other predators (bears, cougars) on elk may confer a competitive edge to bison. Further research should elucidate this relationship (Hebblewhite and Smith 2010).

TROPHIC CASCADES

Vegetation communities can be overutilized by overabundant ungulates in areas without top carnivores such as wolves (Estes 1996; J. Berger et al. 2001). Thus, it has been widely suggested and highly popularized that the restoration of wolves to Yellowstone led to a trophic cascade whereby wolves altered elk behavior and habitat selection to such an extent that feeding on low-growing trees and saplings (aspen, cottonwood *Populus balsamifera,* willow *Salix* spp.) was substantially reduced (Ripple and Larsen 2000; Ripple et al. 2001; Beschta 2003, 2005; Ripple and Beschta 2003). Beyer et al. (2007) reported that willow growth rates were not related to elk abundance but were positively related to wolf presence. Predation risk from wolves evidently contributed to reduced feeding by elk on aspen, cottonwood, and willow at some sites, and this release from browsing contributed to increased plant growth (Ripple and Beschta 2004a, 2004b, 2006, 2007; Beyer et al. 2007). Following the resurgence of willows, beavers *(Castor canadensis)* reoccupied the large river systems and some smaller streams in northern Yellowstone (Ripple and Beschta 2004a, 2004b; Beyer et al.

Twelve different vertebrate scavengers have been documented feeding on wolf kills, with magpies and ravens commonly the first scavengers to appear at carcasses. The grouping behavior and calls of these birds frequently attract other scavenger species, such as this mature bald eagle feeding on a female elk killed by the Slough Creek pack in the Lamar Valley. *Photo by Cindy Goeddel.*

2007). Contrary to expectations, however, Creel and Christianson (2009) found that willow consumption by elk increased in the presence of wolves. Likewise, elk browsing on aspen was not diminished in sites where elk were at higher risk of predation by wolves (Kauffman et al. 2010).

Despite the popularized story of wolves driving a trophic cascade in northern Yellowstone by altering elk behavior and habitat selection, the decrease in elk numbers to between 25 and 50 percent of their abundance prior to wolf restoration is likely more important than decreased willow consumption by individual elk (P. J. White and Garrott 2005b; Creel and Christianson 2009; Kauffman et al. 2010). These decreases in elk numbers were due to the combined effects of multiple predators, human harvest, and drought and snow decreasing maternal condition and recruitment (R. C. Cook et al. 2004; Vucetich et al. 2005; Hamlin et al. 2009). Also, other factors apparently enhanced the growth of riparian vegetation since 1995, including a decrease in moose numbers (Tyers 2003), an increase in the number of days with temperatures suitable for riparian vegetation growth in May through July (D. Despain, U.S. Geological Survey, unpublished data), and mild winters from 1997 to 2010 with low snowpack that allowed elk and other ungulates greater availability to grassland forage and may have reduced feeding on riparian vegetation (Wilmers and Getz 2005). Thus, identifying the specific drivers of the recent increase in woody plant growth and recruitment documented in some parts of Yellowstone's ungulate winter ranges is a complex task. Future research and monitoring may contribute to a better understanding of the relative roles that different-level processes have on vegetation communities.

DISEASE SANITATION

Ungulates that migrate in and out of Yellowstone National Park have been implicated in the transmission of diseases, such as brucellosis *(Brucella abortus)*, sarcoptic mange (scabies), and septicemic pasteurellosis to other wildlife and cattle in the greater Yellowstone area (Aguirre et al. 1995; Wolfe et al. 2002). The restoration of wolves could reduce the commonness of certain diseases in their prey if predation selectively removes animals exhibiting signs of the diseases, a so-called sanitation effect. Conversely, some diseases may be more readily transmitted among elk following wolf restoration if elk concentrate more often due to predation risk and, as a result, increase the likelihood of disease transmission (Barber-Meyer et al. 2007). Diseases that apparently make ungulates more vulnerable to wolf predation include actinomycosis (a severe infection affecting jawbones), skeletal diseases and disorders such as arthritis, severe hydatid tapeworm *(Echinococcus granulosus)*, cyst infestations, and poor body condition, which could be indicative of some serious disease condition in addition to food deprivation (Mech 1970; Mech et al. 1998). For other diseases such as anaplasmosis, bovine viral diarrhea, leptospirosis, and epizootic hemorrhagic disease, wild elk may not exhibit symptoms that would increase their vulnerability to wolf predation, or they may show signs only years after infection (Thorne et al. 2002).

Disease seroprevalences for northern Yellowstone elk were substantially lower 5 to 10 years following wolf restoration for bovine viral diarrhea type 1 and bovine respiratory syncytial virus but similar or higher for *Brucella abortus,* epizootic hemorrhagic disease, *Anaplasma marginale,* and *Leptospira* spp. (Barber-Meyer et al. 2007). Determining whether these apparent changes in disease occurrence resulted from predation, behavioral changes due to predation risk, or other factors not related to predators will require substantially more knowledge of how each disease is transmitted and manifested, as well as whether symptoms can be detected by wolves prior to transmission to other animals (Barber-Meyer et al. 2007).

Chronic wasting disease (CWD) is a contagious, fatal disease of deer, elk, and moose for which there is no vaccine or known treatment. It is transmitted by direct animal-to-animal contact or through carcasses and soil. Deer, elk, and moose in Yellowstone National Park are at moderate risk for infection by CWD because the park is less than 210 kilometers from an infected area, the disease is spreading toward the park, and there are large concentrations of deer and elk in and near the park. If the disease reaches Yellowstone, wolves and other predators could influence CWD prevalence by increasing mortality rates, selectively removing deer and elk with CWD, redistributing deer and elk from areas of high concentration, and removing infected carcasses from the environment (Powers and Wild 2004). Results of preliminary modeling for Rocky Mountain National Park in Colorado predicted that wolves could eliminate CWD within approximately 20 years of wolf reintroduction (Hobbs 2006). The selective removal of CWD-infected animals by predators could slow disease transmission while removing relatively few healthy animals and not substantially slowing population growth. Carcass removal could decrease CWD contamination in the environment to some extent, but the amount of achieved benefits is unclear (Hobbs 2006). Yellowstone supports a healthy large predator and scavenger association that rapidly removes debilitated animals and carcasses, which could dilute environmental contamination.

CONSERVATION IMPLICATIONS

The loss of large native consumers from ecosystems—a process referred to as trophic downgrading—may be one of humankind's most pervasive and disruptive influences on nature (Estes et al. 2011). Recent research indicates that the indirect effects of large predators on their associated ecosystems, many of which were described in this chapter, lead to myriad effects on other species and processes that make these systems more healthy and resilient. Indirect effects of predation have implications for numerous ecosystem processes, including herbivory, the frequency and intensity of wildfires, the prevalence or spread of infectious diseases, the invasion of non-native species, biological diversity, and physical and chemical influences in the atmosphere, soils, and water (Estes et al. 2011). Thus, "disruptions of trophic cascades because of the decline of predation constitute a threat to biological diversity for which the best management solution is likely the restoration of effective predator regimes" (Estes et al. 2011, 306).

The general principles adopted by the National Park Service for managing biological resources include preserving ecosystems, restoring native plants and animals, and minimizing human intervention (National Park Service 2006). Since these principles of ecological process management were adopted in 1969, the conservation and restoration of wildlife resources in Yellowstone National Park has been successful. In turn, the resulting rich wildlife community has sustained a complex and diverse set of interactions, as illustrated for the large mammals described in this chapter. However, there are also a myriad of other less obvious organisms and interactions that are part of the ecological processes sustained in the park and surrounding area. Thus, it is essential that park managers continue to maintain a rich wildlife community and the ecological processes that sustain them.

CHAPTER SEVEN

Omnivory and the Terrestrial Food Web

Yellowstone Grizzly Bear Diets

CHARLES C. SCHWARTZ

MARK A. HAROLDSON

KERRY A. GUNTHER

CHARLES T. ROBBINS

AN OMNIVORE (FROM Latin: *omni* meaning all, everything; *vorare* meaning to devour) is a species that eats both plants and animals. Omnivores are typically opportunistic generalist feeders adapted to eat and digest either animal or plant material. Omnivory provides species with diet flexibility and allows them to adapt to varying environments and adjust their diets accordingly. Humans are omnivores, and our diet diversity worldwide exemplifies this flexibility. Most bear species are likewise omnivorous and select plants and animals that are nutritious, abundant, and available.

Trophic ecology is the study of the structure of feeding relationships among organisms in an ecosystem—often described in terms of food webs or as food chains. Food webs depict trophic links among species in a habitat, whereas food chains simplify this complexity into linear arrays of interactions among trophic levels. Thus, trophic levels such as plants, herbivores, and carnivores are groups of species with similar feeding habits. Omnivores are species that feed on more than one trophic level.

Morphologically and taxonomically, bears possess all the traits of carnivores but, with the exception of the polar bear *(Ursus maritimus)*, have diets composed primarily of plant matter (C. T. Robbins et al. 2004). Many of the bear species evolved with a generalist omnivore strategy that allows them to successfully occupy a broad array of the world's biomes, which are large geographical areas with distinctive plant and animal groups. Today, all bear species have decreased in numbers and distribution as a result of the effects of human activities (Servheen et al. 1999). Understanding the nutritional ecology of the grizzly bears in Yellowstone is of interest in a biological context but is also important when managers are faced with issues of conservation. We cannot plan, implement, and successfully manage conservation programs without solid biological knowledge regarding the needs of the species we are trying to maintain.

Grizzly bears in the greater Yellowstone ecosystem were listed as a threatened species under the federal Endangered Species Act in 1975. Through intensive conservation efforts, this population was restored, and all of the demographic criteria required for removal from this threatened status were attained by 1998. The U.S. Fish and Wildlife Service formally removed grizzly bears in the greater Yellowstone ecosystem from threatened species status in 2007, but bear advocacy groups challenged this change in court. One of the major points of their challenge was that federal biologists had not adequately addressed the potential future impacts of climate change on the abundance of whitebark pine and other bear foods. In 2009, the judge presiding over this case ruled in favor of the bear advocacy groups and ordered grizzly bears returned to threatened species status. The federal government appealed this decision, countering that grizzly bears are omnivores and well adapted to cope with changes in abundance of individual foods. In November 2011, the Ninth Circuit Court of Appeals upheld the lower court decision regarding the potential negative impact of whitebark pine decline on grizzly bear survival and reproduction and sustained the continued threatened status for grizzly bears in the greater Yellowstone ecosystem (Case 09–36100, *Greater Yellowstone Coalition v. Wyoming*). Thus, a thorough understanding of grizzly bear diets and nutrition has important ramifications, and this ruling portends additional studies of the effects of whitebark pine decline on grizzly bear demographics.

Though this book evaluates the effectiveness of management at sustaining essential ecological processes in Yellowstone National Park, we focus on the greater Yellowstone ecosystem as a whole, a spatial scale more relevant to grizzly bears. The greater Yellowstone ecosystem is geographically defined as the Yellowstone Plateau and 14 surrounding mountain ranges above 1,500 meters (Patten 1991). It includes Yellowstone and Grand Teton National Parks, six adjacent national forests (Beaverhead-Deerlodge, Bridger-Teton, Caribou-Targhee, Custer, Gallatin, and Shoshone), and state and private lands in portions of Idaho, Montana, and Wyoming. The ecosystem also contains the headwaters of three major continental-scale river systems: the Missouri-Mississippi, Snake-Columbia, and Green-Colorado. Recent estimates suggest that grizzly bears occupy approximately 37,258 square kilometers in the greater Yellowstone ecosystem, using elevations between 1,580 and 3,660 meters (Schwartz et al. 2002; Schwartz, Haroldson, Gunther, and Moody 2006). A primary component of occupied grizzly bear range within the greater Yellowstone ecosystem is the 23,833-square-kilometer Yellowstone Grizzly Bear Recovery Zone (U.S. Fish and Wildlife Service 1993).

GRIZZLY BEAR DIETS

Early research detailing the diets of the Yellowstone grizzly bear relied on examining undigested food residues in scat or investigating areas where bears were observed foraging (Mealey 1975; R. R. Knight et al. 1982; Mattson et al. 1991). These studies demonstrated that Yellowstone grizzly bears were opportunistic omnivores. They deviate from most other meat-eating carnivores by the volume and variety of vegetative foods in their diet (Mattson et al. 1991). It has been estimated that Yellowstone grizzly bear diets contained more than 120 different plant foods, plus an array of fungi, invertebrates, insects, birds, and mammals (Craighead et al. 1995). Yellowstone grizzly bears commonly consume herbaceous (meaning herblike, not woody) vegetation during spring and early summer (Mattson et al. 1991). In early spring, bears also seek out carcasses of bison and elk that die of starvation during winter. These carcasses provide abundant, easily digested energy and protein at a time when other nutritious foods are limited. As summer progresses, Yellowstone grizzly bears take advantage of other high-quality plant and animal foods that become available. They actively prey on newly born elk calves in late May through June (Mattson 1997; Singer et al. 1997; Barber-Meyer et al. 2008), and ground squirrels (*Spermophilis* spp.), pocket gophers *(Thomomys thalpoides),* and insects are relatively common items in the diet (Figure 3).

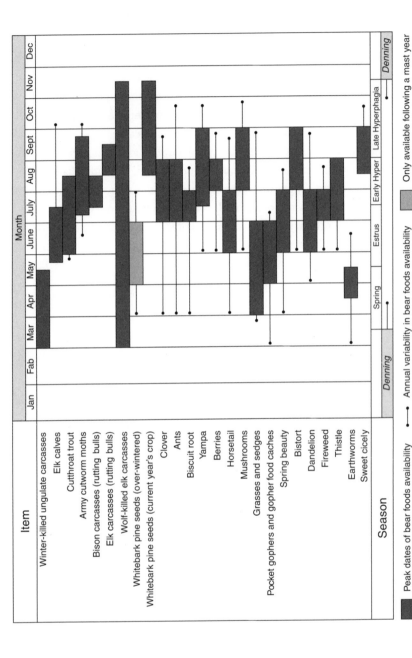

Figure 3. Seasonal food items for grizzly bears in the greater Yellowstone ecosystem. Adapted from Mattson et al. (1991).

In late summer, grizzly bears enter a hyperphagic period (R. A. Nelson et al. 1983), which literally means excessive eating (from Greek: *hyper* meaning over, excessive; *phagia* meaning eating or devouring). Because bears spend nearly half of the year hibernating, they must store large quantities of fat to sustain physiological functions during denning. Females that give birth to cubs in the den require fat stores to maintain themselves, produce young, and nurse their offspring while they remain without food and water until den emergence. In years when foods are abundant in autumn, it is not unusual for bears to enter their dens with more than 30 percent of their body weight as stored fat. To build their fat stores, Yellowstone grizzly bears need to eat large quantities of high-quality foods, including roots and tubers, berries, seeds of whitebark pine, army cutworm moths *(Euxoa auxilaris)* and other insects, and ungulates.

A new technique used to quantify the diets of both living and dead bears is called stable isotope analyses. Isotopes are different forms of the same element, for example, nitrogen isotopes 14 and 15 (^{14}N and ^{15}N). These isotopes are both nitrogen, but the far rarer form, ^{15}N, has one extra neutron, is nonradioactive, and occurs naturally. It is well known that ^{15}N and other stable isotopes tend to concentrate as one moves up trophic levels. When an animal eats plant matter and digests the protein, which contains small amounts of ^{15}N, the body preferentially retains ^{15}N relative to ^{14}N. Thus, bears that have eaten only plants will have ^{15}N concentrations in their hair or bones similar to levels found in other herbivores like deer and elk. However, when a bear eats meat, it is consuming a food at a higher trophic level (above plants) with more ^{15}N because herbivore flesh contains a higher concentration of ^{15}N compared with plants. Consequently, bears that eat meat have ^{15}N levels elevated above those found in herbivores. The ratio of ^{14}N to ^{15}N allows scientists to quantify the proportion of plant and animal matter that a bear ate during the past few weeks, months, or lifetime. By feeding captive bears at Washington State University various diets that included deer, trout, clover (*Trifolium* spp.), grass, and other foods and analyzing the isotope ratios of both the foods and bears, our research team was able to calibrate this technique specifically for grizzly bears (Hilderbrand et al. 1999; Felicetti et al. 2003).

In addition, the Bear Research, Education, and Conservation Program at Washington State University investigated the historical diets of Yellowstone grizzly bears. The oldest grizzly bear bones they analyzed came from a 1,000-year-old packrat (*Neotoma* spp.) midden excavated from the Lamar Cave in Yellowstone National Park. Because of the efforts of this hard-working packrat that had a fetish for bones, scientists were able to show that meat (everything from ants to trout and elk) provided 32 percent of the nourishment for those grizzly bears and that 68 percent came from plants (Jacoby et al. 1999).

During the early part of the twentieth century, when many hotels in Yellowstone National Park were feeding kitchen scraps to attract grizzly bears for tourist entertainment and nearby towns had open-pit garbage dumps, nourishment of Yellowstone grizzly bears switched almost entirely to meat (85 percent meat, 15 percent plants). After all such feeding was stopped by the early 1970s and grizzly bears returned to natural foods again, the diets of all age and sex classes returned to levels of about 40 percent meat and 60 percent plants (Jacoby et al. 1999; Fortin 2011). Large males may use more meat than other classes of bears because they can claim the carcasses of animals that died from other causes (Jacoby et al. 1999). Bears that have been killed for preying on livestock outside Yellowstone National Park had diets that were 85 percent meat and 15 percent plants. These levels of meat consumption contrast with grizzly bears in Montana's Glacier National Park or Alaska's Denali National Park, where plant matter provides 97 percent of their nourishment (Jacoby et al. 1999). Thus, for grizzly bears, the opportunity to consume meat differentiates the Yellowstone ecosystem from many other interior ecosystems where bears must feed primarily on plants.

FOOD WEBS

Because grizzly bears are omnivorous, they consume foods at two trophic levels (plants and animals). This foraging strategy provides flexibility in their diets but also puts them in potential competition with other herbivores and carnivores. Traditionally, community ecologists have assumed that similar species must differ in some aspect of their traits or responses to the environment to coexist in the same habitat (J. M. Chase 2005). This ability for species to coexist depends on the type of interspecific competition, whereby different species compete for the same resource in an ecosystem. Interference competition occurs when individuals of different species directly interact, such as fighting over limited or patchy resources (Case and

Gilpin 1974). Exploitation competition occurs when one species either reduces or more efficiently uses a resource and indirectly depletes the availability of the resource for the other species.

Grizzly bear diets overlap with many herbivores, such as elk, bison, and pronghorn, but bears do not rely solely on vegetation for sustenance. Bears consume the new growth of grasses, sedges, and forbs when they are highly nutritious, rapidly growing, and easily digested. Bears and ungulates are also different in their morphological and physiological adaptations to digest vegetation. Most ungulates (bison, deer, elk, moose) have four-chambered stomachs that contain microbes to efficiently break down forages high in fiber. However, bears have only a single stomach that lacks any specific adaptations to process and digest plant fiber. Hence, the probability for exploitative competition between bears and large herbivores is low—though exceptions may occur in areas where certain plant species are concentrated, highly nutritious, and preferred by both ungulates and bears. A good example is clover, which is an introduced exotic (non-native) plant that grows in dense patches in low-elevation meadows and certain thermally influenced areas within Yellowstone National Park (Mealey 1975; Graham 1978). Bison, elk, and grizzly bears all forage on clover swards in the Pelican and Hayden valleys, and it is not unusual to observe multiple grizzly bears grazing in clover patches (Gunther 1991). Heavy cropping from the intense grazing pressure in these patches maintains persistent regrowth of succulent foliage (Reinhart et al. 2001). Under these circumstances, grizzly bears likely dominate patches when present (interference competition), but bison and elk may limit the availability of clover to foraging bears because of their higher numbers (exploitation competition; Mealey 1975; Graham 1978).

The greater Yellowstone ecosystem is one of a few remaining ecosystems in the continental United States that still contains the full suite of large carnivores and herbivores that existed prior to European colonization of North America. This association of large carnivores includes gray wolves, cougars, grizzly bears, and black bears. There is a much greater opportunity for competition among these large carnivores than in many other ecosystems because of considerable dietary overlap in the prey they consume. However, wolves and cougars are obligate carnivores that must eat meat, whereas grizzly and black bears are opportunistic omnivores. Within the greater Yellowstone ecosystem, the most important large ungulate prey are

Grizzly bears in the greater Yellowstone ecosystem are opportunistic foragers with a diverse diet that seasonally can be composed primarily of various fruits, seeds, and plants. This young female grizzly bear was photographed while she was slowly moving across a meadow digging for pocket gophers and foraging for insects, roots, and grasses. *Photo by Cindy Goeddel.*

elk, bison, and mule deer. Wolves and grizzly bears prey on all three of these ungulates (Schleyer 1983; Mattson 1997; Stahler et al. 2006; Becker, Garrott, White, Gower, et al. 2009), whereas cougars tend to prey on smaller elk and mule deer (Murphy and Ruth 2010). Black bears prey on newborn elk and deer (Irwin and Hammond 1985).

Besides preying on large ungulates, all of the four carnivores scavenge carrion or take kills from one another. Grizzly bears have been documented usurping kills from wolves, black bears, and cougars. Murphy (1998) estimated that the amount of cougar prey stolen by a bear on any given day represented a gain of 71 to 113 percent of the daily caloric requirement for the bear and a loss of 17 to 26 percent of the daily energy requirements of the cougar.

Of all the large carnivores, black bears are the most likely to compete with grizzly bears. Their evolutionary lines diverged more than 3.5 million years ago (Kruten and Anderson 1980; Leonard et al. 2000), but their ranges have overlapped for only about 13,000 years, when the ancestors of current grizzly bears entered North America via the Bering land bridge. Thus, there may have been little time to develop mechanisms to reduce competition (Apps et al. 2006). The diets of grizzly and black bears are similar, though they are rarely compared in the same locations at the same time (Aune 1994; Holm 1998; Apps et al. 2006; Belant et al. 2006). Diet overlap is greatest in the spring, when both species are consuming primarily green vegetation. Diets tend to diverge in summer and are most distinct in autumn, when grizzly bears eat mostly meat, roots, tubers, and seeds of whitebark pine, whereas black bears forage more on berries. Grizzly bears dominate (interference competition) black bears, especially at concentrated food sources, such as ungulate carcasses or garbage dumps (Belant et al. 2006), whereas black bears may outcompete grizzly bears when foods are small and widely dispersed (such as berries) because of their smaller body size and greater foraging efficiency (Welch et al. 1997; Rode et al. 2001; Mattson et al. 2005). Herrero (1978) suggested that aggression, large body size, and long claws for digging gave grizzly bears an advantage in open habitats, whereas smaller body size, recurved claws that allow tree climbing, and the timid behavior of black bears were better adaptations to forested environments.

Intraguild predation involves killing and eating of species that use similar resources and are potential competitors (Polis et al. 1989). Intraguild predation is important in applied ecological problems, such as conservation of mammalian carnivores (Polis and Holt 1992; Linnell and Strand 2000), because it can contribute to the avoidance of larger carnivores in both time and space, reductions in a species density, and changes in demography (Linnell and Strand 2000). Though black and grizzly bears are omnivorous, intraguild predation has been documented with grizzly bears preying on black bears—but the reverse has not been observed (Palomares and Caro 1999; Gunther et al. 2002; Mattson et al. 2005). Both species are also cannibalistic (Tietje et al. 1986; Schwartz and Franzmann 1991; Garshelis 1994; McLellan 1994). Killers are most often larger adult males; victims are most often smaller, dependent young, and female.

DIETARY RESILIENCY: CHANGING WHAT YOU EAT
BASED ON FOOD AVAILABILITY

The greater Yellowstone ecosystem is a dynamic system with environmental variation (both natural and human) causing shifts in the abundance and distribution of natural foods. Variations in temperature and precipitation have major influences over vegetation, including the timing of spring leaf burst, length of the growing season, seed production, and nutrient content. Not all plants respond the same to each environmental perturbation, so in any given year some plant species are more abundant whereas others are scarce. Grizzly bears have adapted to this random process by diet switching. They tend to consume the most nutritious and abundant foods available at the time.

One of the most important plant foods eaten by grizzly bears in the greater Yellowstone ecosystem are the high-fat, energy-rich seeds (nuts) from cones of the whitebark pine. Whitebark pine is a conifer (needle-leaved, cone-bearing plant) that grows at elevations over 2,400 meters. The large size (average = 180 milligrams) of whitebark pine seeds makes them easier to harvest, extract, and consume as well as more energetically rewarding for grizzly bears compared with other pines in the greater Yellowstone ecosystem (Mattson et al. 2001). With a fat content of approximately 50 percent, whitebark pine seeds are also a rich source of dietary fat (Mattson et al. 2001). Dietary fat is important to grizzly bears because it contains concentrated energy that is efficiently converted to body fat. Body fat is important for hibernation and to support lactation (Mattson et al. 2001). The most efficient way to accumulate body fat is to eat fatty foods such as pine seeds. During years when whitebark seeds are abundant, bears feed almost exclusively on them from late August through early November.

Whitebark pine is a masting species that does not produce a seed crop each year but rather produces a large seed crop at irregular intervals of about every two to three years. Thus, animals that use pine nuts as an autumn food source must eat other foods during years without seed production to avoid starvation.

Felicetti et al. (2003) quantified the nutritional value of pine nuts to grizzly bears and discovered that whitebark pines concentrate a rare sulfur isotope (^{34}S) in the seeds' protein. This isotope is absorbed by bears that eat pine seeds and is deposited in their hair. Using isotope analysis similar to what is employed with ^{15}N, these scientists were able to demonstrate that ^{34}S was useful for quantifying pine seed consumption rates in grizzly bears. During years of poor pine seed availability, 28 percent of the bears used pine seeds. During years of abundant seed availability, about 92 percent of the bears used pine seeds and about 67 percent derived over 51 percent of their assimilated sulfur and nitrogen (protein) from pine seeds.

In years following a good crop of whitebark pine seeds, grizzly bear females tend to produce slightly more three-cub litters than one-cub litters (Schwartz, Haroldson, White, Harris, et al. 2006; Schwartz, Harris, and Haroldson 2006). In years of poor seed production, bears in Yellowstone National Park and adjacent wilderness areas shift their diets and their survival rate remains high (Schwartz, Haroldson, and White 2010). However, in areas near the edges of the greater Yellowstone ecosystem, where human activity and developments are located, more bear conflicts occur and mortality rates increase (Mattson et al. 1992; Gunther et al. 2004; Schwartz, Haroldson, and White 2010). Recent work by the Interagency Grizzly Bear Study Team (2009) suggests that autumn grizzly bear mortalities are a function of not only cone production but also population size. Thus, regardless of cone production, more bears die now during autumn since population estimates are three times higher than they were in the mid-1980s.

As an alternative to whitebark pine seeds during years with poor seed production, grizzly bears in the greater Yellowstone ecosystem consume more meat (Mattson 1997; Felicetti et al. 2003), primarily elk, bison, and moose (Mattson et al. 2001), and eat more roots and false truffles (*Rhizopogon* spp.; Fortin 2011). However, the abundance of elk and bison meat as an alternative food to whitebark pine seeds is not consistent from year to year. Since the mid-1990s, counts of northern Yellowstone elk have decreased more than 60 percent from 17,000 to about 5,000 as a result of predation, drought-related effects on elk survival and recruitment, starvation of elk during severe winters, and harvests of elk by human hunters adjacent to the park. Also, Yellowstone bison migrating into Montana were subject to periodic culling to prevent the possible transmission of brucellosis from bison to cattle outside the park. More than 3,600 bison were removed from the population from 2001 through 2011. Because grizzly bears are generalist omnivores exhibiting significant diet plasticity, a link between the abundance of these specific foods and reproduction and survival is difficult to detect (Schwartz, Haroldson, and West 2010; Schwartz, Haroldson, and White 2010). However, meat from elk and bison will likely become more important to the nutritional well-being of grizzly bears in the greater Yellowstone ecosystem if the trend of recent mortality to seed-producing whitebark pine trees by mountain pine beetle *(Dendroctonus ponderosae)* and white pine blister rust *(Cronartium ribicola)* continues.

Whitebark pine is currently under attack by native mountain pine beetles, previous outbreaks of which have resulted in high mortality rates in trees across the western United States. Also, blister rust infection, an exotic fungus, has killed many whitebark pine trees in the Pacific Northwest since it arrived in North America in the late 1920s. Blister rust has been less lethal in Yellowstone but continues to spread, with about 20 percent of the whitebark pine trees in the greater Yellowstone ecosystem being infected (Greater Yellowstone Whitebark Pine Monitoring Working Group 2011a, 2011b).

Estimates of whitebark pine mortality rates suggest that about 52 percent of mature (30+ centimeters in diameter at breast height) cone-producing trees in the greater Yellowstone ecosystem were dead from all causes (wind, fire, beetles, blister rust, lightning) in 2010. Younger trees in smaller size classes had higher survival, with only 5 to 22 percent of trees recorded as dead. Nine percent of marked trees were in the mature size class, whereas 91 percent were in smaller size classes. These data suggest that, although mortality rates have been high for large cone-producing trees, a high proportion of the whitebark pine population remains alive (Greater Yellowstone Whitebark Pine Monitoring Working Group 2011b). How the changes in whitebark cone abundance will affect grizzly bear numbers is not entirely known and cannot be predicted with certainty. Grizzly bears are resourceful, opportunistic omnivore generalists capable of adapting to changes in food resources. Grizzly bears have evolved a life history strategy including large

When grizzly bears emerge from hibernation in early spring, few food sources are available and bears focus on searching for ungulate carrion. A bison that succumbed to winter malnutrition near Round Prairie was discovered in March by this male grizzly bear, which slept and ate on the carcass for several days while chasing away coyotes, ravens, and eagles. *Photo by Cindy Goeddel.*

home range sizes, highly flexible diet switching behavior, and an omnivore generalist diet designed to buffer them against the annual and seasonal vagaries of their foraging environment (Mattson et al. 2001).

In the Northern Continental Divide Ecosystem, whitebark pine seeds were a significant autumn food for grizzly and black bears during the 1950s and 1960s (Tisch 1961; Jonkel and Cowan 1971). After the mid-1970s, whitebark pine use was negligible except on the eastern front of this mountain system (Mace and Jonkel 1986). This decrease in use occurred because most stands of whitebark pine were decimated by blister rust. Grizzly bears adapted to the loss of whitebark pine seeds and switched to other autumn foods, mostly berries.

Whitebark pine seed production has always had boom and bust years in the greater Yellowstone ecosystem. In years with poor whitebark seed production, grizzly bears in the greater Yellowstone ecosystem switch to alternative foods, including ungulates, army cutworm moths, ants, and false truffles. Bears are able to attain adequate body fat levels for denning and reproduction in both good and poor seed years. Though the effects of mountain pine beetle, white pine blister rust, and global climate change on whitebark pine will be hard to manage, the greatest influences on grizzly bear survival are human factors, such as densities of roads, densities of developed sites, the amount of secure habitat available to bears, and big game hunting, because many grizzly bears are killed in self-defense by elk hunters (Schwartz, Haroldson, and White 2010). Land and wildlife managers have more control over these factors.

Humans can also cause systems to change. The cutthroat trout population in Yellowstone Lake is a classic example. Cutthroat trout were once a preferred food resource for grizzly bears living around Yellowstone Lake (Mealey 1975). However, some ill-informed person(s) introduced lake trout to Yellowstone Lake before 1994, and cutthroat trout numbers have decreased precipitously (Koel et al. 2005). Counts of spawning cutthroat trout at Clear Creek decreased from a peak count of more than 70,000 in 1978 to about 500 in 2007.

Early studies of fish use by grizzly bears in the late 1980s relied on detecting fish parts or determining the presence of fish remains in bear scat (Reinhart and Mattson 1990a). In the late 1990s, scientists discovered that mercury in the effluent from thermal vents in Yellowstone Lake could be used to estimate fish consumption rates by bears. Mercury is accumulated up the food chain. When cutthroat trout eat plankton, they concentrate mercury from the plankton in their flesh. Similarly, when a bear eats fish containing mercury, the mercury gets deposited in its hair. Measuring the concentration of mercury in bear hair provides a direct measure of the number of fish consumed by that bear (Felicetti et al. 2004). Coupling mercury concentrations in bear hair with genetic analyses has allowed biologists to estimate how many bears consume fish, how many fish each bear eats, and the sex of the bear. In the late 1990s, male grizzly bears ate over 11 times more fish than female bears, and the total bear population consumed about 2,226 trout (Felicetti et al. 2004). From 2007 to 2010, total trout consumption by grizzly bears had dropped to 332 fish, which was an 85 percent reduction relative to just 10 years earlier and a 98 percent reduction from earlier decades before the introduction of lake trout (Stapp and Hayward 2002b). Bears have apparently adapted to the loss of cut-throat trout by preying on an increasing number of elk calves born in areas around Yellowstone Lake (Fortin 2011).

CONSERVATION IMPLICATIONS

Understanding the basic ecology of grizzly bears provides managers with a strong foundation when making decisions about future management in the greater Yellowstone ecosystem. This work will certainly take on a greater importance as climate change affects regional vegetation, hydrology, insects, and fire regimes. Likely climate change scenarios for the greater Yellowstone ecosystem suggest continued warming, with a 1 to 5 °C increase in average temperature during the twenty-first century (McWethy et al. 2010). Also likely is an increase in annual precipitation, mostly occurring as rain during spring, autumn, and winter (McWethy et al. 2010). Physiologically, grizzly bears in the greater Yellowstone ecosystem should be able to adapt to the changing weather since, until the early twentieth century, they occurred throughout much warmer climates in the southwestern United States and into Mexico (Storer and Tevis 1955; Mattson et al. 1995). However, the changing weather will likely affect some life history patterns, such as the timing of den entry and emergence, and the timing of availability and use of many vegetal foods (Haroldson et al. 2002). For instance, continued warming will bring earlier ripening of berries and whitebark pine cones. Warming temperatures will also increase stream temperatures earlier in the spring, which will bring earlier spawning of cutthroat trout.

Whether these climate-driven changes will adversely affect grizzly bears is unknown, but given the bears' opportunistic nature and diet plasticity, the grizzly bear population in the greater Yellowstone ecosystem will likely find sufficient calories to meet their needs. This was also the conclusion reached by a group of bear experts that gathered in Fernie, British Columbia, in September of 2010 to address the potential impacts of climate change on grizzly bears. The general feeling of the group was that "grizzly bears are opportunistic, omnivorous, and highly adaptable and . . . climate change will not threaten their populations due to ecological threats or constraints; however, climate change may play a significant role in driving grizzly bear/human interactions and conflicts" (Servheen and Cross 2010, 4).

The potential for bear–human conflicts will likely increase as bears adjust to a changing environment, especially if mature, cone-producing whitebark pine trees continue to decrease (Logan et al. 2010; Jean et al. 2011). Increases in grizzly bear–human conflicts will highlight the need to increase human food and garbage security as bears search for alternative foods. Continuing efforts to inform and educate the public on living with bears, such as the Wapiti and Jackson Hole Bear Wise Community projects (Hodges and Bruscino 2008), agency food storage orders, and management of attractants (garbage, livestock, pet food, bird seed, etc.) on both public and private lands will be increasingly important—as will maintaining strong monitoring programs that can identify and work to mitigate negative demographic trends should they occur. Even in the face of likely ecosystem change, prospects for the grizzly bear population in the greater Yellowstone ecosystem remain good because, as this chapter demonstrates, grizzly bears are adaptable opportunistic omnivores, and empirical studies suggest they currently possess traits needed to adapt to change.

communities and
landscape-scale processes

CHAPTER EIGHT

Natural Disturbance Dynamics

Shaping the Yellowstone Landscape

DAVID B. MCWETHY

WYATT F. CROSS

COLDEN V. BAXTER

CATHY WHITLOCK

ROBERT E. GRESSWELL

WHAT WE SEE today in Yellowstone National Park and surrounding eco-systems is largely a result of natural disturbances that shaped, often in dramatic fashion, Yellowstone's physical and biological environments. On September 8, 1988, Yellowstone National Park was closed to the public for the first time in its history following a summer of the largest fires ever recorded since the park was established in 1872. By the end of September, more than 319,700 hectares of primarily lodgepole pine *(Pinus contorta)* forests were burned by over 200 fires fed by a particularly hot, dry, and windy summer. Fires continued to burn into September until cooler temperatures and snow arrived (Schullery 1989). Many considered this event a catastrophe that would forever change the look and feel of the park, and news media worldwide reported that Yellowstone was destroyed by fire (C. Smith 1996). The size and severity of the fires were thought to result from decades of fire suppression, and many people assumed that fires of this magnitude could never have occurred in the past.

The science behind these fires tells a different story. Like other natural disturbances, fires have played an important role in shaping the structure (age, abundance, and type), composition (species), distribution, and natural processes of life in Yellowstone for millennia. Reconstructions of past fires and the response of vegetation indicate that large, severe fire events have occurred in the past and that these events show up in the historical record once every 100 to 300 years (Romme and Despain 1989; Millspaugh et al. 2000; Schoennagel et al. 2004). It is important to note that research suggests that recent large fire events like the Yellowstone fires of 1988 are primarily driven by extreme climatic conditions—not by fire suppression (Romme and Despain 1989; Turner et al. 2003; Westerling et al. 2006). Despite reports of a complete loss of several hundred thousand hectares of forest, the effects of the 1988 fires were highly variable across the landscape, with long strips of forests consumed by high-severity canopy

Fire is a major natural disturbance in the lodgepole pine–dominated forests of Yellowstone National Park and the surrounding area. Most forest fires encompass only small areas; however, in the summer and autumn of 1988, a series of large forest fires burned over 319,000 hectares, creating a mosaic of burned and unburned forest across the landscape, as can been seen in this view looking east from Steamboat Point. *Photo by Cindy Goeddel.*

fires directly adjacent to lightly or unburned forests (Turner et al. 2003; Turner 2010; Romme et al. 2011). The complex mosaic of intensely to lightly burned forests created a highly heterogeneous landscape of forests varying in both structure and composition.

Research over the past 20 years has contributed to a growing understanding that wildfire is a critical ecological process for sustaining the natural structure and function of Yellowstone ecosystems (Reeves et al. 1995; Wallace 2004; Romme et al. 2011), but wildfire is just one process in a rich array of natural disturbances that affect Yellowstone. However, this newfound appreciation for disturbances as important ecological processes must be tempered by the recognition that landscapes like Yellowstone are also responding to human-driven processes, such as climate change and nonnative species invasions, that may lead to novel conditions in the future. In short, scientists are asking how can the past inform the future if future conditions have no analogues to the past? Here we consider the role of disturbance processes that have been operating on decadal to millennial timescales in Yellowstone, a longer-term perspective that is critical for understanding the present-day ecology of Yellowstone and uncertainties facing future management of the Greater Yellowstone Area.

NATURAL DISTURBANCE PROCESSES IN YELLOWSTONE

Natural disturbances are herein defined as discrete events that fundamentally alter the structure, function, and/or composition of an ecosystem, community, or population, or the physical environment (*sensu* Paine and Levin 1981; Pickett and White 1985; Resh et al. 1988). Humans clearly identify fires, floods, and landslides as natural disturbances because they represent discrete events where the consequences, such as tree mortality, mass movement of soil, and destruction of man-made structures, are easily observed. However, organisms perceive disturbances differently depending on their scale of reference and level of biological organization (Figure 4). For example, fires may consume thousands of hectares of forest, but this disturbance event may have minimal effects on local forest bird and fish populations. At large spatial scales (watersheds, basins, or continents) and higher levels of biological organization (multiple populations), organism persistence following disturbance events increases. Disturbance is a normal component in

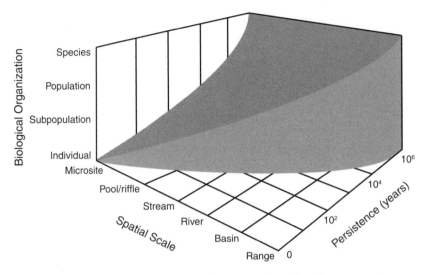

Figure 4. The probability of surviving and recovering from disturbances over time (years) increases with spatial scale and level of biological organization in an aquatic system. Adapted from Currens (1997).

system development, but a disturbance event at one level may be perceived as normal variation at higher levels of organization. Hence, organism response to a disturbance event is contingent on the link between the scale at which the organism and disturbance interact. A more inclusive understanding of disturbance recognizes that organism perception and response to disturbances are hierarchical and scale dependent.

This broader view of natural disturbance includes a number of processes in the greater Yellowstone ecosystem that operate across a hierarchy of scales in space and time (Figure 5). The consequences of a disturbance event vary widely across not only taxa or trophic levels but also in duration and effect. Disturbance events have been classified in three ways: pulse, press, and ramp. Short duration and well-defined events (pulse disturbances), such as floods, rapidly alter environmental conditions, and for most organisms the response occurs quickly following the event (Bender et al. 1984). Rapid but more sustained events (press disturbances), such as the damming of a stream from a landslide, may alter conditions in more lasting ways. Steadily increasing and sustained stresses (ramp disturbances), such as prolonged drought or rising temperatures associated with climate warming,

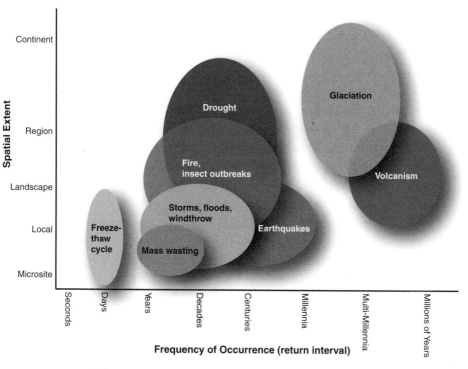

Figure 5. The frequency of occurrence over time and extent of area potentially affected by different types of natural disturbances. The frequency, intensity, and extent of effects from natural disturbances vary both within and across disturbance types.

may fundamentally alter conditions in ways that lead to threshold changes in community structure and biological processes (Lake 2000).

At regional to continental scales, mountain building, supervolcanic eruptions, and glaciation transformed the physical characteristics that serve as the template for present-day life in Yellowstone (Southwood 1977). Though not often thought of as discrete disturbances, these events reorganize nonliving factors that affect living organisms in ways that are comparable to ramp disturbances (C. E. Warren and Liss 1980). An appreciation for these ancient but extreme disturbances is critical for understanding the current distribution of habitats and species. Disturbances that occur more frequently (decades to centuries), such as fire, insect outbreaks, and droughts, cause dynamic changes to the structure and composition of vegetation on shorter timescales. These disturbances often reset forest succession, a process of colonization, establishment, and growth that results in a mosaic of forest patches with different ages and composition. For example, the 1988 fires in Yellowstone consumed thousands of hectares of mature lodgepole pine forests, and contrary to the idea that these forest ecosystems were lost, the fires spurred vigorous new growth of lodgepole pine seedlings, and most pre-fire plant communities returned (Romme et al. 2011). The post-fire heterogeneity provided a variety of new habitat conditions for plant and wildlife species that increased the diversity and structural complexity of Yellowstone landscapes (Bormann and Likens 1979; Romme and Knight 1981; Romme 1982). Similarly, post-fire erosion events often appear to be devastating, but in fact such events are vitally important to the maintenance of long-term habitat complexity and suitable spawning and rearing areas in streams (Reeves et al. 1995; Gresswell 1999). Rather than catastrophes, as many viewed the 1988 fires, natural disturbances are better thought of as events that reallocate resources, whether in the physical or biological environment, and this reallocation provides new opportunities for a set of players within an ecosystem to prosper.

VOLCANISM AND EARTHQUAKES

Yellowstone's geologic history during the last two million years has been punctuated by three large volcanic eruptions related to movement of the earth's plates and the location of magma near the earth's crust under Yellowstone (R. B. Smith and Siegel 2000; Christiansen 2001). These supervolcanic eruptions (greater than 1,000 cubic kilometers of ejected material) occurred 2.05, 1.29, and 0.64 million years ago, respectively, and erased the prior steep mountainous terrain of the region (Christiansen 2001). This reshaping of the mountain landscape by large volcanic eruptions and post-eruption lava flows created the predominantly rhyolite and basalt plateaus of central Yellowstone. Also, the nutrient-poor, silica-rich, and well-drained soils formed on rhyolite are responsible for the extensive lodgepole pine forests that exist at present. Prior to the supervolcanic eruptions and subsequent rhyolitic flows, volcanic activity approximately 50 million years ago ejected less silica-rich andesitic material, and volcanic breccias (rocks composed of broken fragments from explosive eruptions that are cemented together by a finer-grained matrix) form the backbone of the Absaroka Range in eastern and northern Yellowstone (Christiansen 2001). Soils on these andesitic rocks are richer in nutrients such as calcium and potassium, have a greater water-holding capacity than those on rhyolite, and consequently support more productive and diverse forests of spruce *(Picea)*, fir *(Abies)*, and pine *(Pinus)* and higher aquatic productivity in adjacent streams and lakes (Despain 1990).

Humans have not witnessed a supervolcanic eruption similar to the three that occurred in Yellowstone in the last two million years, but their effects obviously dwarf any other disturbance in the region. What is important is that (1) older volcanic activity is responsible for moderately productive andesitic soils and aquatic systems across eastern Yellowstone; (2) more recent eruptions that occurred within the last several million years converted north–south trending mountains into extensive landscapes of dry, nutrient-poor rhyolitic soils and less productive aquatic systems of the Yellowstone Plateau; and (3) the distribution and abundance of the dominant forest types and productivity of streams today is strongly influenced by the rock types that form the physical template of the respective systems.

CLIMATIC FLUCTUATIONS AND YELLOWSTONE'S ICE AGES

Climate fluctuations have occurred for millennia, and for purposes of discussion here, we limit ourselves to those of the last 20,000 years, starting with the Pinedale Glaciation, the most recent of a sequence of glaciations that occurred during the last two million years. During the Pinedale glacial maximum, the Yellowstone ice cap was centered over Yellowstone Lake, with a height of over 1,200 meters above land surface (K. L. Pierce 1979; Good and Pierce 2010). Glacial ice essentially removed all soils and vegetation from the Yellowstone region, setting the stage for primary succession. Most of the glacial landforms in the area relate to the ice recession between 18,000 and 14,000 years ago (Good and Pierce 2010). During this period, streams and lakes formed, soils developed, and plant and animal species recolonized the region (Whitlock 1993). The first plant communities to develop consisted of tundra and steppe vegetation and associated animal forms, but as the climate warmed, tundra was replaced by subalpine parkland approximately 14,000 years ago and then by closed subalpine and montane forest approximately 11,000 years ago. Areas enriched by meltwater deposited fine-grained calcareous sediments in northern Yellowstone and Hayden Valley now contain soils with high moisture-holding capacity and nutrients that support sagebrush steppe and grasslands (K. L. Pierce 1979; J. L. Pierce et al. 2004) as well as aspen parkland. Furthermore, the most productive aquatic systems in Yellowstone are located in this portion of the park. Pollen data suggest that these open vegetation types and surrounding Douglas fir woodland developed in the last 6,000 years (Whitlock 1993; Huerta et al. 2009). This complex habitat supports large populations of elk, bison, predators (including wolves, bears, and cougars), and core populations of Yellowstone cutthroat trout *(Oncorhynchus clarkii bouvieri)*, a keystone aquatic species in this portion of the park.

FIRES

Lake-sediment charcoal records from the past 20,000 years show that fire activity was greatest in the central Plateau region during the early Holocene (6,000 to 11,000 years ago) when fires typically occurred every 75 to 100 years (Millspaugh et al. 2000, 2004). During the last 6,000 years, there has been a shift toward cooler conditions and a longer fire return interval of 200 to 400 years (Millspaugh et al. 2000). Whereas the lodgepole pine forests showed little change on the rhyolite plateaus during the last 11,000 years, vegetation composition on other geologic substrates shifted in response to early-Holocene warming and late-Holocene cooling (Whitlock and Bartlein 1993; Millspaugh et al. 2000, 2004). In the last 2,000 years, major fire events in Yellowstone occurred about every 100 to 300 years, with drought and a warming climate playing important roles in fire frequency and extent (Millspaugh and Whitlock 1995; Whitlock, Marlon, et al. 2008; Higuera et al. 2010). The unique vegetation and fire histories in different parts of Yellowstone reflect the strong control of substrate in shaping vegetation's response to climate change.

Today the fire season in Yellowstone generally extends from early July to mid-October, when soil and fuel moisture are low, lightning frequency is high, and weather patterns are conducive to fire spread (Renkin and Despain 1992). An average of 17 fires (range = 2 to 58) have been started by lightning each year since 1930. Furthermore, most fires in Yellowstone start when lightning strikes a tree in an old-growth stand, when there is a sufficient amount and distribution of fine fuel in the vicinity, and when weather conditions are dry and windy (Despain 1990).

From 1972 through 1987, under a natural burn policy, 235 fires burned 16,187 hectares without human intervention (Despain 1990). Contrary to assertions that intentional fire suppression and an unprecedented accumulation of fuels were directly responsible for the large size and intensity of the 1988 fires, subsequent research shows that they resulted from extreme climatic and weather conditions, including sustained winds greater than 30 miles per hour associated with six cold fronts, rainfall that was 32 percent of the 30-year average, and a prolonged period of low fuel moisture less than 13 percent (Balling et al. 1992; Renkin and Despain 1992). High-resolution tree-ring reconstructions and charcoal studies from the central Plateau point to decades of large-area burns in both the seventeenth and eighteenth centuries (Romme and Despain 1989) but confirm that the 1988 fires were unprecedented over the last 800 years in the amount of area burned in a single year (Higuera et al. 2010).

Revisiting conditions in Yellowstone 20 years after the 1988 fires, scientists found that while the post-fire composition of the plant community generally resembles pre-fire conditions, there was a dramatic increase in the structural variability of forests, especially related to tree densities (Turner et al. 2003, 2004; Romme et al. 2011). New seedling aspen stands established in regions where they did not previously occur (Kay 1993; Romme et al. 1997, 2005; Forester et al. 2007). Ecosystem processes such as nutrient cycling and primary productivity responded rapidly to the 1988 fires, but in general, pre-fire conditions had returned within years to decades following the fires. For example, in the immediate years following the fires, many locations showed rapid release and microbial immobilization of nutrients that were previously locked up in plant tissues (Turner et al. 2007). Elk populations decreased and then increased in the first years following the fires, but the long-term response to changes in forage and habitat is still uncertain.

The effects of fire on aquatic systems include changes in productivity, nutrient cycling, spatial complexity of stream channels, and diversity and biomass of aquatic invertebrates, and all these effects are intricately linked with terrestrial ecosystem responses to fire (Gresswell 1999). The 1988 fires initially had stronger effects on smaller streams than on larger rivers, largely related to the proportion of watershed area burned in the smaller streams. For instance, channels of some small streams were substantially altered and were incised for several years following the fires. However, as dead wood moved into these small stream channels approximately 5 to 10 years post-fire, more sediments were stored and channels began to fill and stabilize (Minshall et al. 1997, 1998). Post-fire increases in suspended sediments in streams in the northern portion of Yellowstone National Park were sustained for three to four years following the fires, but dissolved nutrient concentrations were elevated for only about one year. Both algae and leaf litter, important resources that fuel aquatic food webs, increased rapidly following the fires in response to elevated light, nutrients, and rapid growth of streamside shrubs. Stream invertebrate communities showed elevated biomass and shifts toward dominance of disturbance-adapted taxa, such as midges and some mayflies (Minshall et al. 2001), changes that have now been sustained for at least 20 years (Romme et al. 2011). Fish mortality appeared to be minimal and limited to extreme-severity burns in small streams. In addition, there was no measurable effect of the fires on growth of cutthroat trout in Yellowstone Lake (Gresswell 1999, 2004), perhaps because large-water bodies are buffered from changes in temperature and nutrient availability. Though small streams appeared most affected in the decade immediately following the fire, the geomorphology and ecology of some streams and rivers may be affected over a much longer time frame. For instance, in streams whose channel widths exceed the length of most logs, dynamic large wood jams are now more common, shifting locations of major habitat features and providing important structure for fish (G. W. Minshall, C. V. Baxter and T. B. Mihuc, unpublished data).

The severity and size of the 1988 fires will certainly shape the park's ecological future for decades as the mosaic of burn severity leaves a legacy of new patterns of young to old vegetation, new seedling aspen stands, stand density variation across burned areas, and dynamic changes to stream geomorphology and the amount of in-stream woody material (Turner et al. 2004; Landhäusser et al. 2010; Marcus et al. 2011; Romme et al. 2011).

FOREST INSECT INFESTATIONS AND PATHOGENS

Yellowstone supports a diverse assemblage of insects that feed on plant tissues or plant products and represent important agents of natural disturbance in Yellowstone's past. Insects that have historically had the greatest effect on Yellowstone ecosystems include mountain pine beetle, spruce beetle *(Dendroctonus rufipennis)*, western spruce budworm *(Choristoneura occidentalis)*, and the Douglas fir bark beetle *(Dendroctonus pseudotsugae Hopk.;* Logan et al. 2010). Bark beetles (Curculionidae: Scolytinae) invade the woody tissues of living trees and deposit eggs in the living tissue that carries organic nutrients. This process inoculates trees with fungi that cause diseases that can weaken and eventually kill the tree. Beetle movements through the bark and the introduction of pathogens such as the blue-stain fungus *(Grosmannia clavigera)* combine to prevent movement of water and nutrients throughout the tree. The effects of defoliating insects, like the western spruce budworm and their larvae, are less acute, stressing trees by removing photosynthesizing needles but typically only resulting in mortality when other factors undermine the tree's vigor. The effects of insect outbreaks are highly variable, but large outbreaks can cause widespread and dynamic restructuring of communities.

Today we are seeing broad-scale effects of one insect in particular on Yellowstone's ecosystem, the mountain pine beetle. The mountain pine beetle is a small insect that has led to widespread mortality of pine trees in Yellowstone. In the greater Yellowstone ecosystem, large population outbreaks of mountain pine beetles have led to near 95 percent mortality where relatively homogenous stands of mature pines were once present (Logan et al. 2010). Mountain pine beetles feed on the inner bark of host trees and prefer mature, large-diameter pine trees. Beetle feeding activities introduce a fungus, which, combined with beetle activity in the cambium, results in high rates of tree mortality. Mountain pine beetle outbreaks are cyclical, lasting between several to over 10 years, and are thought to be influenced by extreme cool temperatures and the availability and vigor of host trees across the landscape among other factors (Raffa et al. 2008). During the early stages of an outbreak, large populations of beetles disperse from brood trees and seek out larger trees under stress from drought, injury, poor site conditions, overcrowding, or disease. High levels of mountain pine beetle mortality are not uncommon, but even during severe attacks generally less than 30 percent of standing trees are killed in any one year (W. E. Cole and Amman 1980).

Large-scale outbreaks of mountain pine beetle occurred in Yellowstone during droughts in the 1930s and 1970s and ended with the onset of wet periods (Despain 1990; Gibson et al. 2008). Today outbreaks from northern Mexico to Alaska have affected hundreds of thousands of hectares of forest and continue to cause widespread mortality where host trees are still available (Raffa et al. 2008). The intensity, extent, and synchronicity of the current outbreak on whitebark pine is thought to be unprecedented in the last century, caused by the presence of extensive homogenous stands of host trees, recent decades of drought and warm temperatures, and the synergistic effects of mountain pine beetles and non-native white pine blister rust (Kaufmann et al. 2008; Raffa et al. 2008; Bentz et al. 2009; Cudmore et al. 2010; Logan et al. 2010; Tomback and Achuff 2010).

WINDSTORMS, MASS WASTING, FLOODING, AND DROUGHTS

Windstorms, mass wasting (soil slumps, landslides), and flooding events often occur at decadal frequencies on any given landscape, and their effects are typically localized. These disturbances play an important role in maintaining the heterogeneity of biological communities across the landscape, allowing early-successional tree and shrub species such as aspen, willow, and cottonwood to persist. Powerful wind events have blown down hundreds of hectares of trees in the greater Yellowstone ecosystem, contributing to variation in forest age-class and stand development (Fujita 1989). Homogenous mature stands of lodgepole pine are especially susceptible to extreme winds because of their shallow root development. Tornadoes and microbursts have produced extreme winds in Yellowstone sufficient to blow over or break off all of the large trees in a stand (Despain 1990). For example, in 1984 a severe wind event left a series of blowdowns across Yellowstone ranging from 1 to 40 hectares, including a 57-square-kilometer swath of lodgepole pine between Canyon and Norris. In 1987, over 6,000 hectares of forest were leveled by a tornado in the Teton-Yellowstone wilderness (Fujita 1989). Susceptibility of trees to windthrow (uprooting or breakage by wind) in Yellowstone is also heightened in areas of shallow poorly drained soils, uniform-age stands, and steeper slopes (Alexander 1964). As with fire, succession after a blowdown is contingent on the presence of species that regenerate by sprouting and the nature of the seed bank. One of the important differences between fire and blowdown events is that fires often create thermal or mineral soil conditions that promote establishment of serotinous species, such as lodgepole pine, whereas post-blowdown soil conditions may allow other species to increase in dominance (Despain 1990; Romme et al. 2011).

Mass-wasting events such as severe erosion or landslides have been triggered by the combination of severe fire followed by extreme rain events in the past. Meyer et al. (1995) examined and radiocarbon-dated debris flows and other geologic deposits in northern Yellowstone and found a close relationship between severe fires, mass wasting, and climate. Fire-triggered erosion events were associated with dry periods, such as the Medieval Climate Anomaly, when severe fires were frequent in forests of spruce and fir. Mass-wasting events likely occurred when heavy rains destabilized newly burned slopes. During wet periods, the absence of fire-triggered debris flows led to fluvial terrace development in the main stem and tributaries of Soda Butte and Slough creeks (Meyer et al. 1995).

Large floods in each of the years from 1995 through 1997 affected fluvial geomorphology and created a highly variable distribution and density of in-stream coarse woody debris, an important structural attribute supporting aquatic biota (Marcus et al. 2001, 2011; Leigleiter et al. 2003). The severity of these floods is thought to be related to runoff from slopes that were deforested by the 1988 fires, where the absence of vegetation caused rapid melt of record or near-record snowpacks (Marcus et al. 2011). Large floods that result from storms or above-average snowmelt have led to redistribution and reworking of stream and river sediments, which maintain the dynamic character of riverine habitats important to their ecological integrity (Meyer et al. 1995; Reeves et al. 1995; Marcus et al. 2011). These events are also critical for recruitment and regeneration of key riparian species, including willow and cottonwood, and the many terrestrial wildlife species that depend on these riparian corridors (Rood et al. 2003).

Multiyear droughts and associated water stress are linked to greater vulnerability of vegetation to fire, insect outbreaks, and disease, as well as mass wasting. The period between A.D. 650 and 1300 featured extremely dry years in the Yellowstone ecosystem that were associated with more frequent fires, less beaver activity, and reduced flowering (S. T. Gray et al. 2007; Whitlock, Dean, et al. 2008; Huerta et al. 2009; Persico and Meyer 2009). Individual trees or large patches of forest in vulnerable locations, such as thin soils, drier aspects, and higher exposure, may have died from severe droughts and particularly long (multiyear or decadal) and severe drought events (van Mantgem and Stephenson 2007). Drought stress would have also contributed to the overall effects of multiple disturbances (Raffa et al. 2008; Logan et al. 2010). In aquatic systems, low flow periods are a part of the natural flow regime (A. M. Berger and Gresswell 2009), but extended droughts and subsequent reduced stream flows and elevated water temperatures during late summer can negatively affect fish and other aquatic organisms. As with other disturbances, river and stream reaches where populations are depleted are recolonized by individuals surviving these events, in refugia habitats of streams and lakes less affected by drought conditions.

YELLOWSTONE IN TRANSITION

Natural disturbances operating over a range of temporal and spatial scales have created the Yellowstone landscape that we see today, and much of the heterogeneity in physical and biological environments of Yellowstone is directly tied to past disturbances and the resulting mosaic of habitats and conditions. The ecological response to the 1988 fires showed an important dichotomy: species that were important before the fires retained dominance on the post-fire landscape, but increases in structural heterogeneity and dynamic changes to some systems illustrate how postdisturbance response is both contingent on pre-fire trajectories and new conditions created by disturbances.

Dominance of lodgepole pine in central Yellowstone's landscape for over 10,000 years suggests that elements of the Yellowstone ecosystem are remarkably resilient to disturbance occurring on timescales of years to millennia. Similarly, most species we see today have survived multiple disturbances over the duration of their existence and likely survived and recovered by a combination of adaptation, migration, and reestablishment via surviving populations. Today, however, some species may be pushed to the brink of extinction or regional extirpation because the rates of climate warming are of a higher magnitude than witnessed in the past (Shafer et al. 2005; van Mantgem et al. 2009). As a result of the rapid pace of climate warming now and into the future, these survival strategies may become more difficult given fragmented landscapes within and outside Yellowstone National Park that will impede migration. Hence, the effects of disturbance events in Yellowstone landscapes are dynamic, contingent on current trajectories of change, and consequently unpredictable—especially when multiple disturbances are interacting. We are witnessing new and complex interactions between human-caused change and natural disturbances that enhanced diversity in Yellowstone in the past but may have different consequences in the future (Logan et al. 2010; Westerling et al. 2011). These multiple interacting factors are leading to important, possibly unprecedented changes to Yellowstone's species and ecosystems and are worth considering in more detail.

NEW SYNERGIES BETWEEN NATURAL AND HUMAN-CAUSED DISTURBANCES

In the greater Yellowstone ecosystem, mountain pine beetle outbreaks are acting together with non-native white pine blister rust, drought stress, and a general warming trend to undermine the viability of high-elevation whitebark pine forests. Whitebark pine populations have endured disturbances such as insect outbreaks and fire for millennia (Figure 6), but current research suggests that a combination of stresses on whitebark pine populations and subsequent mortality will eventually end their role as dominant forests in the region (Logan et al. 2010). Why are whitebark pine forests more vulnerable to disturbance effects than they were in the past; what are the conditions contributing to their decline; and what can and should managers do to support or maintain these forests?

Whitebark pine occurs at elevations over 2,400 meters and has been periodically infected by native mountain pine beetles. Since the early 2000s, the mountain pine beetle has caused substantial mortality to whitebark pine trees, affecting hundreds to thousands of hectares of pine forests (Logan et al. 2010). Adding to and interacting with the current mountain pine beetle outbreak, white pine blister rust, a non-native fungus originating in Asia, is also infecting both limber *(Pinus flexilis)* and whitebark pines. Though white pine blister rust affects white pines throughout North America, the most negative consequences arise when trees are simultaneously exposed to multiple stressors (Tomback and Achuff 2010). Recent research suggests that mountain pine beetles preferentially attack trees infected with white pine blister rust, and the synergistic effects of the non-native fungus and drought are exacerbating the effects of the pine beetle outbreaks across Yellowstone (Logan et al. 2010). Additionally, less severe winters and

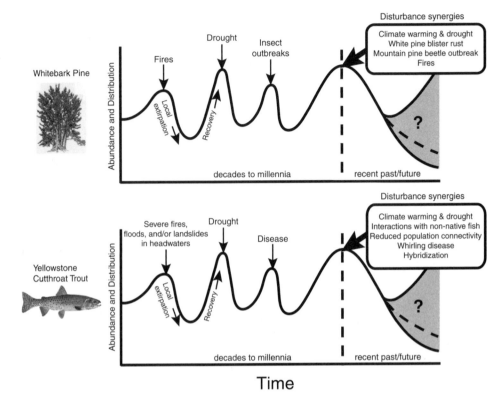

Figure 6. Conceptual framework illustrating how natural disturbances have influenced, and might influence, the abundance and distribution of whitebark pine and Yellowstone cutthroat trout in the greater Yellowstone ecosystem. Dynamics in the distant past emphasize historic resilience to perturbations and show short-term local reductions in abundance and/or distribution followed by recovery via regional dispersal, establishment, and persistence of surviving populations. Dynamics in the recent past suggest a loss of resilience related to synergies among multiple human-caused and natural disturbances and stressors. The dotted line and the gray shaded area indicate large uncertainty in future trajectories for these populations, and management that explicitly acknowledges these synergies through adaptive management and monitoring will be most effective. Images provided by Joseph Tomelleri (trout) and Katherine C. Kendall at the Whitebark Pine Ecosystem Foundation (pine).

a generally warming climate have allowed mountain pine beetle populations to increase substantially, swamping the defenses of whitebark pine trees. Natural disturbances that have operated for millennia are now acting simultaneously to lead to the possible loss of this keystone species across the greater Yellowstone ecosystem.

Another example is the Yellowstone cutthroat trout, which entered the Yellowstone region and large portions of the west sometime after the last glaciation 12,000 years ago, coming up the Snake River and crossing into the region via the Two Ocean Plateau (Behnke 1992; Gresswell 2011). In Yellowstone National Park, cutthroat trout played key roles in aquatic and terrestrial food webs as predators (principally of aquatic invertebrates) and as prey for important wildlife species, such as grizzly bears and river otters (Felicetti et al. 2004; Crait and Ben-David 2006). Throughout their long evolutionary history, Yellowstone cutthroat trout have persisted in the face of broad-scale disturbances by recolonization of disturbed areas through movement of surviving individuals from regional strongholds (Gresswell et al. 1994; Gresswell 1999). In fact, disturbances like floods and fires have likely played a positive role in maintaining the shifting habitat mosaic important to Yellowstone cutthroat trout (Gresswell et al. 1994; Gresswell 1999; Stanford et al. 2005) and in supporting terrestrial fauna that capitalize on postdisturbance pulses of aquatic prey from streams (Reeves et al. 1995; Malison and Baxter 2010). Conversely, more recent human disturbances have substantially affected populations of Yellowstone cutthroat trout both inside and outside the park.

Throughout its history, Yellowstone National Park has served as an important stronghold of the Yellowstone cutthroat trout (Gresswell and Liss 1995; Gresswell 2011). Even as numbers started to decrease in areas outside the park as a result of hybridization with non-native rainbow trout *(Oncorhynchus mykiss)*, habitat degradation, and overharvest by anglers, the park represented one of the last great core populations of the once widespread Yellowstone cutthroat trout (Gresswell and Varley 1988; Gresswell and Liss 1995; Varley and Schullery 1998). Over the past century, there has been a substantial reduction in the distribution and abundance of this iconic subspecies outside the park, and many unique populations have disappeared (Gresswell 2011). In fact, Yellowstone cutthroat trout are currently found in only 42 percent of their historical range, and only about 28 percent of the historic range is still occupied by genetically unaltered populations. Most of the remaining populations are found in small headwater streams where there are fewer than 94 Yellowstone cutthroat trout per kilometer. Furthermore, over half of stream-dwelling populations are found with non-native species (Gresswell 2011). In contrast, 65 percent of the historic range of Yellowstone cutthroat trout in the park (about 12 percent of the total range of the subspecies) is still occupied by genetically unaltered populations, and two lakes in the park (Yellowstone Lake and Heart Lake) represent almost 62 percent of the lake habitat where the subspecies was historically found (May et al. 2007; Gresswell 2009).

In the past several decades, however, the introduction of non-native lake trout to Yellowstone Lake has substantially reduced the abundance of Yellowstone cutthroat trout. Effects of the introduced predator have extended far beyond a reduction in cutthroat trout abundance, and consequences have been documented in the aquatic and terrestrial food web dynamics of the Yellowstone Lake ecosystem (Tronstad et al. 2010). For example, terrestrial predators such as grizzly bears that used to forage on Yellowstone cutthroat trout during their spawning migrations into adjacent streams must now search elsewhere to meet their food requirements (Haroldson et al. 2005; Tronstad et al. 2010).

Furthermore, this core population of Yellowstone cutthroat trout is now facing the threat of multiple interacting stressors associated with humans. For example, whirling disease, caused by a parasite originating in farmed trout imported from Europe, has recently been added to the list of threats to Yellowstone cutthroat trout (Koel et al. 2005, 2006). In addition, recent climate warming has led to increased stream temperatures and altered flow regimes of many rivers in the west, and this may have important consequences for Yellowstone cutthroat trout distribution and life history characteristics, as well as their susceptibility to the effects of exotic disease or non-native fishes (Kaeding 2010; Gresswell 2011; Wenger et al. 2011). Clearly, multiple interacting stressors significantly compromise the Yellowstone cutthroat trout's ability to persist. How might managers prevent the extinction of species like the whitebark pine and Yellowstone cutthroat trout and maintain natural aquatic food web and ecosystem processes?

CONSERVATION IMPLICATIONS

Natural disturbances are directly responsible for shaping the geology and soils that serve as the template for life in Yellowstone, the distribution and amount of early to late successional vegetation, and the wildlife and ecosystem processes that these habitats support. Understanding how natural disturbances shape Yellowstone ecosystems is essential for effective management of the natural resources that humans value. As one of the largest protected and intact ecosystems in the continental United States, Yellowstone is a valuable natural laboratory for exploring the effects of disturbance processes across multiple spatial and temporal scales. We now know that attempts to control and limit disturbances such as fire ultimately prove unsuccessful. How can managers be effective in supporting the diversity of life in Yellowstone in light of what we know about past disturbances and new synergies between natural and human-caused disturbances?

The answer may lie in what we already know about fundamental principles of conservation, including (1) acknowledging the importance of disturbance processes by protecting landscapes large enough to support recolonization following disturbances that occur at multiple scales; (2) providing buffer habitat around protected areas and connections among critical habitats; (3) mediating human effects within and around public lands; and (4) using adaptive management to better integrate research and management. Implementing management that fulfills these principles, however, has proved elusive, partly because of a strongly held perception that landscapes are relatively static and that life in Yellowstone can be sustained through protection of resources within park boundaries. The idea that the consequences of fire are catastrophic illustrates the lack of appreciation for disturbances as important ecological processes that are largely responsible for Yellowstone's diversity and unique assemblages of wildlife. Any strategy for supporting landscapes and keystone species that are experiencing multiple stresses today will require greater focus on combining adaptive management with efforts to improve public understanding of the dynamic history of change in Yellowstone.

Difficult decisions will have to be made in the future with respect to mitigating the effects of human-caused stresses on keystone species in Yellowstone, and clearly articulating management goals will require education that increases public appreciation for disturbance processes. The examples of whitebark pine and Yellowstone cutthroat trout underscore the notion that human intervention may be necessary to support species persistence threatened by human activities, and public support for interventions (especially outside park boundaries) will likely be possible only through greater ecological understanding among stakeholders that help set management goals and objectives. Supporting populations in the face of dynamic change will likely require continuing efforts to protect a full suite of biophysical and environmental gradients, maintaining or increasing habitat connections, and minimizing human effects in landscapes surrounding protected areas. For Yellowstone, this means we need to think of protecting the park's resources as a part of a larger effort to acknowledge the important role of disturbance in shaping the Yellowstone landscape and to adapt to change by protecting environments beyond the borders of Yellowstone National Park.

CHAPTER NINE

Climate and Vegetation Phenology

Predicting the Effects of Warming Temperatures

CHRISTOPHER C. WILMERS

KARTHIK RAM

FRED G. R. WATSON

P. J. WHITE

DOUGLAS W. SMITH

TAAL LEVI

WITH THE MELTING of winter snow, plant growth commences in the greater Yellowstone ecosystem, resulting in a greening of the spring landscape. Plants grow in predictable sequences of leaf, bud, flower, and seed formation, which together are referred to as the phenological stages of growth. The timing of phenological events is highly variable and depends on a suite of climatic and biotic factors. At the landscape scale, this phenological development has been characterized as a green wave that travels up the elevation gradient as snow recedes from late winter to summer (Frank and McNaughton 1992). This green wave is crucial to the area's migratory ungulates, such as elk, bison, mule deer, pronghorn, and bighorn sheep, which travel up to 550 kilometers round-trip (J. Berger 2004) as they track this highly nutritious young green forage to feed their young and accumulate fat reserves needed to survive the following winter and reproduce. In turn, a diverse suite of carnivore species preys and scavenges on these herbivores.

As around much of the globe, temperatures in and near Yellowstone National Park have been increasing over the last 50 years, leading to more rapid snowmelt on low-elevation winter ranges (Wilmers and Getz 2005). Understanding how climate change will influence plant phenological development at both local and landscape scales, and its impacts on wildlife, requires a comprehensive understanding of the park's climate system and how it interacts with plants and animals. We review what is currently known about the park's climate system and assess how the green wave dynamics may be transitioning in a changing climate.

Yellowstone National Park experiences two distinct climatic regimes (Whitlock and Bartlein 1993; Curtis and Grimes 2004). A winter-wet/summer-dry regime is driven from the west by the jet stream bringing moist air and precipitation from the eastern Pacific in winter and is blocked by subtropical highs in the eastern Pacific as the jet stream moves poleward in summer. A separate summer-wet/winter-dry regime is brought about by the North American monsoon, which brings precipitation from the gulfs of California and Mexico as far north as Yellowstone National Park in late spring and early summer.

Winter precipitation arrives primarily as snow, leading to a seasonal snowpack that is the dominant climatic control on wildlife. Several different physical characteristics of the snowpack are important. The total depth of the snow is perhaps the most universal metric of the impact of snow on wildlife, but snow density is also important in both beneficial and adverse ways depending on the species and the time of year. Density measures the mass of snow within a given volume (kilograms per liter; usually stated as a percentage). Thus, it impacts the weight of snow that must be moved by wildlife to forage beneath the snow or travel through it. Mid-winter in Yellowstone is often cold (Newman and Watson 2009), leading to relatively low-density snow (less than 20 percent) in the surface layers after heavy snowfalls. With increasing time since substantial snowfall, and as warmer temperatures arrive, the density of the snowpack gradually increases to 30 to 40 percent because of such processes as vapor exchange between snow crystals and mechanical compaction under its own weight (Watson, Anderson, Kramer, et al. 2009; Watson, Anderson, Newman, et al. 2009). The combined effects of density and depth are reflected in the snow-water equivalent or total depth of liquid water that would be obtained if the snowpack were completely melted. This measure is widely used to quantify the net impacts of snowpack on wildlife at landscape scales since it is easily measured and its spatial distribution can be estimated over large areas

(Watson, Anderson, Newman, et al. 2009). Of particular importance to many wildlife situations is snowpack hardness, which is measured as the force that must be exerted over a certain distance to penetrate each layer of the snowpack. Hardness is difficult to measure and difficult to predict (Pielmeier and Schneebeli 2003). To some extent, it can be approximated by snow-water equivalent but not always. Strong winds and daily surface melt-thaw cycles can lead to a hard crust on an otherwise low-density snowpack, creating precarious situations for ungulates and opportunities for predators (Telfer and Kelsall 1984; Crête and Larivière 2003; Tucker et al. 2010).

For about half of each year, most of Yellowstone National Park is covered by snow. In an average year, the snowpack begins to accumulate in October through most of the central region at elevations from 2,000 to 2,300 meters (Newman and Watson 2009). Heavy snowfalls typically occur between December and February, and the snowpack continues to deepen until a maximum depth of about 1.1 to 1.4 meters is reached usually by late March on the central plateau (Watson, Anderson, Newman, et al. 2009; National Climatic Data Center cooperative weather station data). Melting is relatively rapid, with most of the snowpack melting within about one month of the date of maximum depth and complete melt-out occurring before the end of May. Thus, the total snowpack season is about seven months long. However, on Parker Peak, Wyoming (2,865 meters elevation), the season length averages over nine months, while at Gardiner, Montana (1,608 meters elevation), it averages only three months—with most of the interannual variation occurring at the end of the season (National Resources Conservation Service snow telemetry station summary data; Western Regional Climate Center and Cooperative Observer Network data).

The snowpack varies from year to year in a manner that depends on location within the park. The winter of 1996–1997 was severe throughout the central and southern parts of the park but not at lower elevations in the north. Snow depth peaked at 2.4 meters at the Snake River on the southern boundary (2,098 meters elevation)—the deepest snow there in 87 winters of useable records (National Climatic Data Center cooperative weather station data). There were near-record snow depths of 1.9 meters at Lake Yellowstone (2,368 meters elevation) and 1.7 meters at Old Faithful (2,243 meters elevation)—about 40 to 50 percent above average. In the north at Mammoth Hot Springs (1,902 meters elevation), snow depth peaked at only 0.31 meter, and this was actually 24 percent below average.

The snowpack melted completely by mid-May at Snake River, Lake Yellowstone, and Old Faithful. At Mammoth, snowpack melted by mid-March, leaving a six- to eight-week period between mid-March and mid-May with an exceptional contrast between the snowbound south and interior and the snow-free northern valleys.

In contrast, a universally mild winter occurred in 1986–1987. Snake River, Lake Yellowstone, and Old Faithful peaked at 28 to 39 percent below their average snow depths of 0.84 to 1.04 meters. Snake River experienced its third lowest peak on record, Old Faithful its fifth lowest, and Lake Yellowstone its tenth lowest in 102 years of useable records. Mammoth experienced its second lowest peak, 62 percent below its mean, at a snow depth of only 0.15 meters. The snowpack came and went several times, and snow was on the ground for a total of only about two months. These interannual variations have a substantial impact on the population dynamics of wildlife, such as elk (P. J. White et al. 2010), bison (Geremia et al. 2009), and bighorn sheep (P. J. White et al. 2008).

Snowpack characteristics vary markedly throughout Yellowstone National Park at various scales and due to a variety of influences. There is more snow on the high-elevation peaks and plateaus, an intermediate amount of snow below these on the central plateau, and less snow in the valleys, particularly the Lamar and Yellowstone valleys (Watson, Anderson, Newman, et al. 2009; Watson and Newman 2009). At a given elevation, there tends to be more snow in the west and south, since the major winter storms tend to arrive from these directions, fully laden after traveling up the Snake River Valley in Idaho and becoming depleted as they move eastward and northeastward over the park (Watson, Anderson, Newman, et al. 2009; Watson and Newman 2009). More snow tends to accumulate in the meadows than in the surrounding forests, but the same tree cover that reduces snow accumulation can slow down snowmelt, so the snowpack often persists longer into the spring in the forests (Watson, Anderson, et al. 2006; Watson, Newman, et al. 2006). In the open country, there is less snow on exposed ridges and hilltops and more snow in the sheltered hollows— particularly on the downwind side of windblown ridges (Watson, Anderson, Kramer, et al. 2009). Some meadow complexes are characterized by hummocky terrain and exhibit large areas of both reduced and increased snowpack relative to flatter meadows where the snow is more homogeneous (Watson, Anderson, Kramer, et al. 2009). During the spring melt, slopes that face the sun are more rapidly depleted of the snow and become snow-free well before surrounding areas (Watson, Anderson, et al. 2006; Watson, Newman, et al. 2006). There is no snow on any of the intense geothermal features and a substantially reduced snowpack on the large low-intensity geothermally influenced patches that might otherwise go unnoticed (Watson, Newman, et al. 2009). The combined effect of these influences is greatest in spring, when the park becomes a receding mosaic of snowbound and snow-free areas (Watson, Anderson, Newman, et al. 2009). In an average year, the melt-out works its way up the lowest valleys in March, reaching the meadows on the central plateau in late April and the highest plateaus in June (Watson, Anderson, Newman, et al. 2009; Watson and Newman 2009). At smaller scales, melt-out on sunny, exposed, and non-wooded patches precedes the large-scale pattern by a few weeks, while shaded, sheltered, and wooded patches hold snow for a few weeks after the large-scale melt sequence has passed.

Monsoonal precipitation arrives in late spring and early summer with a somewhat different spatial pattern than winter precipitation (Whitlock and Bartlein 1993; Mock 1996). Monsoonal airflow moves northward along the Great Plains to the east of the park, bringing afternoon thunderstorms to areas along the eastern front of the Rocky Mountains (Curtis and Grimes 2004). The northern valleys and northeastern passes in the park are most exposed to these systems (Mock 1996; S. T. Gray et al. 2004), and up to 42 percent of their annual precipitation occurs during May through July (Western Regional Climate Center and Cooperative Observer Network data). Higher-elevation areas in the central and southern portions of the park also receive precipitation peaks in May and June, but the contrast with the winter months is much less than in the north. In January, precipitation increases about 6.9 millimeters with every additional 100 meters in elevation, but in June the corresponding increase is only about 2.5 millimeters (Western Regional Climate Center and Cooperative Observer Network data). By about late July, the monsoonal influence subsides and the entire park moves into a relatively dry period. In some years, the monsoon fails to reach as far north as Wyoming and drought conditions ensue (Curtis and Grimes 2004).

The climate and snowpack of Yellowstone National Park have changed in the past and are expected to change in the future. The details depend heavily on the temporal and spatial scales under consideration. During the past century in the western United States, temperatures have generally warmed by 1 to 2 °C, precipitation has increased, the fraction of

In winter, moose feed extensively on the twig tips of woody plants such as willow that generally protrude above the snowpack and allow feeding without the energetically costly need to displace snow to uncover grasses, sedges, and forbs at ground level. This moose was feeding in deep snow on willows growing along Crystal Creek. *Photo by Cindy Goeddel.*

precipitation falling as snow versus rain has decreased, annual peak snow-water equivalent has decreased, the snowpack has melted earlier, and the peak in spring runoff has occurred earlier (Mote et al. 2005; Regonda et al. 2005; I. T. Stewart et al. 2005; Knowles et al. 2006; I. T. Stewart 2009; McWethy et al. 2010; Ray et al. 2010; Littell et al. 2011). In this regional context, Yellowstone is typical with respect to a warming in temperature but not necessarily an increase in precipitation (Newman and Watson 2009). The past century of variation in Yellowstone's precipitation lies within the range observed during the past millennium (S. T. Gray et al. 2007; Whitlock, Dean, et al. 2008). Despite substantial swings at decadal and multidecadal timescales, there has been no clearly observable longer-term precipitation trend throughout Yellowstone National Park (S. T. Gray et al. 2007; Newman and Watson 2009). Complicating matters further, the changes that have been observed vary throughout the park because it straddles the boundary between several climate regimes (Whitlock and Bartlein 1993; Mote et al. 2005; L. R. Stevens and Dean 2008; McWethy et al. 2010). Predictions of future change reflect historic change, with temperatures predicted to increase substantially during the next century, but precipitation predictions are uncertain and variable (Ray et al. 2008; I. T. Stewart 2009). There is weak agreement among models that winter precipitation will increase and minimal agreement about whether summer precipitation will increase or decrease (Ray et al. 2008).

The response of snowpack to increasing temperature and precipitation depends on the absolute temperatures involved. If the temperature is near freezing, as is the case throughout much of the western United States during winter, then an increase in temperature will cause a decrease in snow-water equivalent because of a state change in precipitation from snow to rainfall and enhanced warming of remaining snowpack. At high elevations where the temperatures are well below freezing, however, a change in temperature has minimal effect, and any increase in precipitation will translate to an increase in snowpack. This dichotomy is borne out throughout the western United States in historical observations (Mote et al. 2005; Knowles et al. 2006) and future predictions (Littell et al. 2011). The elevation range of Yellowstone National Park straddles this dichotomy (Regonda et al. 2005; Newman and Watson 2009). Low-elevation areas experienced decreases in snow-water equivalent from 1948 through 2003 (Wilmers and Getz 2005), and these decreases are predicted to continue (Littell et al.

2011). However, increased snow-water equivalent was observed in some higher-elevation locations in and near the park between 1950 and 1997 (Mote et al. 2005), and continued increases in snow-water equivalent at high elevations are not inconsistent with the varied coarse-scale model predictions that are available (Ray et al. 2010; Littell et al. 2011).

There are four moderately high-elevation snow courses at Yellowstone National Park where snow conditions have been measured since the 1930s (Crevice Mountain, 2,512 meters elevation; Lake Camp, 2,371 meters; Thumb Divide, 2,432 meters; and Aster Creek, 2,362 meters). The immediate environs of the Thumb Divide and Aster Creek sites were burned in 1988 (Watson and Newman 2009), and a fire-related decrease in snow-water equivalent at these sites began in 1989 (analysis of Natural Resources Conservation Service data measured on March 1st). Crevice Mountain had no trend in snow-water equivalent from 1939 through 2010, and snow-water equivalent at Lake Camp trended upward at a rate of 34 millimeters per century from 1936 through 2010. The annual peak snow depth at the moderately high elevation of the Lake Yellowstone climate station (2,368 meters elevation) also trended upward at 16 millimeters per century from 1904 through 2010 (National Climatic Data Center cooperative weather station data), while the lower-elevation Snake River and Old Faithful sites trended downward from 1905 through 2010 at 53 and 71 millimeters per century, respectively. Mammoth is at the lowest elevation of the long-term sites, and its peak snow depth trended downward at 182 millimeters per century. Thus, snowpack depth and water equivalent appear to be decreasing at locations below about 2,000 meters elevation but stable or increasing at locations above 2,300 to 2,500 meters (about 40 percent of the park and adjacent mountains; Romme and Turner 1991). The integrated runoff response from the Yellowstone River has been toward earlier spring runoff peaks (Regonda et al. 2005; I. T. Stewart et al. 2005), which suggests that the majority of the park is experiencing shorter winters and longer summers as a result of snowpack changes.

PLANT PHENOLOGY AND CLIMATE CHANGE

Yellowstone's climate influences plant phenology at both the plant and landscape scale. At the plant scale, the phenological development of individual plants from first growth to fruit set and senescence (aging past maturity) are influenced by a number of environmental conditions, such as snow depth, soil moisture, photoperiod, and temperature. The melting of the snowpack initiates the first growth of low-lying vegetation. The interaction between soil moisture, temperature, plant genetics, and photoperiod determines the timing of later phenological stages as well as the amount of biomass accumulated during the growing season.

Studies of plant reproductive phenology elsewhere suggest that flowering date has advanced in response to climate change (Abu-Asab et al. 2001; Fitter and Fitter 2002; Primack et al. 2004; Miller-Rushing et al. 2007), though this effect can be highly species dependent (Tooke and Battey 2010). While some species have considerably advanced their flowering date (Abu-Asab et al. 2001), others have not (N. L. Bradley 1999). Species whose life cycles are strongly synchronized with the start of snow melt may experience a significant increase in fitness if earlier onset of vegetative activity also allows them to benefit from a longer growing season (Galen and Stanton 1995; Dunne et al. 2003; Wipf 2009). For species where snowmelt has little direct impact on their life cycle, other cues such as stable summer temperatures and increased photoperiod may be advantageous (Wipf 2009). Longer day lengths and warm summer temperatures allow them to benefit from peak pollinator densities.

Plants that are adapted to flowering early in the season are most likely to show a strong response to changing environmental drivers (Fitter and Fitter 2002). Evidence from high-latitude systems suggests that flowering and pollinator activity initiate soon after snowmelt (Hoye and Forchhammer 2008; Forrest and Miller-Rushing 2010). However, it is unclear whether organisms are responding to snowmelt as an immediate cue or if the relationship can be explained by associated environmental changes. For example, earlier snowmelt allows for a greater number of days over which warming temperatures can influence both flowering time and insect pollinator development (Thorhallsdottir 1998). Thus, rapid snowmelt facilitated by lower snow accumulation in a previous winter may not advance phenology without a concomitant rise in spring temperatures (Forrest and Miller-Rushing 2010).

Increasing temperatures are not only responsible for triggering an earlier onset of spring but also have other positive impacts on plant growth. Warmer temperatures alter nitrogen mineralization rates, which allow plants greater access to nitrogen. The resulting increase in foliar nitrogen can then trigger higher rates of photosynthetic activity over the duration of the longer growing season (Richardson et al. 2009). Further, earlier spring onset is also known to increase area of leaf exposure to sunlight, resulting in a greater degree of photosynthetic activity (Jolly et al. 2005; Luyssaert et al. 2007). If soil moisture levels are low, however, increased temperature can lead to desiccation-induced reductions in biomass and reproduction.

Unlike strong evidence documenting an earlier start of spring onset, evidence for changes in autumn phenology have been less compelling (Cleland et al. 2007; Hülber et al. 2010). The end of the growing season can be initiated by changes in photoperiod, temperature, genetic sensitivity of plants, and nutrient requirements (Shaver and Kummerov 1992; Oberbauer et al. 1998). This suggests that the majority of effects observed on plant phenology are mediated through changes arising from spring advancement rather than from a delay in autumn phenology. Studies of experimentally warmed plots in Greenland showed that elevated temperatures resulted in earlier phenologies but also compressed life histories (Post et al. 2008a). Whereas some species respond to warming by shifting the date of only one phenological stage (such as leaf emergence), others respond by accelerating their entire life history. This results in fewer days between leaf emergence and seed set under warming scenarios. From an herbivore's perspective, accelerated life histories by plants mean fewer days that they are available as high-quality forage. The Greenland study also revealed that warming synchronized plant phenology across space, resulting in a more narrow time window over which high-quality forage is available to herbivores (Post et al. 2008b). As a consequence, caribou calf survival was much lower during warm years than cool years.

At the landscape scale, snow recedes from low to high elevations with the onset of spring warming. Grasses and shrubs initiate vegetation growth and quickly progress through various phenological stages resulting in a burst of vegetative activity known as the green wave (Frank and McNaughton 1992; Merrill et al. 1993). As the growing season progresses, green biomass concentration peaks early and then decreases as dead tissue begins to accumulate. When grasses enter later phenological states, nutritional values decrease because of lower crude protein content, lower nitrogen, and a concurrent decrease in digestibility resulting from accumulation of indigestible fiber (R. G. White 1983). The peak of the green wave occurs during early phenological states when green biomass concentration is maximized. Herbivores preferentially select plants in this state because they provide the most efficient nutritional yield (Frank and McNaughton 1992). In Yellowstone National Park, this peak occurs on average 28 days after snowmelt (Frank and McNaughton 1992).

The elevational gradient of the green wave in Yellowstone National Park is driven by the influence of topography on temperature and precipitation. At higher elevations, more snow accumulates and lower temperatures prolong the growing season by delaying melt-out. As temperatures warm with the progressing summer, the gradual melt-out is responsible for the distinct elevational gradient of the green wave (Thein et al. 2009). The total snow accumulated in a season can also impact total vegetation productivity because prolonged snow cover leads to an increase in biomass in the subsequent growing season (D. H. Knight et al. 1979). However, this effect was significant only in dry meadows and had little effect in wet meadows, suggesting that increased snowfall and subsequent melt creates a deeper water table that allows grass roots to penetrate further, acquire more nutrients, and produce a greater biomass (Thein et al. 2009).

The green wave can be measured using either field- or satellite-based methods. Frank and McNaughton (1992) measured aboveground net primary productivity by using the canopy intercept method, whereby standing biomass is related to the number of hits a pin makes when it is passed at an angle through vegetation. They measured productivity at several sites across an elevational gradient multiple times through the growing season. Fences were used to exclude ungulates so that they could test the influence of herbivory on rates of productivity among sites. Satellite methods use remotely sensed imagery to indirectly measure plant phenology and biomass. The most commonly used satellite product is the Normalized Difference Vegetation Index (NDVI). Plants absorb light in the visible spectrum to conduct photosynthesis, which results in a low visible spectrum score, and scatter light in the near-infrared spectrum to avoid overheating, which results in a high near-infrared spectrum score. NDVI measures the difference in near-infrared and visible spectrums. Thus, the more leaves a plant has, the higher the NDVI value. Once snow disappears, NDVI values increase as plants begin to grow. NDVI peaks later in the growing season when plants reach maximum biomass and then decreases as plants senesce. The period during which NDVI is increasing most rapidly corresponds with when plants are at their peak nutritive value. This has been validated by studies comparing vegetation reference plots (Hebblewhite et al. 2008; Thein et al. 2009) and studies correlating peak fecal crude protein levels in ungulates (Christianson and Creel 2009) with satellite-derived NDVI values. Unlike field-based methods, satellite methods allow exploration of green wave dynamics over large spatial and temporal scales but cannot control for factors such as herbivory.

In mid-winter (December through February), most grassland areas have negative NDVI values due to snow cover. In early spring (April), large expanses of the northern grassland and lower- to middle-elevation portions of the central range exhibit sharp increases in NDVI values that reflect the combined effects of snowmelt and new vegetation growth. By late spring or early summer (May–June), most grassland areas in Yellowstone National Park have increasing NDVI values, including the high-elevation summer ranges for ungulates, while vegetation growth in some low- and middle-elevation grasslands has already peaked. Yellowstone National Park is at its greenest in mid-summer (July), after which NDVI in most patches slowly decreases in August. Rapid decreases in NDVI during October and November signal the beginning of winter and the end of the growing season (Thein et al. 2009). In some years, autumn rains can initiate a second smaller-scale green-up that may be important to ungulate nutrition heading into winter.

Specific patterns of snowmelt and the onset of new vegetation growth in northern Yellowstone are typically earliest in the Gardiner basin of Montana and progress upslope along the valleys to the south and east, with the high meadows on the Mirror Plateau having the latest onset dates. The patterns of new vegetation growth in central Yellowstone are more complex because the first patches to green up are generally in the mid-elevation meadows (Lower Geyser basin, Nez Perce Creek, Madison Canyon, and Norris) where geothermal features facilitate snowmelt. This is followed shortly thereafter by the lower-elevation Madison Valley region to the west, and later the higher-elevation Hayden and Pelican valleys begin to green. Meadows along the upper reaches of Pelican and Raven creeks are consistently the last areas in the central range to green up. There is considerable local variation in vegetation growth leading to significant overlap in green-up between meadow complexes. Thus, patches just starting to green up can be found in a number of meadows spanning a range of elevations. However, in lower areas onset of the growing season typically occurs approximately 100 days earlier than in the highest areas (Thein et al. 2009). The date of peak biomass is also correlated with elevation, with lower elevations peaking approximately 30 days before higher elevations. Peak biomass dates tend to occur approximately 70 days after onset at low elevations and 40 days after onset at high elevations, which is consistent with the periods of measurable vegetation production reported by Frank and McNaughton (1992). Peak biomass is relatively low in the Gardiner basin (where a lack of precipitation limits production), near geothermal features, and in a few areas in the Madison Valley. High maximum NDVI values indicating dense vegetation biomass occur in the Pelican Valley, despite a relatively short growing season (Thein et al. 2009).

Temperature, through its interactions with snowmelt and precipitation, influences spatial and temporal patterns of the summer green wave in Yellowstone National Park. An analysis of NDVI data for growing seasons between 2000 and 2010 suggests that the peak green wave occurs at higher elevations during years with warmer temperatures. The peak green wave also spans a wider elevation gradient in warmer years. Within elevational bands, a strong interaction between snowmelt and temperature lead to an increase in distance over which green wave is synchronized, with larger patches more pronounced in lower elevations. Thus, climate change is likely to speed the progression of the green wave up the elevational gradient and decrease spatial variability within elevation bands—though there is much uncertainty in this prediction.

Interactions between moisture levels and fire regime could greatly expand or contract the geographic extent of Yellowstone's grasslands. Also, certain warming scenarios may favor the spread of invasive species, which might be less palatable to grazers. Invasive species often have short generation times, high dispersal ability, and high fecundity. Thus, they may expand more quickly than native species into newly suitable habitat as the climate changes (B. A. Bradley et al. 2010). Because of their short generation times, invasive species may also be better able to evolve and adapt to climate change. In addition, carbon dioxide fertilization favors a range of invasive plants, with elevated carbon dioxide increasing the invasion of annual grasses into perennial shrub lands, perennial shrubs into forests, and annual forbs into grasslands (B. A. Bradley et al. 2010).

CONSERVATION IMPLICATIONS

Many ungulates undertake seasonal migrations across distinct geographical ranges to maximize energy intake, and such strategies are believed to have evolved in response to spatial heterogeneity in resource availability (Fryxell and Sinclair 1988; J. Berger 2004). In fact, seasonal fluctuations in vegetation green-up as derived from NDVI data suggest that migratory routes strongly track seasonal changes in vegetation (Boone et al. 2006). As ungulates migrate across the landscape, they preferentially forage on early plant phenological stages to maximize weight gain (Mysterud and Langvatn 2001).

Both snowpack and forage green-up appear to influence the timing of spring migration (10 to 140 kilometers) by northern Yellowstone elk to higher-elevation summer ranges, with elk delaying migration after winters with high snowpack but migrating earlier in years with lower snowpack and earlier vegetation green-up (P. J. White et al. 2010). Elk spending winter at lower elevations are generally able to begin tracking the retreat of snowpack and phenology of vegetation green-up earlier than elk spending winter at higher elevations (P. J. White et al. 2010). Likewise, the 15- to 50-kilometer migrations of Yellowstone pronghorn from their winter range to various summer ranges occur as snow melts and vegetation green-up progresses. Autumn migrations occur before snow covers migrant summer ranges (P. J. White, Davis, et al. 2007). In addition, bison move from higher-elevation summer ranges to lower elevations during autumn through winter, until returning to the summer ranges in June (Meagher 1989b; Bjornlie and Garrott 2001; Bruggeman et al. 2009). As winter progresses and snow depths increase on the summer range, the available foraging areas for bison are reduced to increasingly limited areas at lower elevations and on thermally warmed ground, even though many geothermal areas contain relatively little and poor-quality forage (Meagher 1989b; Bruggeman 2006; Bruggeman et al. 2009). Also, snow melts earlier at lower elevations, so there are earlier green-up and energy-efficient foraging opportunities while summer ranges are still covered with snow (Bjornlie and Garrott 2001; Bruggeman et al. 2006). Thus, there are often mass movements of bison to lower-elevation areas along the western boundary of the park and nearby areas of Montana during mid-April through mid-May that coincide with the onset of vegetation green-up and growth (Thein et al. 2009). The return

Snow is a major factor that influences nearly all the animals in the Yellowstone area. While some species such as small mammals may benefit from a thick layer of snow that protects them from predators, snowpack also makes it difficult for Yellowstone's large ungulates to obtain forage. A bull bison rests after plowing the snow with his massive head to graze. *Photo by Cindy Goeddel.*

migration of bison from lower-elevation winter ranges aligns with temporal and spatial patterns of new vegetation growth, which progresses at the
rate of approximately 10 days for every 300 meters of elevation gained (Despain 1990; Thein et al. 2009).

Ungulates that inhabit seasonal environments typically time their reproductive events around periods of peak vegetation green-up (Rutberg 1987)
to enable temporal overlap with high-quality plant phases during development, thereby allowing juveniles to build up fat reserves (Gaillard et al.
2000). The impacts of climate change might positively or negatively impact
newborn ungulates. Advancement of spring can lead to a longer snow-free
season where forage access and movement are not encumbered by snow. An
increase in growing season length without a corresponding increase in access to high-quality forage, however, might negatively impact herbivores.
Juvenile survival in Alpine ibex *(Capra ibex),* bighorn sheep, and mountain
goats was lower when warming increased the rate of green-up and made
high-quality forage available for a shorter period of time (Pettorelli et al.
2007). Also, earlier plant phenologies as a result of warming temperatures
resulted in a mismatch in the timing of caribou calving and the date of
peak forage resources, resulting in high mortality of newborn calves (Post
et al. 2008b).

Changes in the spatial heterogeneity of the green wave might also im-
portantly impact ungulate weight gain and newborn survival. When green-
up is highly synchronized in space, as is predicted under warming scenarios, ungulates are limited to a shorter period of time during which early
phenological plant stages are available as forage. This can lead to lower neo-
nate survival (Post and Forchhammer 2008) and lower body weights entering winter. A longer growing season, however, allows ungulates access to
snow-free forage for a longer period of time, which can result in higher
body masses (Pettorelli et al. 2005). Warming might also result in milder
winters, leading to increased winter survival of ungulates. These opposing
forces of warming on ungulate foraging dynamics make predicting the ul-
timate consequences of climate change on ungulate populations a difficult
task (Alonzo et al. 2003).

As climate change increases the speed of the phenological development
of plants, resulting in a narrower window of time during which ungulates
have access to high-quality forage, this could tip the predator–prey balance,
causing ungulates to adopt alternate behavioral strategies. In the Clark's

Fork elk herd northeast of the park, some elk are now resident year-round near their historic winter range, while others continue to migrate during summer into the high-elevation meadows of Yellowstone National Park in search of forage. Research suggests that a decrease in forage availability on the summer range for migrant elk due to a contraction of the growing season, combined with increased predation, is leading to a decrease in the number of migratory elk (Milius 2010). This pattern of decreasing migratory elk compared with resident elk is also occurring elsewhere, such as in Banff National Park (Hebblewhite et al. 2006), where supplemental feeding of resident elk is also a factor influencing elk behavior.

Within Yellowstone National Park, wolf prey selection of elk appears to be driven by phenological development on the summer range. In the early years after wolf reintroduction, wolves killed bull elk primarily in late winter after the cumulative effects of the autumn rut and long winter weakened them. This weakened state was reflected by poor bone marrow condition of bulls killed by wolves in late winter. However, wolves began to prey on bulls in early winter during the drought years that followed. These bulls already had poor marrow condition in early winter, possibly due to below average snowfalls in the prior winter contributing to an abbreviated and lower magnitude green wave during spring and summer. This resulted in bulls entering the rut with low fat reserves such that by early winter they were already in poor condition and susceptible to wolf predation. In recent years, above average snowfalls have led to a resumption of the winter predation pattern whereby wolves kill bull elk primarily in late winter.

The dynamics of winter snowpack and summer phenological development of plants are primary drivers in animal movement, physiology, behavior, interactions with other species, and, ultimately, population dynamics. Understanding the links between climate change and these drivers will be critical to informing the ecology and management of Yellowstone's pronghorn, bighorn sheep, mountain goats, deer, elk, bison, and their predators in the years to come. Future work needs to elaborate on mechanisms driving the summer green wave and the resulting influence on wildlife.

CHAPTER TEN

Migration and Dispersal

Key Processes for Conserving National Parks

P. J. WHITE

GLENN E. PLUMB

RICK L. WALLEN

LISA M. BARIL

TWO PROCESSES THAT typify the life histories of many animals worldwide and illustrate the importance of ecological process management are migration and dispersal. Migration has been defined as a seasonal, round-trip movement between separate areas not used at other times of year (J. Berger 2004). Seasonal migrations are the norm for many species to obtain access to greater food supplies, make more efficient use of resources by tracking vegetation growth, and reduce the risk of predation (Fryxell et al. 1988). Dispersal is essentially movement from one area to another without return (at least in the short term; Stenseth and Lidicker 1992), and the term *range expansion* is often used to describe the outward dispersal of animals beyond the limits of the traditional distribution for a population (Gates et al. 2005). Dispersal can regulate populations below the food-limited capacity of the environment to support them by reducing local densities in the source population (Coughenour 2008).

Several authors have documented the worldwide demise of these endangered processes as habitats have been degraded and fragmented and wildlife have been compressed into relatively small, disconnected areas that threaten species, communities, and ecosystems (Brower and Malcolm 1991). For example, more than 75 percent of migration routes for bison, elk, and pronghorn in the greater Yellowstone region of the western United States have already been lost, and many of the remaining routes are shortened or disrupted (J. Berger 2004). The greater Yellowstone ecosystem has been "less altered by human activities than many areas in the United States because about 64 percent of its area is publicly owned" (Hansen 2009, 27). Thus, the opportunity remains to sustain ecological processes and native organisms into the future. However, "critical resources and habitats are underrepresented within the protected lands, and livestock grazing is extensive on the national forests in this region" (Hansen 2009, 27). Also, valley bottoms and flood plains with higher plant productivity and more moderate winter conditions are primarily located on private lands (Hansen 2009). Thus, habitat modification, destruction, and fragmentation have differentially affected areas outside protected lands that are crucial for the migration and seasonal use by many species in this mountain environment (Hansen and DeFries 2007).

There are many consequences to migrant animals and the ecosystems that sustain them if animals cannot move across the landscape (Coughenour 2008). For example, migratory ungulates on grasslands in Yellowstone National Park have strong influences on plants and soil microbes that can increase the availability of nutrients for plants and stimulate the production of leaves, roots, and stems (Frank and McNaughton 1993; Frank and Groffman 1998; Frank et al. 2002). Thus, impaired migration by ungulates could diminish the transfer of nutrients across the landscape and spatially reorganize energy and nutrient dynamics in ecosystems (Singer and Schoenecker 2003; Holdo et al. 2006). Likewise, barriers to migration, such as fencing and urbanization, or incentives not to migrate, such as supplementary feeding, could result in ungulates becoming more sedentary and concentrating grazing in smaller areas, with negative consequences for rangelands (Coughenour 2008).

Understanding the consequences of impaired migration and dispersal is essential for preserving species, communities, and ecosystems. However, these relationships have not been well documented, and past and present land-use changes have rarely considered the importance of migratory and dispersing animals as "agents of change" in ecosystems (Coughenour 2008, 33). This chapter provides examples of the importance of sustaining or restoring migration and dispersal by wildlife in Yellowstone National Park that are integral components of the greater Yellowstone ecosystem.

BISON

Bison historically occupied about 20,000 square kilometers in the headwaters of the Yellowstone and Madison rivers (Schullery and Whittlesey 2006). They were nearly extirpated in the early twentieth century, with Yellowstone National Park providing sanctuary to the only remaining wild and free-ranging population (Plumb and Sucec 2006). Intensive husbandry, protection, and relocation were used to restore the population (Meagher 1973), and in summer 2012, about 4,200 bison were an integral part of the ecosystem.

Bison seasonally migrate and began to expand their winter range from the interior of Yellowstone National Park onto lower-elevation areas along the park boundary and into Montana as numbers increased during the 1980s and bison began to experience nutritional deficiencies (Meagher 1989b; Taper et al. 2000; DelGiudice et al. 2001; Coughenour 2005). However, these movement processes occurred well before undernutrition became sufficient to decrease survival and recruitment, suggesting that migration and dispersal served to limit the numbers of bison feeding in the interior of the park below the food-limited capacity of the environment (Owen-Smith 1983; Coughenour 2005, 2008). During winter, deep snow accumulates on much of the bison range inside the park and restricts their access to forage. Thus, the numbers and timing of bison moving to lower-elevation winter range in or outside the park increases as snow builds up in the park interior (Bruggeman et al. 2009; Geremia et al. 2011).

Although Yellowstone National Park provides a large amount of habitat for bison, it does not provide sufficient habitat for the population during some winters when deep snows limit access to forage at higher elevations. As a result, some bison migrate to low-elevation habitat outside the park in search of forage during winter and spring, similar to deer, elk, and pronghorn. For example, severe weather with deep, prolonged snow induced about one-half of the 3,900 bison counted in summer 2010 to exit the park in search of food. If bison are forced to remain in the park by humans, as was largely the case until recent years because of the risk of disease (brucellosis) transmission to cattle in Montana, then the bison population would be limited by the amount of food available in the park. Under these circumstances, bison numbers could increase during a series of milder winters and cause significant deterioration to other park resources (such as vegetation, soils, and other ungulates) and processes as they exceed the capacity of the environment to support them. At some point, bison would then die in large numbers from starvation during drought conditions or a winter with deep snowpack that limited available forage (Coughenour 2005).

Alternatively, management culls (removals) could be implemented to keep bison numbers in the park well below the food- and snow-limited capacity of the environment to support them. Either way, ecological processes within the park could be diminished by a lack of human tolerance for the natural processes of migration and dispersal by bison outside the park, and the suitability of the park to serve as an ecological baseline or benchmark for assessing the effects of human activities outside the park could be diminished (Boyce 1998; Sinclair 1998; Coughenour 2008).

Human-imposed restrictions on migration and dispersal resulted in culls of more than 1,000 bison in some winters (1997, 2006, 2008) and perpetuated irruptive (boom-and-bust) dynamics in the bison population that could have consequences to the age, breeding herd, and sex structure of the population that persist for decades (P. J. White, Wallen, et al. 2011). Thus, a coalition of federal, state, and tribal managers recently agreed to management practices that would decrease the need for large culls of bison and maintain more population stability. These practices included increased tolerance for bison migrating to habitat outside the northern and western boundaries of Yellowstone National Park in Montana (U.S. Department of the Interior et al. 2008; Interagency Bison Management Plan Agencies 2011). From 2008 through 2011, up to 700 bison migrated beyond the western boundary of the park and accessed suitable habitat in the Hebgen basin of Montana (P. J. White, Wallen, et al. 2011). In addition, during 2011 more than 300 bison migrated north of the park boundary onto habitat in the Gardiner basin of Montana.

PRONGHORN

Yellowstone pronghorn were once numerous (1,000 to 1,500 animals) and migrated 80 to 130 kilometers down the Yellowstone River from higher-elevation summer ranges in what is now Yellowstone National Park to lower-elevation winter ranges in Montana (Skinner 1922). However, human settlement reduced pronghorn numbers and effectively eliminated their migration outside the park sometime before 1920 (Skinner 1922). Feeding, irrigation, and fencing efforts until 1934 apparently reinforced the tendency for some pronghorn to remain on the winter range year-round (Skinner 1922; Keating 2002). By the 1970s, there were serious concerns about the long-term viability of the population because low abundance (fewer than 150) increased their susceptibility to random, naturally occurring catastrophes (D. B. Houston 1982), and their migration still effectively terminated at the park boundary (Caslick 1998).

Historically pronghorn inhabiting the northern portion of Yellowstone National Park in summer migrated down the Yellowstone River valley to spend winter at lower elevations outside the park. As a result of a number of anthropogenic factors, today's migrations are less extensive but still involve animals seasonally moving in and out of the park, like this small herd photographed near the famous Roosevelt Arch that greets visitors to the park at the northwestern entrance near the town of Gardiner, Montana. *Photo by Cindy Goeddel.*

The pronghorn population appears to be limited by the amount of food that is available on its truncated winter range near the northern boundary of the park (P. J. White, Davis, et al. 2007). This winter range may no longer support populations of more than 500 pronghorn for sustained periods because of habitat fragmentation and degradation of sagebrush caused, in part, by historic farming and ranching in the area, followed by intense browsing from congregated ungulates, such as elk, during winter (Boccadori et al. 2008). Thus, pronghorn numbers can decrease rapidly in response to decreased food availability in this key resource area, especially during periods of extended drought or severe winters (P. J. White, Bruggeman, and Garrott 2007). As a result, impaired migration and dispersal in this pronghorn population have contributed to irruptive dynamics rather than a tendency toward more gradual fluctuations (P. J. White, Bruggeman, and Garrott 2007).

In spring and summer, most pronghorn (80 percent) follow the green wave of vegetation growth as snow melts at progressively higher elevations in the park (P. J. White, Davis, et al. 2007). These movements enable pronghorn to use nutritious food when it is available and release the lower-elevation winter range from intensive use for a portion of the year. Also, migrating pronghorn appear to have somewhat higher survival rates among adults (93 to 97 percent) and fawns (13 to 25 percent) through summer than among nonmigrants (adults: 78 to 98 percent; fawns: 2 to 14 percent) that remain on the winter range year-round (Barnowe-Meyer 2009). Coyotes account for about 63 percent of predation on adult pronghorn and up to 81 percent on fawns (Barnowe-Meyer et al. 2009). However, high numbers of wolves on the summer ranges of migrating pronghorn apparently increased the survival rates of adult female pronghorn and their fawns by altering the behavior, spatial distribution, and predation risk from coyotes compared with the pronghorn remaining on the winter range where wolf densities were quite low (Barnowe-Meyer et al. 2010).

During the past several decades, the proportion of pronghorn migrating and dispersing apparently increased (P. J. White, Davis, et al. 2007) as the greater Yellowstone ecosystem experienced a severe and extended drought and food on the pronghorn winter range became even lower quality and quite patchy (Boccadori et al. 2008). In 2000, a small herd of pronghorn was detected approximately 30 kilometers north of Yellowstone National Park in the southern portion of the Paradise Valley in Montana. This herd increased to 82 animals by 2009 and represents the first significant return of pronghorn to the southern Paradise Valley since the early twentieth century (Lemke 2009). Genetic and telemetric data suggest that this population was likely started or supplemented by Yellowstone pronghorn and is still maintained by frequent dispersal from the park (K. Barnowe-Meyer, University of Idaho, unpublished data). The persistence of this recently formed population, with dispersal and gene flow between the two populations, would improve the long-term viability of the Yellowstone population.

ELK

The largest elk population in Yellowstone National Park spends the winter on grasslands and shrub steppes along the northern boundary and nearby areas of Montana (D. B. Houston 1982; Lemke et al. 1998). Northern Yellowstone elk migrate seasonally, moving to higher-elevation ranges throughout the park during summer and returning to the winter range in autumn (Skinner 1925). Their summer range includes the majority of Yellowstone National Park, and elk that spend winter at lower elevations in or outside the park tend to spend summer in the western part of the park, while elk that spend winter at higher elevations in the park generally spend summer primarily in the eastern and northern parts of the park (Craighead et al. 1972; P. J. White et al. 2010). However, when the park was established in 1872, portions of the winter range outside the park at lower elevations were intensely hunted and converted to domestic livestock grazing and irrigated hayfields (D. B. Houston 1982). Thus, a large proportion of elk were essentially forced to spend winter inside the park, which is generally at higher elevation than the surrounding area and receives more snowfall. Also, a combination of hunting and management culls from the 1930s to 1968 eliminated several migration routes and population segments (Craighead et al. 1972). However, the population reestablished winter range north of the park boundary and into the Paradise Valley of Montana during the late 1970s in response to increasing elk abundance, changes in the structure and timing of hunter harvests, and protection of winter ranges outside the park (Coughenour and Singer 1996; Lemke et al. 1998). This range expansion provided the population with increased access to food and spurred rapid growth in abundance (Eberhardt et al. 2007).

Migration in summer increases access to nutritious food (Merrill and Boyce 1991) and distances elk from the most concentrated area of wolf activity in the northern portion of the park, thereby decreasing mortality (P. J. White et al. 2010). Migration also dilutes the risk of predation for individual elk because they mix with other elk from about seven other discrete populations that winter outside the park but occupy portions of it during summer. During winter, one segment of the northern Yellowstone elk population primarily occupies middle to upper elevations in the park, another herd segment occupies middle to lower elevations in and outside the park, and some elk move across the northern range in response to variations in the snowpack (Craighead et al. 1972; Hamlin 2006). This spatial structuring exposes elk to different risks of mortality because elk that spend winters at higher elevations are killed by wolves at a higher rate (D. W. Smith 2005; Kauffman et al. 2007), while elk that spend winters at lower elevations outside the park avoid high densities of wolves but are exposed to hunter harvest (P. J. White and Garrott 2005a). Since wolf restoration between 1995 and 1997 and substantial reductions in hunter harvest since 2002, elk spending winter inside the park, where predator densities are quite high, have had lower survival rates (85 to 89 percent) and recruitment (11 to 14 calves per 100 adult females) compared with elk spending winter outside the park (survival: 91 to 99 percent; recruitment: 17 to 22 calves per 100 adult females), where predator densities are lower (Evans et al. 2006; Cunningham et al. 2009).

These differences in recruitment and survival have contributed to a higher proportion of the elk population now spending winter outside the park and migrating to summer areas in the western portion of the park (P. J. White et al. 2010, 2012). The diversity and flexibility of migration tendencies in northern Yellowstone elk have enhanced the stability of this population living with some of the highest wolf densities reported for North America (D. W. Smith 2005). Numbers of northern Yellowstone elk that spend winter in upper elevations of the park have decreased by about 80 percent since wolf restoration, similar to decreases observed in the nonmigratory Madison headwaters elk population in the west-central portion of the park (Garrott et al. 2009b). Conversely, numbers of northern Yellowstone elk that spend winter at lower elevations outside the park initially decreased following wolf restoration but then restabilized after hunter harvests were greatly reduced (Hamlin 2006).

TRUMPETER SWANS

Yellowstone National Park supports resident, nonmigratory trumpeter swans through the year as well as regional migrants from the greater Yellowstone ecosystem and longer-distance migrants from Canada and elsewhere during winter. During spring and summer, swans occupy small lake, river, and wetland habitats throughout the park. However, nesting habitat used by trumpeter swans in Yellowstone National Park is marginal because nesting lakes are small, shoreline complexity is low, shorelines are commonly timbered, habitat for feeding and nesting is often discontinuous, and feeding areas are generally only at the outside edges of wetlands because of deeper water at the center of lakes (Gale et al. 1987). During winter, aggre-

Many of the wildlife species that inhabit the Yellowstone region are seasonal migrants that routinely move between distinct summer and winter areas. Although trumpeter swans are year-round residents of the greater Yellowstone area, large numbers of migrants, like these birds flying across Swan Lake in December, arrive every year from Canada to winter along the unfrozen waterways of the Yellowstone area. *Photo by Cindy Goeddel.*

gations of trumpeter swans within the park occupy ice-free waters along the shores of lakes and along rivers.

Counts of resident adult trumpeter swans in Yellowstone National Park decreased from a high of 69 in 1961 to 4 in 2009, while the abundance of the overall Rocky Mountain population increased from fewer than 1,000 to more than 5,000 swans (Proffitt, McEneaney, et al. 2009; Baril et al. 2010). Evidence suggests that dispersal of swans from the larger subpopulation in the Centennial Valley of Montana and elsewhere may be an important factor for maintaining resident swans in Yellowstone National Park by filling vacant territories or pairing with single adult birds (McEneaney 2006). Decreases in the numbers of resident swans in the park became more dramatic after supplemental feeding of grain outside the park was terminated in the winter of 1992–1993 (Proffitt, McEneaney, et al. 2009). Thus, Yellowstone National Park may be reliant on swans dispersing from more productive areas within the region, with the dynamics of resident swans being influenced by management actions outside the park. In addition, the high-elevation habitat in Yellowstone National Park provides marginal conditions for nesting, which results in chronically low numbers of nesting pairs and fledglings (Proffitt, McEneaney, et al. 2010). This effect has been compounded over the last several decades by natural changes in habitat, such as decreased wetlands due to long-term drought or chronic warming, and the recovery of predator populations. Thus, the low number of resident swans in Yellowstone National Park will likely continue unless there is an increase in swans dispersing from nearby areas into the park.

BALD EAGLES

Because of a population decline caused by pesticides and other factors, the bald eagle was protected as an endangered species in the United States in 1967. However, habitat protection, management actions, and reduction in levels of certain types of pesticides (DDT) resulted in significant increases in the breeding population, and the bald eagle was declared recovered in 2007 (U.S. Fish and Wildlife Service 2007b). Numbers of nesting and fledgling bald eagles in Yellowstone National Park also increased incrementally since 1984, with 31 to 34 nest attempts per year since 2001 (Baril et al. 2010). Resident and migrating bald eagles are now found throughout the park, with nesting sites located primarily along the shorelines of lakes and larger rivers.

Yellowstone National Park supports resident, nonmigratory bald eagles through the year as well as regional migrants from the greater Yellowstone ecosystem and longer-distance migrants from Colorado and elsewhere during summer (Swenson et al. 1986). Bald eagles move to their nests in March or April and remain nearby through July, after which they disperse through the park. By November, they are generally on their winter range, which in Yellowstone consists of stretches of open water. Bald eagles in Yellowstone National Park have highly variable but generally low nest success (34 percent) and productivity (one young per occupied nest) because of severe spring weather and late ice breakup at elevations higher than 2,100 meters (Swenson et al. 1986). Thus, recruitment of eagles outside the park, with subsequent migration and dispersal into the park, has probably maintained the population in Yellowstone National Park (Swenson et al. 1986). Dispersal into the park is also important to provide mates for eagles on territories that have lost their mate or failed at reproductive attempts previously. In addition, the limited availability of food in Yellowstone during winter (ice-up) results in more than 80 percent of adults and essentially all subadults moving out of the area (Swenson et al. 1986). These movement dynamics may become more important in the near future because of the collapse of cutthroat trout numbers in and near Yellowstone Lake as a result primarily of the introduction of non-native lake trout. Numbers of bald eagles spending summer in the park and their migration tendencies may be altered due to decreased food and attrition of eagles on nesting territories. Thus, migration and dispersal of eagles into the park will become even more important for population stability.

SONGBIRDS

Approximately one-half of the 70 species of songbirds breeding in Yellowstone National Park migrate long distances to the southern United States (north temperate migrants) or western Mexico and northern Central America (neotropical migrants; Carlisle et al. 2009). The remaining species are residents, with at least some portion of each population migrating short distances to lower-elevation wintering areas within the greater Yellowstone ecosystem (altitudinal migrants). An additional handful of songbirds migrate to the greater Yellowstone ecosystem from their breeding areas in Canada (Sibley 2003). Riparian areas dominated by cottonwoods and willows support a greater diversity and abundance of birds during migration than any other habitat type in the region (Skagen et al. 2005; Carlisle et al. 2009). Diversity during the breeding season is also high, with nearly one-half of all bird species in the region nesting only within this habitat type (Skagen et al. 2005).

Deciduous woody vegetation represents roughly 3 percent of the landscape in the greater Yellowstone ecosystem and occurs mostly in productive valley bottoms at lower elevations overlapping extensively with private land (Parmenter et al. 2003). Land use in the greater Yellowstone ecosystem has traditionally been a mix of agriculture and ranching, but more recently development outside towns and cities has emerged as a rapidly expanding land-use type (P. H. Gude et al. 2007). Maintaining riparian corridors is crucial to preserving bird migration as an ecological process within the greater Yellowstone ecosystem, but the condition of riparian corridors for migrating birds has been degraded by these land uses. For example, cattle grazing can reduce vegetation structure and increase erosion, irrigation and damming can limit cottonwood and willow recruitment and increase die-off, and human settlement adjacent to riparian areas can increase predation risk by household pets and predators attracted to developments (Ammon and Stacey 1997; Stromberg 2001; Hansen and Rotella 2002).

Though approximately 68 percent of the greater Yellowstone ecosystem is publicly owned, the majority of protected areas occur in less productive regions located at relatively high elevations (P. H. Gude et al. 2007) where summer is shorter and bird species diversity is lower (Finch 1989). Thus, lower-elevation riparian areas, which may be increasing in extent following wolf restoration and significant decreases in the abundance and distribution of browsing elk, tend to be hotspots for bird diversity and may serve as sources of birds that can disperse to less productive, higher-elevation regions (Hansen and Rotella 2002). However, development in rural areas can reduce nesting success of songbirds by attracting higher densities of nest predators, such as skunks *(Mephitis mephitis)*, raccoons *(Procyon lotor)*, and black-billed magpies *(Pica hudsonia)* or brood parasites such as brown-headed cowbirds *(Molothrus ater)* than would otherwise occur (Ammon and Stacey 1997; Hansen and Rotella 2002). If this occurs, then fewer birds will be available for dispersal into Yellowstone National Park and other high-elevation protected areas, and the populations of songbird species will likely be reduced in these regions despite their protected status (Hansen and Rotella 2002).

CONSERVATION IMPLICATIONS

These examples indicate that the ecological process of migration is essential for the long-term conservation of many species occupying protected areas such as parks in temperate mountain environments. Many bald eagles, bison, elk, pronghorn, songbirds, trumpeter swans, and other species in Yellowstone National Park seasonally migrate across the park boundary to obtain access to greater food supplies and make more efficient use of available resources. Migration was also important for reducing predation risk for elk and pronghorn, which in turn contributed to population stability by increasing survival and recruitment in some segments of the populations. Likewise, dispersal appears essential for the persistence and health of some wildlife populations. Bison, elk, and pronghorn expanded their winter ranges to lower elevations in and outside Yellowstone National Park as local densities increased and additional resources were needed. Also, the dispersal of wolves and grizzlies outside the park, despite the higher mortality incurred as a result of humans, has been a great success and increased the range of these two threatened species. Conversely, bald eagles and trumpeter swans relied on dispersal into the park to provide a source of mates.

Though the migration and dispersal of numerous species within and outside Yellowstone has improved substantially since the implementation of ecological process management, much remains to be done to sustain this progress and ensure it is not fleeting. For example, the National Park Service principles for managing biological resources provide scant mention of key processes such as migration and dispersal (National Park Service 2006). Thus, there is an immediate need for the agency to acknowledge that the protection of species inside park units is insufficient for preserving biodiversity and ecosystems into the future and to promote initiatives that focus on effectively maintaining, protecting, and restoring key ecological processes across jurisdictions and paradigms (Boyce 1998). Ideally, these efforts would involve strong partnerships among state and federal agencies,

American Indian tribes, and other stakeholders, such as nongovernmental organizations and private landowners, to assess risks (including uncertainty), develop a collaborative vision and agenda, monitor the resiliency of the system and how it evolves in response to change, discuss alternate management approaches, and implement proactive measures to ensure that unacceptable changes do not occur (Peterson et al. 2003). Management of migratory wildlife will require linking ecological understanding of populations and their movements with social and economic concerns related to human–wildlife interactions so that communities can develop effective ways of managing them (Gordon et al. 2004; du Toit et al. 2004). Also, the management of animal–human conflicts will be essential for maintaining species such as bison, grizzly bears, and wolves outside preserves, where they are currently most limited by interactions with people and livestock.

Ecosystem management would enhance connections between park systems and lands under other jurisdictions (Yung et al. 2010). Unfortunately, "the transition to ecosystem-based management in the greater Yellowstone area has been a piecemeal, faltering process" (Keiter and Boyce 1991, 405). Federal and state agencies have succeeded in establishing cooperative, interagency, institutional structures, such as regional committees with public involvement, to break down boundary-based management traditions and coordinate the use of science in natural resource management decisions (Keiter and Boyce 1991; P. J. White, Garrott, and Olliff 2009). There is also wide recognition that ecological processes are dynamic and do not end at the boundaries of protected areas (National Research Council 2002). However, the continuation of wildlife migration across park boundaries and tolerance of wildlife on human-dominated systems outside Yellowstone National Park and other protected areas is the prerogative of surrounding states that manage wildlife and private citizens in these areas (Plumb et al. 2009). Many state and local officials have strong, opposing values from federal land managers on how to integrate humans and nature in this system, and as a result, private land owners have not been convinced that the maintenance of healthy ecosystems and the processes that sustain them is in their interests (Keiter and Boyce 1991). Scrutiny by opposing groups and congressional representatives has served to stalemate rather than facilitate the process of ecosystem-based management. Thus, comprehensive natural resource policies have not been defined or implemented (Keiter and Boyce 1991), and coordinated management of the greater Yellowstone ecosystem is unlikely to become reality in the near future. Rather, it is likely that the extent to which processes such as migration and dispersal are preserved in the greater Yellowstone ecosystem will be forged one issue or species at a time in the political arena, either through legislation or litigation, rather than comprehensively (Keiter and Boyce 1991; Schullery 2004).

Regardless, the National Park Service has a responsibility to promote management priorities for ecosystems or watersheds to ensure that essential ecological processes to preserve park resources are sustained into the future (Keiter and Boyce 1991). What happens in Yellowstone, the world's first national park, has influenced the debate on conservation approaches throughout the world for over a century and will continue to do so in the future. Thus, the National Park Service should continue to lead efforts to preserve processes that sustain ecosystems, even if recommendations sometimes conflict with human activities inside the park or on public and private lands outside park unit boundaries. Over time, ecosystem process management will enhance the lives of people living in areas near reserves and other public lands by providing sustainable resources, such as clean water, wildlife populations, and vegetation communities. It will also enable federal and state agencies to better address new challenges that complicate the management of protected areas, including changing climate, massive forest die-offs, and diseases that can be transmitted among wildlife, livestock, and people.

CHAPTER ELEVEN

Have Wolves Restored Riparian Willows in Northern Yellowstone?

N. THOMPSON HOBBS

DAVID J. COOPER

THE REINTRODUCTION OF wolves to Yellowstone represents one of the most important natural experiments in the history of ecology, offering an unparalleled opportunity to understand how predators change the operation of fundamentally important ecological processes. In particular, wolf reintroduction created the chance to understand how a top predator influences its plant-eating prey and how changes in those prey influence plants. Predators can reduce the numbers of herbivores or change their behavior so that plants benefit from the presence of predators. Ecologists call this series of effects a trophic cascade. Trophic means feeding or nutrition, and the term *trophic cascade* describes how the effects of predators on herbivores modify the effects of herbivores on plants. The idea is essentially the same as the proverb that has originated in many cultures throughout the world— the enemy of your enemy is your friend. That is, by reducing the abundance of herbivores or changing their feeding behavior, predators release plants from the effects of being eaten.

A related concept, fundamentally important to managing ecological processes worldwide, is what ecologists call *alternative states*. The state of an ecosystem is shorthand for the configuration of plants, animals, and the physical environment that characterize an ecosystem at any given time. Ecologists have long recognized that the same ecosystem can exist in dramatically different states and that stresses, particularly stresses resulting from human management of ecological processes, can cause these states to change rapidly, sometimes without warning. For example, excessive harvest can cause the collapse of fisheries, where desirable species suddenly become so rare in the ecosystem that they can no longer support commercial fishing. A desirable state of this ecosystem includes abundant, commercially valuable fish, while in the alternative, degraded state, fish become so sparse and small in size that they cannot support commercial harvest.

In many cases, eliminating the stress that caused the original state change cannot quickly reverse the changes to an undesirable ecosystem state. For example, overgrazing by livestock can cause perennial grasslands to change to a state dominated by annual plants and shrubs. Because overgrazing causes changes in the fertility of soils and their ability to retain water, simply removing grazing does not allow the return of grasslands on dry, infertile soils. Alternative states that resist change to their original condition, even when the stressor causing the state change is removed, are called alternative stable states. Understanding the forces that stabilize alternative states can provide useful information for the practice of ecosystem restoration (Suding et al. 2004).

The concepts of trophic cascades and alternative stable states frame the discussion of ecological processes offered in this chapter. We focus on two questions. First, did the extirpation of wolves from Yellowstone during the early twentieth century cause a state change in the riparian plant communities of northern Yellowstone? Second, is this new state stable, or has the reintroduction of wolves restored the riparian zone to its historical condition as a result of a trophic cascade?

Historic photographs and monitoring data show that during the late 1800s and early 1900s, communities of tall willows dominated riparian zones along small streams throughout northern Yellowstone, extending up to 40 meters laterally from stream margins. The life histories of willows are tightly tied to streams. Willow seeds are aerially dispersed in early summer, and the seeds require moist, bare mineral soil to germinate. These are the kinds of soils that occur along streams and on their floodplains but are not found in the uplands above the reach of floodwaters. Moreover, willows are water-loving plants—they thrive when their roots can reach groundwater, ensuring a steady supply of water to support growth.

Woody plants like willows are critical parts of riparian ecosystems throughout the world (Hughes 1997). Their sturdy upright stems and strong perennial roots prevent erosion of stream banks, increase deposition of nutrient-rich sediment during floods, maintain stream water quality, and contribute large woody debris and fine particles to streams (Naiman and Decamps 1997). They provide habitat for a wide range of plants and animals.

In particular, abundant and tall willows are a critical component of habitat for beaver along small streams, providing food and materials for dam building. Long, flexible stems from willows serve as reinforcements that strengthen the wood, rock, and mud dams that beavers build to impound water. Beavers live in lodges in the ponds that form behind their dams, lodges that are made safe from predators by the surrounding water. A large biomass of willows or aspens near streams is needed as a food source and to provide materials for dams. Willows are vital to beaver.

Dam building by beavers is also vital to willow. Beaver dams raise local water tables and encourage the deposition of fine-grained sediment along streams (Meentemeyer and Butler 1999), producing conditions that are especially favorable for willow establishment and growth (Bigler et al. 2001; Bilyeu et al. 2008). Even after beaver abandon a dam, it may impound water for decades, thereby continuing to provide ideal habitat for willows. Once a beaver pond drains, bare sediment is exposed, providing sites for willow establishment (Read 1958). Thus, the presence of willows benefit beaver, and the presence of beaver benefits willows, thereby forming a mutualistic relationship where both species benefit (Baker et al. 2005).

We refer to the historic condition of the landscape where riparian zones were dominated by tall willows and where stream networks were punctuated by beaver dams as the *willow state*. Comparisons of modern and historic photographs illustrate that striking reductions in the abundance and stature of willow occurred in northern Yellowstone during the twentieth century (Kay 1990; National Research Council 2002). Willows were replaced by grasses and herbaceous flowering plants (dicots) in many areas (D. B. Houston 1982; Engstrom et al. 1991; Singer et al. 1994), leading to an alternative condition in the riparian zone that we will call the *grassland state*. In this state, willow communities are limited to small isolated

Beaver generally require extensive stands of aspen or willow in close association with streams and rivers for both food and as materials for constructing lodges and dams. This beaver has pulled a willow branch out from a cache under the ice of the Lamar River and clambered out onto the ice to eat. *Photo by Cindy Goeddel.*

fragments, active beaver dams are absent, and individual willow plants are short in stature.

Coincident with the loss of willows, beavers abandoned the small streams of northern Yellowstone (E. R. Warren 1926; Jonas 1955). Almost a third of mainstream reaches historically experienced beaver-related aggradation or deposition of sediment (Persico and Meyer 2009), and the absence of beaver dams transformed many areas of the riparian zone. Competing explanations exist for the disappearance of beaver from northern Yellowstone. One idea is that increasing elk populations in the early twentieth century excluded beaver by outcompeting them for food (Kay 1990). Another possibility is that warmer temperatures and lower stream flows in recent decades may have initiated beaver abandonment (Persico and Meyer 2009). Irrespective of its cause, the loss of beaver dams from the landscape created feedbacks that increased the stability of the alternative grassland state: the absence of willow removed an important resource for beavers, and the absence of engineering by beavers degraded habitat for willow (Wolf et al. 2007).

DID THE EXTIRPATION OF WOLVES CAUSE A STATE CHANGE?

Accumulating evidence suggests that the loss of wolves from the food web of northern Yellowstone led to the loss of willows and other woody deciduous plants from the ecosystem by allowing excessive browsing on willows by an abundant elk population (National Research Council 2002). Although it is impossible to absolutely establish cause, many scientists have concluded that the shift in ecosystem state from willows to grasslands resulted from elevated numbers of elk after wolves were exterminated during the early 1900s. Here we review the evidence supporting this conclusion.

Northern Yellowstone historically provided winter habitat for the largest migratory elk herd in North America. The area where elk spend the winter is known as the northern range. Count data for northern Yellowstone elk extend from the 1930s to 2011 (Figure 7; Taper and Gogan 2002; Eberhardt et al. 2007; Yellowstone National Park, unpublished data). Data on the numbers of elk spending winter on the northern range before the elimination of wolves are not available, but it is likely that the elk population grew rapidly after the elimination of market hunting when the park was established in 1872. Park records show that there was concern over the impacts of wintering elk on shrubs within the park during the early twentieth century, and as a result, culling was initiated in the 1920s to maintain the population at levels that would not overbrowse vegetation.

Although thousands of animals were removed from the park during the period from 1920 through 1967, it was unknown what level of reduction was needed to prevent overbrowsing. Thus, there is no way to know whether numbers of elk during this period were excessive relative to their food supply or what the elk population size would have been if wolves had been part of the food web. During 1968, all culling of elk ceased as Yellowstone National Park began a program of ecological process management known as natural regulation, where elk numbers within the park were allowed to be regulated entirely by their access to food during winter and the effects of severe winter weather (National Research Council 2002).

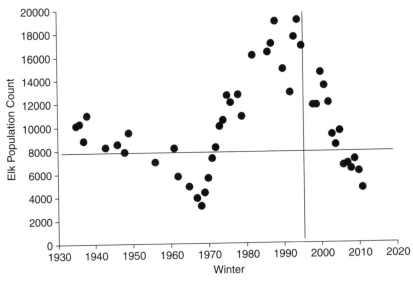

Figure 7. Counts of northern Yellowstone elk from 1930 through 2011 (D. B. Houston 1982; Taper and Gogan 2002; National Park Service, unpublished data). The horizontal line shows the mean population size (7,300) from 1930 through 1970, the approximate time interval when state change occurred. The vertical line indicates when wolves were reintroduced (1995).

From 1930 to 1990, dramatic changes occurred in willow communities on the northern range. One of the ways that scientists are able to understand past events in plant communities is through dendrochronology—the study of characteristic patterns of annual growth rings in woody plants. By counting rings in cross-sections of plant tissue, each ring representing one year of growth, it is possible to determine the year when an individual plant was established. Dendrochronological studies provide the best objective evidence that change from the willow to the grassland state occurred after the elimination of wolves from the food web of the northern range.

Two sets of dendrochronological observations are particularly compelling (Wolf et al. 2007). First, establishment of willows appears to have decreased dramatically after the 1930s, essentially producing missing generations from the age distribution of willows (Figure 8). Older-age willows were absent from the dendrochronological record, indicating that plant establishment or survival was reduced from the 1930s through the 1990s (Wolf et al. 2007). In addition, by aging plants and studying the character-

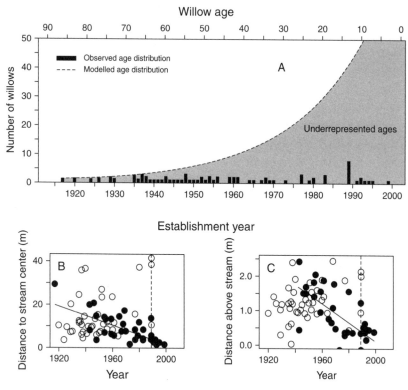

Figure 8. (A) Dendrochronological studies (Wolf et al. 2007) reveal that there were entire generations of willows that failed to show up in the historic record obtained by aging plants. After 1930, the age distribution suggests reduced rates of establishment and/or survival. (B) After 1940, willows established primarily on gravely soils (closed circles) adjacent to streams. Before that period, plants established on beaver pond margins (open circles) that extended up to 40 meters from the current stream center. The line fit through points for establishment near streams (B) shows trends in area of establishment habitat and (C) stream down cutting. Exposed, fine-grained sediment following the fires of 1988 stimulated seedling establishment (dashed vertical lines).

istics of soils where they germinated, scientists were able to show that the locations on the landscape where willows germinated shifted from fine-particle soils on margins of beaver ponds, extending up to 40 meters from the center of streams, to areas with gravelly soils immediately adjacent to the stream. This pattern indicated that the width of riparian habitat establishment along stream corridors was 20 to 40 times greater in 1930 than in 1990.

These same studies showed that in the 1930s, willows established on floodplain locations that were meters above the current stream channel. How could this be? The data on change in elevation of the willow germination surface indicates that a change in the streambed's elevation has occurred. That is, stream channels in the 1990s were more deeply cut into their flood plain relative to their location in the 1930s, a process known as incision. Radiocarbon dating of woody materials in eroded stream banks indicated that the stream incisions that occurred during the last century were greater than any seen for at least hundreds of years (Wolf et al. 2007), although they may not be unprecedented in the geologic record of the past few thousand years (Persico and Meyer 2009).

The loss of beaver dams from the northern range can explain the narrowing of habitat for willows as well as stream incision. Narrowing of habitat likely occurred as a result of lowered water tables and reductions in the area of exposed substrate and moist soils following the loss of dams from steam channels (Wolf et al. 2007). Stream incision can be explained by increased velocity of stream flows. Beaver dams slow the flow rates of streams, reducing their ability to erode their banks. Unimpeded by beaver dams, stream flows could accelerate, causing locally significant incision that effectively disconnects stream channels from their historic floodplains. For reasons that are not well understood, stream incision also occurred in areas of the northern range that were not associated with historic activity of beaver (Persico and Meyer 2009). This could be due to the loss of stream bank stability provided by roots of willows.

Historic photographs also provide evidence of a state change from willows to grassland on the northern range. This evidence is based on comparisons of pictures of willows early in the twentieth century to pictures from later decades (D. B. Houston 1982; Kay 1990; Meagher and Houston 1998; National Research Council 2002). It is hard to argue that the photographic record shows a state change in riparian zones from 1930 to 1990, though there are problems with interpreting them (as will be discussed in the next section).

These data strongly implicate the loss of wolves from the northern range as a cause for the change from the willow state from 1930 to 1990 (National Research Council 2002). Changes in climate and the absence of beaver are also implicated (Wolf et al. 2007; Persico and Meyer 2009). This evidence is consistent with the idea that during the late 1800s and early

1900s, a trophic cascade extending from wolves to elk to willows maintained a state on the northern range that differed markedly from the state during the 1990s.

It has been hypothesized that the prevailing grassland state along stream corridors has recently reorganized to the willow state as a result of a trophic cascade caused by the successful restoration of wolves to Yellowstone in 1995. The central idea in the trophic cascade hypothesis is that the indirect effects on elk foraging behavior have released willows from top-down control by herbivory and have promoted a rapid recovery in willow biomass, distribution, and stature (Ripple and Beschta 2004a, 2006; Beschta and Ripple 2007, 2010; Beyer et al. 2007). In the subsequent sections, we examine evidence consistent with the trophic cascade hypothesis and evidence arguing against it.

HAS THE REINTRODUCTION OF WOLVES REVERSED THE STATE CHANGE?

There are four types of data that have been used to support the trophic cascade hypothesis: (1) comparisons of photographs before and after the restoration of wolves, (2) measurements of willow height after the introduction of wolves, (3) dendrochronological studies of annual variation in willow growth rate, and (4) measurements of willow heights in relation to spatial locations assumed to coincide with differences in predation risk. We examine each of these lines of evidence in turn.

Some of the original evidence for behavioral effects of wolves on willows showed pictures of lush, healthy-looking willows taken during 2002 compared with willows at the same location in 1996 (Ripple and Beschta 2004a, 764). The caption to the figure showing these photographs reads, "In 2002, after 7 years of wolf recovery, willows show evidence of release from browsing pressure (increases in density and height)." Similar comparisons of photographs taken before and after wolf restoration seem to imply that willows were doing much better after wolf introduction than during the decades when willows seemed to vanish from the northern range (Ripple and Beschta 2004b, 2006).

This photographic evidence was reinforced by measurements of willow height during the years following wolf introduction. Increased willow heights observed from 1997 through 2003 and decreasing trends in the intensity of browsing were taken as evidence that wolves had initiated a trophic cascade in 1997 by influencing elk foraging behavior (Beschta and Ripple 2007). A similar conclusion was reached based, in part, on observations of increased heights from 2001 through 2006 (Beschta and Ripple 2010). Of course, this evidence would be improved by data on willow growth before the introduction of wolves. Dendrochronological studies provided this before and after comparison (Beyer et al. 2007). Analysis of the width of rings on willow stems from 1990 to 2000 indicated that growth rates of willows after wolves were introduced were more rapid than during the years before introduction. Because the effects of weather could not explain the change in growth rate that occurred during this decade, it was concluded that the effect must be due to wolves reducing elk herbivory on willows (Beyer et al. 2007).

Evidence for a behaviorally mediated trophic cascade also comes from examining willow growth along assumed spatial gradients in predation risk from wolves in the Gallatin Valley, north and west of Yellowstone National Park (Ripple and Beschta 2004b). Areas where the valley was narrow were assumed to be relatively risky habitats, whereas more open areas were assumed to be relatively safe habitats. It was shown that willows were taller in the narrow valleys relative to the open areas, and this observation was concluded to be consistent with a behaviorally mediated trophic cascade. Other spatial comparisons examined willow heights between valley bottom sites close to streams with upland sites distant from streams (Ripple and Beschta 2006). It was assumed that valley bottoms were perceived by elk as being riskier places to forage than the surrounding, more open uplands. Willows grew taller in valley bottoms than on the uplands, and there were no observed differences in groundwater depth between the two locations. Willow heights increased with decreasing view distance and distance to impediments, which were assumed to be indicators of predation risk. Again, it was concluded that these data were consistent with the trophic cascade hypothesis.

In summary, all of the evidence supporting the conclusion that the reintroduction of wolves has released willows from control by excessive browsing by elk is what scientists call correlative. Correlative studies show that two variables change simultaneously when we examine them over time in the same place or we compare them at different places at the same time. Thus, willows have been seen to increase over a period of time following the restoration of wolves to Yellowstone and have been observed to thrive in areas where predation risk is presumed to be high. Many ecological studies depend on correlation, and it is a widely used approach that has provided important insights. However, though correlation might seem to imply that one variable causes the other to change, we cannot infer causation from correlation. As an example, there is a strong correlation between people's net financial worth and their age. However, this does not mean that getting older makes you rich. It simply means that age and wealth are correlated to something else (for example, how long you have worked).

Thus, although it may be true that many previous observations are consistent with the trophic cascade hypothesis, many of these explanations are inconsistent. For example, photographic evidence showing release of willows after the introduction of wolves can be easily explained by the season when the photos were taken. Before-wolf photographs were taken at the end of winter, after plants had been heavily browsed. After-wolf pictures were taken in the late summer, after plants had leafed out and grown (Ripple and Beschta 2004a, 2004b). Comparing photographs from different seasons can produce erroneous conclusions. Bilyeu et al. (2008) showed that differences in the appearance of willows ostensibly caused by their release by wolves from browsing can be more simply explained by differences in the appearance of plants between seasons. Willows that look lush and healthy during summer look bedraggled at the end of winter regardless of the presence or absence of wolves.

None of the studies reporting increases in willow height after the restoration of wolves accounted for the fact that the years when wolves were reintroduced to the northern range coincided with exceptionally deep snow. The three winters of 1995 through 1997 had the heaviest accumulation of snow during the previous 28 years, while snow depths were lower during the following decade (Creel and Christianson 2009). Elk are known to consume more browse when snow is deep (Christianson and Creel 2007). Thus, it is entirely possible that willows were browsed particularly heavily during the initial years after wolves were introduced and were at low abundance but were browsed more lightly after wolves became abundant, allowing willows to appear to grow tall in response to wolves. However, this temporal trend can also be explained well by elk response to snow. Thus, the apparent wolf effect may be coincidental to the effect of temporal patterns of winter severity.

A similar problem exists for the spatial comparisons used as evidence to support the trophic cascade hypothesis. For example, the idea that uplands must be riskier than valley bottoms was justified because willow water table depth could not be correlated with willow height (Ripple and Beschta 2006). However, subsequent studies using stable isotopes to measure groundwater use by willows showed that water-table depth is not necessarily a good indicator of willow water-acquisition patterns (Johnston et al. 2011). When groundwater use was properly measured, it accounted for much of the variation in willow growth (Tercek et al. 2010; Johnston et al. 2011). This highlights a fundamental weakness of the studies of spatial predation risk referenced above. In all of these studies, landscape characteristics were assumed to be risky, and when willows grew well in these locations, the conclusion was that the data were consistent with the original assumption. However, there is little scientific basis for the original assumption, other than, perhaps, a rough-hewn plausibility (Creel and Christianson 2009).

There is evidence to show that the behavioral effects thought to be responsible for releasing willows from browsing do not exist. Two studies have examined browsing by elk in response to direct measurements of wolf behavior as indicators of predation risk. In the first study, fecal samples were collected from elk that were known to be browsing in areas where wolves were nearby, as revealed by radiotelemetry. These samples were compared with fecal samples of elk browsing in areas where wolves were absent. Microscopic analysis of the samples for fragments of willow showed that willows contributed a greater proportion of elk diets when wolves were nearby than when they were far away, a result diametrically opposed to the predictions of the trophic cascade hypothesis (Creel and Christianson 2009).

A separate study examined browsing by elk on aspen, which also are ostensibly protected from browsing by increased risk of predation by wolves (Kauffman et al. 2010). A map of predation risk was developed using an extensive, detailed data set on locations where elk were killed by wolves. Browsing intensity on aspen was measured in stands chosen at random locations across the map. The intensity of browsing on aspen increased in direct proportion to the risk of predation (Kauffman et al. 2010). Again, this result was the opposite of what is predicted by the trophic cascade hypothesis (but also see Beschta and Ripple 2011).

Although correlative studies like the ones discussed above can be useful for developing hypotheses, these hypotheses must be carefully tested. Evidence for causation, as opposed to correlation, depends on properly designed experiments, where treatments are randomly assigned to individuals and some individuals are left untreated as controls. To be most useful, experiments must be replicated, that is, they must be repeated in different places, to be sure that the results obtained in one place match the results obtained elsewhere. Replicated, randomized, and controlled experiments are the gold standard of science.

A replicated, randomized, and controlled experiment cast doubt on the operation of a simple trophic cascade on the northern range after the reintroduction of wolves (Bilyeu et al. 2008). The experiment sought to understand whether willows could recover quickly if they were released from browsing or, alternatively, if the absence of beaver dams might prevent willow recovery even when browsing was removed. This work was designed to test the hypothesis that reducing browsing might be necessary for willow recovery, but it might not be sufficient to allow the rapid recovery that has been suggested by the trophic cascade hypothesis. It might be that the grasslands state is stable because the loss of beavers from the system degraded habitat for willows to the extent that they could not recover even when they were not browsed. The experiment manipulated browsing by building fences such that some willows were exposed to browsing by elk and others were completely unbrowsed. Depth of groundwater was manipulated by raising water tables with dams constructed to resemble those built by beavers.

Five years of study revealed several surprising results (Bilyeu et al. 2008). The most surprising outcome was that browsing on willows in the control plots, that is, in the unfenced areas that were ostensibly protected from browsing by the trophic effects of wolves, was consistently high, exceeding 60 percent of the growth every year and, at least qualitatively, exceeding browsing levels observed before the reintroduction of wolves (Singer et al. 1994). Willows in plots that were browsed and had access to elevated water tables were able to grow rapidly. In two of three willow species, they grew as rapidly as willows that were totally unbrowsed but had ambient water tables. Removing browsing from plants on historic floodplains slowed willow growth by increasing moisture stress (Johnston et al. 2007). After five years of protection from browsing, the growth of unbrowsed willows was slowing

Riparian plant communities dominated by tall willows, like this beaver-modified wetland along Thorofare Creek just outside the south boundary of Yellowstone National Park, are becoming more prominent after a long period of decline during the past century. Experimental studies have demonstrated that the recovery of willow communities is complex and involves changes in browsing pressure by elk, whose numbers have decreased following wolf reintroduction; physical processes controlling the availability of water; disturbances by beaver; and changes in climate. *Photo by Cindy Goeddel.*

down, and their heights were far below the 200-centimeter-tall plant heights that historically characterized the willow state (Bilyeu et al. 2008). The only plants that recovered quickly to the imposed treatments and that approached historic heights were those that were unbrowsed and had access to elevated water tables (Bilyeu et al. 2008). Other studies have confirmed the importance of access to groundwater in willow growth on the northern range (Tercek et al. 2010; Johnston et al. 2011).

These experimental results showed that a composite of forces control the transition from the grassland to the willow state. A trophic cascade can occur only if the "top-down" effects of herbivores on plants exceed the "bottom-up" effects of limitation by resources needed by plants to grow (for example, the availability of water). If willows have insufficient water to grow, then moderating browsing by elk will not promote growth. Although willows responded to removal of browsing, their response was slow unless they had access to elevated water tables. Moreover, it was clear from the experiment that willows with adequate water could tolerate high levels of browsing. This experiment implied that the loss of beavers may dramatically slow the recovery of willows. It follows that if willows are required by beavers, and beavers require willow, then the reintroduction of wolves will not rapidly restore willows to the conditions that prevailed before wolves were extirpated.

Even though this experiment meets high scientific standards for evidence, it is not without its limitations. It applies only to areas of the landscape that were historically occupied by beavers. It is a relatively brief snapshot in time (five years) and needs to be extended for an additional decade or more to fully understand the interplay of browsing and water availability. The experiment is limited in spatial extent, and it is important to know how other areas, randomly selected from areas formerly occupied by beaver, behave relative to the experiment.

CONSERVATION IMPLICATIONS

The fundamental implication of the work we reviewed is that controls on the state of the willow riparian zone along small streams in northern Yellowstone remain only partially understood. However, it is clear that a composite of forces will determine the future of the riparian ecosystem: browsing by elk, physical processes controlling availability of water, disturbance by beaver, and changes in climate. Change in the establishment locations of willows was occurring as late as the 1980s and 1990s. These changes reflected modifications of the physical environment—down cutting of streams and loss of establishment habitat—that are not likely to quickly change to the conditions that prevailed at the turn of the nineteenth to twentieth century. The grassland state required the better part of a century to develop. Why would we think that it would be reversed fifteen years after the reintroduction of wolves?

Predation by bears and wolves on elk, human hunting outside Yellowstone National Park, and drought have caused precipitous decreases in elk counts (Vucetich et al. 2005; P. J. White and Garrott 2005a; Eberhardt et al. 2007; Barber-Meyer et al. 2008), from almost 20,000 animals in 1968 to fewer than 5,000 in 2011. However, it is imperative to understand that elk numbers have only recently dropped below the population levels that were able to maintain the alternative grassland state for seven decades. It is entirely possible that the effects of reduced elk numbers on willow communities have simply not had sufficient time to be manifest, let alone observed.

The restoration of wolves to the landscapes of Yellowstone National Park during 1995 and 1996 represents one the most important natural experiments in the history of ecology. It is vital to the discipline of ecology and the practice of conservation that this story be told fully and accurately. There is overwhelming evidence from managed areas of the world that the removal of top predators from food webs degrades the structure and function of ecosystems (Estes et al. 2011), and the work we review here adds to that body of evidence. It is tempting to believe that the converse is true, that restoring predators to ecosystems will quickly and easily fix the problems caused by their extirpation. However, it has been shown repeatedly that degraded, alternative states of ecosystems may resist restoration by simply fixing the original errors in the management of ecological processes (Scheffer 2009). It is often much easier to conserve states and processes in ecosystems than to restore them.

The practice of conservation is politically charged, and this contentious atmosphere means that it is especially important that those who support conservation with science do so objectively, fully revealing the strengths and weaknesses of the evidence they present. The emerging Yellowstone story amplifies the fundamental value of conserving predators by revealing that failing to maintain intact food webs can lead to degraded, alternative states that are not restored simply, quickly, and easily. A substantial body of science from Yellowstone emphasizes the urgency of conserving intact food webs and shows the challenges of restoring simplified ones.

CHAPTER TWELVE

Assessing the Effects of Climate Change and Wolf Restoration on Grassland Processes

DOUGLAS A. FRANK

RICK L. WALLEN

P. J. WHITE

AS HERBIVORES, UNGULATES are a highly integrated and interactive component in many ecosystems that impart strong influences on the turnover of soil nutrients, competitive interactions among plant species, and the composition of plant species over time (Hobbs 1996; Pastor et al. 1997; Kie et al. 2003). Herds of grazing ungulates can stimulate the flow of energy and nutrients in grassland ecosystems with high nutrient and moisture availability, with the classic example being the Serengeti in northwestern Tanzania and southwestern Kenya (McNaughton 1976, 1985, 1990; McNaughton et al. 1988, 1997). Grazing intensity in grassland ecosystems is typically much higher (40 to 65 percent) than in other terrestrial ecosystems (9 percent) and often has substantial effects on plant and soil processes (Hobbs 1996; Frank et al. 1998; Knapp et al. 1999).

Ungulates increase or decrease plant production depending on several interacting factors, such as grazing intensity, site fertility, and the history of the system (Hobbs 1996; Pastor and Cohen 1997; Bardgett and Wardle 2003; Kie et al. 2003). In fertile systems where plants and ungulates share a long history, soil nutrients are available to plants that are adapted to regrow after being grazed. The facilitating influence of ungulates on their forage production in these systems will increase from low to moderate grazing intensities and decrease at higher grazing intensities until a point at which the amount of leaf tissue that is removed is great enough to reduce carbon assimilation and plant production. The response of these grasslands to grazing in which plant production is maximized under intermediate grazing intensities has been coined the *grazing optimization function* (McNaughton 1979; Hilbert et al. 1981; K. M. Stewart et al. 2006).

Grassland production is limited by a single resource (light) aboveground and primarily the availability of two resources (water and nitrogen) belowground (Vitousek and Howarth 1991; Frank and Groffman 1998). Grazing opens up the grassland canopy, which increases the light intensity at the soil surface where regrowth occurs in the form of new stems or shoots produced at the base of grazed grasses (Parsons and Penning 1988; Wallace 1990; Knapp et al. 1999). By reducing transpiring leaf surface area, grazing also reduces water loss from these plants and increases the moisture available to support plant regrowth (McNaughton 1985). At the same time that ungulates reduce light and water limitation for grazed plants, they also increase soil nitrogen availability by several pathways. First, ungulates consume plant tissue and deposit fertilizers in forms (feces and urine with nitrogen) that are more accessible to soil microbes than in the original plant material (Ruess 1984; Ruess and McNaughton 1987). Second, by improving the quality of leaf and root material flowing to the soil (Risser and Parton 1982; Holland et al. 1992; Shariff et al. 1994), ungulates stimulate rates of litter breakdown and mineralization of soil nitrogen, which is the conversion of nitrogen from decaying plant material into ammonium that can be used by microbes and living plants (Frank and Groffman 1998). Third, defoliation (removal of leaves) increases the secretion of small organic compounds from the roots of grazed plants into the rhizosphere, which is the narrow zone of soil immediately surrounding the root system (Hamilton and Frank 2001). These compounds stimulate the population growth and activity of microbes and increase the availability of nitrogen to grazed plants (Hamilton and Frank 2001; Hamilton et al. 2008). Fourth, grazing can induce a shift in the community composition of Mycorrhizae, which are fungi that colonize the roots of host plants and supplement plant water and nutrient uptake, to facilitate plant growth (Frank et al. 2003).

In fertile grasslands, low to moderate grazing levels increase the availability of light, moisture, and nitrogen to defoliated plants, which stimulates the growth of new tissue and the overall productivity of the system. The removal of older, lower-quality material and increase in younger, nutrient-rich plant tissue feeds back positively on the diet quality of ungulates. Conversely, grazing in infertile systems or by dense populations of ungulates can cause reductions in the cycling of nutrients, plant production, and forage availability and quality. At infertile sites, the selective removal of preferred, nitrogen-rich forage species can shift community composition toward plant species that produce litter of relatively poor quality, which can slow nitrogen cycling and reduce soil nitrogen availability and plant production (Pastor et al. 1993; Ritchie et al. 1998; Díaz et al. 2007). High grazing intensities can lead to the degradation of plant communities by removing vegetation, compacting soils, and reducing the diversity of plants (Terborgh et al. 2001; Coté et al. 2004). High densities of ungulates can also shift the composition and structure of plant communities (Augustine and McNaughton 1998; Olff and Ritchie 1998; Kie et al. 2003). K. M. Stewart et al. (2006) found that the quality and consumption of some forages decreased at high densities of elk (20 to 35 per square kilometer), which corresponded with reductions in body condition and pregnancy rates. Decreased maternal body condition also reduces the survival of young (Singer et al. 1997).

Grasses are an important component of the diet of most ungulates in Yellowstone National Park, including bighorn sheep (58 to 65 percent of diet), bison (53 to 54 percent), elk (75 to 79 percent), mule deer (19 to 32 percent), and pronghorn (4 to 10 percent; Singer and Norland 1994). Ungulates in this temperate mountain environment migrate seasonally and move selectively over the landscape, feeding preferentially on grasses in early stages of development when forage is most nutritious (Senft et al. 1987; Frank and McNaughton 1992; Wallace et al. 1995; Frank et al. 1998). Ungulates consume green forage in the spring on the lower-elevation winter ranges before most of them migrate with the wave of young, nutritious vegetation as it moves up the elevation gradient during the growing season. Ungulates then return to the winter range in late autumn to access mostly dead, low-quality forage while deep snow covers the higher-elevation summer range. Ungulate migration in and near Yellowstone from low to

The grasslands of Yellowstone National Park and the surrounding low-elevation river valleys support a diverse community of ungulates, including bighorn sheep, which were decimated throughout the region by market hunting, competition with domestic livestock, and the introduction of exotic diseases from livestock. *Photo by Cindy Goeddel.*

high elevations is also positively associated with forage concentration and foraging efficiency (amount of plant material removed per bite; Frank and McNaughton 1992; Frank et al. 1998). Thus, by migrating, ungulates increase foraging efficiency while at the same time improving their diet quality.

Prior to wolf restoration, some areas in northern Yellowstone were intensively grazed (up to 55 percent), but there was near total recovery or compensation of the grazed plants due to the facilitating effects of grazers on the availability of plant resources and animal movements to other patches (Frank and McNaughton 1993; Despain 1996; Wallace 1996). Seasonal migration allows grazed vegetation an extended period to recover while soil conditions are suitable for plant growth. Many grasses in Yellowstone are generally adapted to grazing and respond positively to moderate grazing levels (Huff and Varley 1999). Wallace (1996) found that small, previously grazed plants were highly preferred by bison and elk in northern Yellowstone, similar to the grazing lawns found in the savanna systems of the Serengeti, where regrazing of the same individuals or groups of plants occurred (McNaughton 1976, 1979). Grazing ungulates stimulate soil nutrient availability by several previously described pathways that provide readily available nutrients essential for the rapid regrowth of grazed plants and the production of dense, short grass stands (Frank and McNaughton 1992; Wallace 1996). Thus, grazing tends to reduce leaf and seed height (the classic grazing lawn effect). Diversity of northern range grassland has been shown to increase at moist, higher productive sites and remain the same or decrease at drier, lower productive sites in response to grazing (Frank 2005). Plants respond positively to patchy grazing by bison and more isolated grazing by elk through increased growth and photosynthesis (conversion of carbon dioxide into sugars using the energy from sunlight), suggesting that preferred forage species are retained in the system (Wallace 1996).

CLIMATIC AND PREDATION EFFECTS ON GRAZED GRASSLANDS

Grasslands occur in areas with the greatest year-to-year variation in weather among the earth's land ecosystems (Frank et al. 1994). Because moisture is the principal factor limiting grassland production, high between-year variation in weather should produce high between-year variation in grassland processes (Knapp and Smith 2001). Frank and McNaughton (1992) reported a 19 percent reduction in the amount of green plant material on northern Yellowstone grasslands from 1988 to 1989 as a result of the death or injury of plants during a severe drought. This drought also contributed to substantial starvation of bison and elk during the ensuing winter, with subsequent decreases in consumption and feces deposition that contributed to effects that persisted for several years after the drought event (Singer et al. 1989). In addition, the positive feedback of moderate grazing on shoot and root production during an average moisture year in 1999 was not observed during a two-year drought in 2000 and 2001 (Frank 2007), suggesting that the stimulating effect of ungulates on plant production was modulated by soil moisture conditions.

During the 1980s and 1990s, migratory ungulates on the northern grassland of Yellowstone National Park had tight chemical, physical, and biological linkages with plants and soil microbes that doubled the rate of nitrogen mineralization, stimulated aboveground plant production of leaves and stems by as much as 43 percent, and increased belowground production of roots by 35 percent (Frank and McNaughton 1993; Frank and Groffman 1998; Frank et al. 2002). These biogeochemical linkages were largely driven by high densities of elk (12 to 17 per square kilometer) that influenced the availability of water and soil nutrients to plants (Maschinski and Whitham 1989; Frank et al. 1998). The facilitation of nitrogen availability and the related increase in nitrogen assimilation (uptake and conversion) by plants promoted aboveground plant production, which demonstrated the importance of large herds of migratory ungulates in controlling soil nitrogen cycling in this grassland system (Frank and McNaughton 1993; Frank and Evans 1997; Frank and Groffman 1998). However, rates of ungulate grazing and grassland nitrogen cycling were reduced by 25 to 53 percent by 1999 through 2001, largely a result of fewer elk (10 to 11 per square kilometer) due to hunter harvest, the addition of wolf predation, and severe starvation of elk during the winter of 1997 (Frank 2008). These decreases in consumption and nitrogen cycling suggest that the strength of the ungulate stimulation of ecosystem processes was reduced, resulting in overall less nitrogen cycling and probably lower production.

Wolf predation also changed the distribution and extent of grazing and microbial activity in the northern Yellowstone grasslands (Frank 2008). Prior to wolf restoration, sites that supported the most forage were grazed more by ungulates, primarily elk. After wolf restoration, however, the relationship between consumption and production changed, with grazing decreasing more at higher productive sites than at lower productive sites (Frank 2008). Thus, by shifting the location that ungulates grazed Yellowstone grassland, wolves contributed to widespread effects on elk, plants, and soil microbes that likely spatially reorganized energy and nutrient dynamics in this system. During this same period, however, the number of bison in northern Yellowstone doubled, which likely concentrated grazing and animal-induced effects on nitrogen cycling compared with the likely more diffuse effects on grassland processes that occurred when elk numbers were high. Scientists are monitoring the potential effects of this relatively recent (15 years) change in the system.

The seeds of grasses are an important food for a wide variety of birds and mammals, including this least chipmunk, stretching to reach the seed head on top of a tall stem of grass. *Photo by Cindy Goeddel.*

CONSERVATION IMPLICATIONS

Prior to the restoration of wolves, there was a long debate regarding whether too many ungulates were causing rangeland overgrazing and deterioration of the Yellowstone system. As ungulate numbers approach the food-limited capacity of the environment to support them, their facilitating influence on plant productivity (grazing optimization) decreases because higher grazing intensities remove enough leaf tissue to reduce plant production. Thus, ungulate populations must be held well below their carrying capacity or be highly migratory to maintain high plant productivity through grazing optimization (K. M. Stewart et al. 2006). Ungulates in Yellowstone are highly migratory and follow the green-up of vegetation to higher elevations during the spring and summer. They concentrate foraging on highly nutritious, young vegetative growth at a location for a short interval before moving upslope, which gives plants a sufficient period of time to recover while conditions for growth are suitable. Thus, the National Research Council (2002) concluded that grazing by abundant ungulates had not irreversibly damaged the grasslands of northern Yellowstone because plant and nutrient dynamics are currently within their expected long-term ranges, with no indication that processes or the system are near some critical undesirable threshold.

Diverse predator associations typically limit ungulate densities well below carrying capacity (Kie et al. 2003; K. M. Stewart et al. 2006), and the restoration of wolves and the reestablishment of a complete native predator community in Yellowstone may keep elk well below the ecological carrying capacity of the environment. In fact, there is some indication that the dynamic northern grassland system of Yellowstone National Park and nearby Montana is in a state of flux. The density of northern Yellowstone elk decreased to approximately three to five per square kilometer from 2006 through 2011 from a high of 12 to 17 per square kilometer in the late 1980s and early 1990s, before wolf reintroduction. Fewer elk have resulted in less forage consumed and less intense feedbacks by elk on soil and plant processes, which likely has contributed to lower plant production and forage quality (K. M. Stewart et al. 2006; Frank 2008). As mentioned previously, however, the effects of an increasing bison population on soil and plant processes remain to be determined.

A warming climate could also influence the diet, nutrition, and condition of Yellowstone ungulates. Likely climate change scenarios for Yellowstone suggest a 1 to 3 °C increase in average temperature during the twenty-first century, with a corresponding increase in annual rainfall—though it is unknown precisely how precipitation patterns will change and how those changes will affect the Yellowstone system (McWethy et al. 2010). Merrill and Boyce (1991) noted that fewer green grasses are available in middle to late summer when green-up occurs early in spring and grasses are cured by early summer. Temperatures in northern Yellowstone have shown a pronounced warming over the past 50 years and contributed to decreased snow levels below elevations of 2,000 meters, earlier snowmelt and warming of the soil, earlier peak in the growing season followed by earlier death of grasses, and increased frequency of drought, such as the severe, extended drought from 2000 through 2006 (Romme and Turner 1991; Wilmers and Getz 2005; McMenamin et al. 2008). Early spring green-up could improve the diet quality for ungulates during the latter stages of gestation and lactation, which would contribute to increased birth weight, growth, and survival of calves (B. L. Smith et al. 2006). Conversely, an early decrease in food quality and quantity could result in lactating adult females entering autumn and winter with marginal fat reserves for pregnancy and survival. For example, DelGiudice et al. (2001) reported that bison and elk in Yellowstone were severely nutritionally restricted and had low fat reserves during the winter of 1989 following the severe drought and fires the previous summer. Clutton-Brock and Albon (1989) reported that a smaller portion of adult female red deer (*Cervus elaphus*) in the Scottish Highlands conceived, and overwinter survival of calves was lower, during years when September was dry.

Climate change likely will influence Yellowstone elk herds through weather effects on forage availability and quality. Levels of body fat in adult female elk in northern Yellowstone during the winters of 2000 through 2002, in the midst of a severe drought, were sufficiently high to preclude appreciable overwinter mortality but marginal enough that R. C. Cook et al. (2004) suggested the possibility of nutritional limitations occurring on the summer range. The nutritional demands of lactating females and their offspring are somewhat higher during summer and autumn than winter (J. G. Cook 2002). Thus, small to moderate nutritional limitation in summer and autumn could decrease reproduction and survival, and reduce calf growth on summer range (Merrill and Boyce 1991; R. C. Cook et al. 2004; K. M. Stewart et al. 2005).

Bison and elk have a substantial degree of dietary and habitat overlap (Singer and Norland 1994), which may result in competition between bison and elk for food resources (Coughenour 2005). Thus, P. J. White and Garrott (2005a) speculated that wolf recovery would contribute to increased bison abundance by decreasing elk numbers, and indeed, bison numbers on the northern range increased from 866 bison in 1996 to approximately 2,600 in 2012, after wolf reintroduction. Model simulations suggest that the number of bison the range can support in northern Yellowstone could increase from a range of 1,800 to 3,600 animals with greater than 20,000 northern Yellowstone elk, to a range of 1,900 to 4,900 animals as elk numbers decrease toward 5,000 (Coughenour 2005). This prediction is consistent with the observed decrease in northern Yellowstone elk numbers and concurrent increase in bison numbers from 1995 through 2011.

The feeding habits of bison are strongly influenced by social interactions and dominated by grasses, sedges, and rushes throughout the entire growing season, while elk feed more on forbs and browse (Plumb 1991). Thus, bison tend to create distinct grazing patches, unlike elk, which feed more on isolated individual plants (Wallace 1996; Bruggeman 2006). Also, the majority of the northern bison herd remains on low-elevation northern grassland through summer, whereas most elk migrate to higher-elevation summer ranges throughout the park after some early spring grazing (Wallace 1996; P. J. White et al. 2010). Thus, grazing by bison could subject grasses to repeated tissue loss throughout the growing season. In addition, bison are much more formidable prey for wolves because of their large size and tendency to use group defenses (Carbyn and Trottier 1987; D. W. Smith et al. 2000; Becker, Garrott, White, Gower, et al. 2009). Thus, bison comprise less than 5 percent of wolf kills on Yellowstone's northern range, and there is no evidence that wolves have modified the abundance, distribution, and movements of bison in the park (D. W. Smith 2005). As a result, large concentrated groups of bison that repeatedly graze sites throughout the summer could potentially have quite different effects on the dynamics of Yellowstone's northern range than herds of elk that graze sites for relatively short periods during their migration to high-elevation summer range.

The larger body size of bison, combined with strong herding tendencies and selective consumption of dominant grasses and sedges in moist, relatively productive sites, could contribute to an increase in grazing-tolerant plants that are repeatedly grazed more intensely through the summer (Plumb 1991; Knapp et al. 1999; Person et al. 2003). Bison grazing alters the composition of plant communities by converting grass-dominated sites to sites of higher diversity with more forbs and higher spatial variation (Knapp et al. 1999). Potential long-term changes could include a reduction in the dominance of grasses; an increase in other species, and increased spatial variation (including wallows; Knapp et al. 1999). However, the nature of these relationships likely will vary as the abundance of bison increases or decreases relative to the food-limited capacity of the environment to support bison in response to competition, management removals, predation by wolves, and weather (Kie et al. 2003; K. M. Stewart et al. 2009). Wolf packs concentrate predation on bison in the central Yellowstone valleys and live off bison in Canada's Wood Buffalo National Park (Carbyn and Trottier 1987; Becker, Garrott, White, Gower, et al. 2009). Thus eventually, wolves in northern Yellowstone may start to kill more bison as the ungulate community becomes dominated by bison. Research has been initiated to elucidate the influence of recent changes in the Yellowstone grazing community on ecosystem processes such as the spatial pattern and intensity of ungulate grazing and grassland energy and nutrient dynamics. We predict that the transition to an increasing population of bison will influence the grasslands in northern Yellowstone much differently than elk.

invasive, non-native species

Altered Processes and the Demise of Yellowstone Cutthroat Trout in Yellowstone Lake

ROBERT E. GRESSWELL

LUSHA M. TRONSTAD

A FOOD WEB describes the trophic relationships among the species in a community (Pianka 1978). In most cases, a food web has numerous food chains that represent a single pathway between trophic levels. In essence, a food web depicts energy flow from primary producers (lowest trophic level) to consumers and decomposers (organisms like bacteria that break down dead organisms). Consumers include herbivores that feed directly on producers and carnivores that feed on herbivores or other carnivores. Omnivores are also consumers that occur in many food webs, but they are more difficult to place at any specific trophic level (Pianka 1978).

At about the same time wolves were reintroduced to Yellowstone National Park, another predator, the non-native lake trout, was discovered for the first time in Yellowstone Lake. Though the effects of both predators on the complex food webs of the respective ecosystems might have been predicted, wolves were reintroduced to a system in which they had existed for millennia and among an assemblage of animals and plants with which they had evolved. In contrast, there were no predatory fishes in Yellowstone Lake historically, but there were numerous birds and mammals that depended on the native Yellowstone cutthroat as an energy source, especially during the spring when the cutthroat trout were vulnerable to predation as they ascended tributary streams to spawn.

In the intervening decades, both systems have changed. The current assemblage of ungulate and predator species and the processes operating in the system reflect the natural capacity of the terrestrial ecosystem as it continues to evolve. In Yellowstone Lake, however, the cutthroat trout population is on the verge of collapse, and the emerging system is evolutionarily novel for the lake. In this chapter, we review the history of the Yellowstone Lake ecosystem and examine the changes that have occurred through time, especially since lake trout became established in the lake during the latter part of the twentieth century.

Yellowstone Lake is a large (surface area = 34,108 hectares), high-elevation (2,357 meters) mountain lake with almost 240 kilometers of shoreline (Kaplinski 1991). It is a deep lake, with an average depth of 48.5 meters and a maximum depth of 107 meters (Kaplinski 1991). There are 124 tributaries entering the lake, but 81 of these tributaries are small, with drainage areas that are less than 100 hectares, collectively comprising only about one percent of the entire lake drainage (Gresswell et al. 1997).

Geologically, the drainage is dominated by volcanic bedrock, except in the southern portion of the basin, where sedimentary deposits are common in some tributary drainages (U.S. Geological Survey 1972). During the last glaciation, the Yellowstone Lake watershed was under about 1,600 meters of ice. As the ice melted 12,000–14,000 years ago, the lake level rose more than 24 meters above the current lake surface (Richmond 1976). Currently, lower elevations in the watershed reflect this history, and postglacial lake and river deposits are common (Richmond 1976).

Lodgepole pine is the most common tree in the forested portion of the drainage, but Engelmann spruce *(Abies engelmannii)*, subalpine fir *(Abies lasioicarpa)*, and whitebark pine are locally abundant. Deciduous (plants that lose their leaves annually) shrubs found in some of the tributary corridors are primarily willows. Subalpine meadows occur in the watersheds of some of the larger tributaries and at higher elevations.

In the lake, the growing season is short, and an average of 215 ice-free days occurred annually from 1951 through 2010 (U.S. Geological Survey, unpublished data). Historically, the phytoplankton (microscopic aquatic plants) community was dominated by diatoms except in periods of thermal stratification, when blooms of *Anabaena flos-aquae* (a blue-green alga that can fix nitrogen) occurred regularly (J. C. Knight 1975). Common zooplankton (tiny invertebrates that float through the water) included *Diaptomus shoshone, Daphnia schødleri,* and *Conochilus unicornis* (Benson 1961). In the part of the lake close to shore, amphipods (specifically *Gammarus lacustris* and *Hyallela azteca*) and aquatic insect larvae were abundant (Benson 1961).

Only two fishes, Yellowstone cutthroat trout and longnose dace *(Rhinichthys cataractae),* were native to Yellowstone Lake (Simon 1962). By the 1950s, three non-native fishes, the redside shiner *(Richardsonius balteatus),* lake chub *(Couesius plumbeus),* and longnose sucker *(Catostomus catostomus),* had become established (Gresswell and Varley 1988). During the latter part of the twentieth century, longnose suckers were distributed throughout the lake, but redside shiners and lake chubs were generally limited to shallow shoreline areas and lagoons that are common in the West Thumb area of Yellowstone Lake (Biesinger 1961; Dean 1972). Longnose dace were most often associated with the mouths of tributaries.

Prior to 1994, there were no other salmonid fishes in Yellowstone Lake, but an introduced population of brook trout was discovered in one of the larger tributaries and subsequently removed (Gresswell 1991). Fortunately, early attempts to introduce mountain whitefish *(Prosopium williamsoni),* rainbow trout, and landlocked Atlantic salmon *(Salmo salar)* to the lake and/or the Yellowstone River downstream of the lake were not successful (Gresswell and Varley 1988).

Though policies of the National Park Service provided substantial protection from pollution and land-use practices that often degrade habitat, native Yellowstone cutthroat trout were subjected to the effects of non-native fish introductions, egg-taking operations, commercial fishing, and intensive sport-fishery harvest through the middle of the twentieth century (Gresswell and Varley 1988; Gresswell et al. 1994). Cessation of commercial harvest in the 1920s and egg-taking operations in the 1950s helped sustain a robust population for decades, but by the late 1960s, overharvest resulted in decreases in abundance, size, and age structure of the population. In order to reverse this trend and reduce angler harvest, the National Park Service implemented special angling regulations in the 1970s. Numbers and average size increased, and by the mid-1980s, the assemblage of Yellowstone cutthroat trout in Yellowstone Lake appeared to be relatively secure (Gresswell et al. 1994). Most important, despite the fluctuations in population abundance, it appeared to be the largest genetically unaltered inland population of cutthroat trout in the world (Gresswell and Varley 1988). In fact, Yellowstone Lake is believed to represent about 80 percent of the remaining lake habitat (by surface area) for the Yellowstone cutthroat trout (Gresswell et al. 1994).

Prior to the introduction of lake trout, Yellowstone cutthroat trout were the top predator of a food web composed of three trophic levels: fish, zooplankton, and phytoplankton (Tronstad et al. 2010). Yellowstone cutthroat trout consumed invertebrates (organisms without a backbone), including those that lived on the bottom of the lake and zooplankton. Consumption of zooplankton intensified as cutthroat trout densities increased. Gut contents of cutthroat trout from Yellowstone Lake suggested that most of them consumed small crustaceans (R. D. Jones et al. 1990), and even large Yellowstone cutthroat trout fed on zooplankton. Because cutthroat trout primarily eat zooplankton in Yellowstone Lake, changes in cutthroat trout density may alter lower trophic levels (primary producers) in the lake's food web. It is likely that Yellowstone cutthroat trout altered portions of the food web associated with the bottom of the lake and open water, though historical information is limited to the open-water portion of the food web.

Fish rarely occur in cutthroat trout stomach contents even where the four species (cutthroat trout, longnose dace, redside shiner, and lake chub) occupy the same area, such as lagoons and spawning streams. Though Dean and Varley (1974) documented consumption of redside shiner by Yellowstone cutthroat trout during the annual spawning migration at a fish trap on Pelican Creek, Gresswell (1995) suggested that this unusual observation was an artifact of trap operation.

Historically, there were 42 birds and mammals in the Yellowstone Lake watershed that used Yellowstone cutthroat trout for food, including two endangered species, the bald eagle and the grizzly bear (Schullery and Varley 1995). Prior to the introduction of lake trout to Yellowstone Lake, fish-eating birds probably had the greatest effect on cutthroat trout (Gresswell 1995; Stapp and Hayward 2002b). Fish consumption (size and weight) per day varied among 20 or more bird species (Swenson 1978; Swenson et al. 1986; Schullery and Varley 1995), but total biomass (weight) of Yellowstone cutthroat trout eaten by birds may have exceeded 100,000 kilograms annually (Davenport 1974). Between 1972 and 1982, Yellowstone cutthroat trout comprised up to 23 percent of the diet of bald eagles during the breeding season (April to August) in the Yellowstone Lake area (Swenson et al. 1986), and eagles consumed Yellowstone cutthroat trout almost exclusively during the spawning period (May to July); (Ball and Cope 1961; Gresswell et al. 1997).

Other fish-eating birds included American white pelican *(Pelecanus erythrorhynchos)*, osprey *(Pandion haliaetus)*, great blue heron *(Ardea herodias)*, common merganser *(Mergus merganser)*, California gull *(Larus californicus)*, common loon *(Gavia immer)*, Caspian tern *(Hydroprogne caspia)*, Barrow's goldeneye *(Bucephala islandica)*, bufflehead *(Bucephala albeola)*, belted kingfisher *(Megaceryle alcyon)*, and double-crested cormorant *(Phalacrocorax auritus)*. All of these birds bred in the Yellowstone Lake area and were dependent on Yellowstone cutthroat trout spawners, eggs, and offspring as a food source. Most of these birds focused on fish in shallow water near shore and tributaries where the Yellowstone cutthroat trout were abundant (Schullery and Varley 1995; McEneaney 2002).

Stapp and Hayward (2002b) suggested that approximately 7 percent of the Yellowstone cutthroat trout population was consumed annually by mammalian predators before lake trout became established in the Yellowstone Lake ecosystem. Yellowstone cutthroat trout were seasonally important in the diet of grizzly bears in the lake area (Mealey 1980; Mattson and Reinhart 1995; Haroldson et al. 2005). In contrast, river otters *(Lontra canadensis)* were believed to be dependent on Yellowstone cutthroat trout throughout the year (Crait and Ben-David 2006), and Yellowstone cutthroat trout were the most frequent prey consumed by river otters near the spawning tributaries and the lake itself during the summer. Furthermore, river otters affected the prevalence and growth of plants living along the edge of streams and lakes by transferring lake-derived nutrients into them (Crait 2005). After angling regulations led to an increase in Yellowstone cutthroat trout abundance in the 1970s, grizzly bear activity increased along spawning streams. For example, the number of streams frequented by bears during the spawning season was higher from 1985 to 1987 than from 1974 to 1975 (Reinhart and Mattson 1990a).

In addition to the important ecological role of the Yellowstone cutthroat trout in Yellowstone Lake, this assemblage supported a popular fishery that attracted anglers from around the world. Despite decades of overharvest and an egg-taking operation that removed 800 million eggs during the first

The osprey is one of more than 20 bird species that forage on Yellowstone cutthroat trout, with estimates of the total biomass (weight) of trout eaten by birds potentially exceeding 100,000 kilograms annually when trout populations were abundant in Yellowstone Lake. *Photo by Cindy Goeddel.*

half of the twentieth century, the Yellowstone cutthroat trout population in the lake was robust by the early 1990s. The size and age structure of the population at that time closely resembled historic proportions (Gresswell et al. 1994). The economic value of the fishery in the lake for 1994 was estimated to be over $36 million (Varley and Schullery 1995a). In addition to the ecological, recreational, and economic values, a substantial alteration of the fish assemblage in Yellowstone Lake would have negative repercussions for the aesthetic, or nonconsumptive, values (such as fish watching) associated with the Yellowstone cutthroat trout in the ecosystem (Gresswell and Liss 1995; Varley and Schullery 1995a).

CHANGES IN THE FISH COMMUNITY SINCE THE LATTER PART OF THE TWENTIETH CENTURY

The first documented evidence of non-native lake trout in Yellowstone Lake occurred in 1994 (Kaeding et al. 1996), but subsequent research suggests that lake trout were established in the lake at least a decade prior to their discovery (Munro et al. 2005). The effects of predation on Yellowstone cutthroat trout have been documented over the past decade in Yellowstone Lake, and current evidence suggests that non-native lake trout are directly linked to the observed decreases in the abundance of Yellowstone cutthroat trout (Ruzycki et al. 2003; Koel et al. 2005). Because cutthroat trout in the lake evolved without large fish-eating predators (Gresswell 2011), it appears that adaptive behaviors to reduce predation had not arisen. Lake trout can consume cutthroat trout approximately 27 to 33 percent of their body length, and therefore, juvenile cutthroat trout were especially vulnerable (Ruzycki et al. 2003). Estimates from the late 1990s suggest that a single lake trout can consume about 41 cutthroat trout annually (Ruzycki et al. 2003).

It was during the late 1990s that *Myxobolus cerebralis* (the causative agent of whirling disease) was first discovered in Yellowstone Lake (Koel et al. 2005). Though up to 20 percent of juvenile and adult Yellowstone cutthroat trout in Yellowstone Lake may be infected with this non-native parasite, infection does not appear to be uniform in the watershed (Koel et al. 2006). For example, *Myxobolus cerebralis* has been detected in Pelican Creek, Clear Creek, and the Yellowstone River downstream from the lake.

However, the parasite was not found in the Yellowstone River upstream of the lake inlet and 13 other spawning tributaries (Koel et al. 2006). Rate of infection was highest in Pelican Creek and the Yellowstone River below the lake (Koel et al. 2006). Evidence suggests that more than 90 percent of the Yellowstone cutthroat fry from Pelican Creek are infected with *Myxobolus cerebralis,* and since 2001, few young cutthroat trout have been observed in lower sections of the stream (Koel et al. 2005). Conversely, nonmigratory Yellowstone cutthroat trout are still prevalent in the headwaters of Pelican Creek despite high densities of *Myxobolus cerebralis.*

Numerous years of below-average precipitation in the Yellowstone Lake drainage (six of nine years between 1996 and 2005) represent another environmental factor influencing the Yellowstone Lake ecosystem in the late 1990s and early 2000s (Koel et al. 2005). Though droughts probably have occurred frequently over the past 12,000 years (Gresswell et al. 1994), decreased water availability and increased water temperatures may have negatively affected Yellowstone cutthroat trout recruitment for years (Kaeding 2010). Futhermore, the combined effects of lake trout predation and whirling disease may have worsened the effects of drought on the Yellowstone cutthroat trout population. On the other hand, precipitation has been at or above normal for the past six years, and it is assumed that climatic variables are not currently having negative effects on the native cutthroat trout.

There is substantial evidence of change in the community composition of Yellowstone Lake since the early 1990s, and monitoring programs suggest substantial decreases in the abundance of Yellowstone cutthroat trout. For example, the number of spawning Yellowstone cutthroat trout that entered Clear Creek from Yellowstone Lake decreased from an average of 43,580 between 1977 and 1992 (Gresswell et al. 1994) to 3,828 between 2001 and 2004 (Koel et al. 2005). From the mid-1970s to the early 1990s, the size structure of the spawning population shifted to larger individuals, apparently in relation to reductions in angler harvest or delayed maturity. The average cutthroat trout spawner was about 400 millimeters in length at a time when the number of spawners was consistently above 20,000 (Gresswell et al. 1994). From 2003 through 2007, the number of spawners was at historic lows (about 500 Yellowstone cutthroat trout spawners during 2006 and 2007; Koel et al. 2008), but the shift to larger individuals was even more dramatic. Few fish less than 380 millimeters in length entered Clear Creek, and the proportion of the spawning population greater than 450 millimeters was historically high. In previous decades, Yellowstone cutthroat trout greater than 480 millimeters in length were uncommon, but by 2003, almost half of the spawners entering Clear Creek were in this size group. These more recent changes in length of spawning cutthroat trout appear to be directly related to drastic decreases in the abundance of Yellowstone cutthroat trout that have made food more available to those that remain.

In Pelican Creek, the second largest tributary to Yellowstone Lake, the number of spawning Yellowstone cutthroat trout averaged almost 24,300 between 1980 and 1983. The fish trap is no longer operational in Pelican Creek, but in 2003, sampling with nets at the historical site suggested that Yellowstone cutthroat trout from the lake are no longer spawning in this tributary (Koel et al. 2005). Annual visual surveys conducted on 9 to 11 tributary streams in West Thumb and along the west shore of the lake since 1989 provide additional estimates of the relative abundance of spawning Yellowstone cutthroat trout (Reinhart and Matson 1990a; Haroldson et al. 2005; Koel et al. 2005). Trends in spawner abundance have been similar to Clear Creek in most of the streams that were monitored, and since 2004, counts are much lower than in the mid-1980s, when the assessment was initiated.

Angler use and harvest data have been collected annually since 1950 (Moore et al. 1952; R. D. Jones et al. 1990). Surveys were conducted by creel clerks until 1974, but beginning in 1975, all angler information has been collected through a voluntary postal survey. The mean landing rate for Yellowstone cutthroat trout in Yellowstone Lake for the 15 years prior to the discovery of lake trout was 1.5 fish per hour. Since the late 1990s, the landing rate for anglers steadily decreased to a low of less than 0.5 fish per hour in 2006, and despite increases since that time, the landing rate has remained less than 1.0 fish per hour since 2002. Concomitantly, the mean length of Yellowstone cutthroat captured by anglers has been at historic highs (greater than 400 millimeters) since 2000. Again, this increase in length of Yellowstone cutthroat trout is assumed to be related to the decrease in cutthroat trout abundance and associated increase in food availability.

A monitoring program using experimental gill nets was initiated in 1969, and since 1978 nets have been set in late September at the same 11 sites (five nets per site) in Yellowstone Lake. Thus, it is possible to compare the relative abundance, size, and age structure of Yellowstone cutthroat trout through time (Gresswell 2004). The number of Yellowstone cutthroat trout captured per net since 1998 has remained below historic lows recorded during the previous 19 years (1977 to 1997). Since the late 1990s, length structure has also changed substantially. The greatest changes occurred between 2000 and 2010, when the proportion of Yellowstone cutthroat trout between 330 and 450 millimeters substantially decreased, and those greater than 460 millimeters greatly exceeded historic proportions (Koel et al 2005; National Park Service, unpublished data). These results are consistent with other Yellowstone cutthroat trout monitoring information.

EFFECTS OF LAKE TROUT ON THE YELLOWSTONE LAKE FOOD WEB

The food web of Yellowstone Lake was redefined after lake trout were introduced, and because these fish are apex predators (D. M. Post et al. 2000), a fourth trophic level was added. Trophic cascade theory predicts that moving from three to four trophic levels would result in lower biomass of Yellowstone cutthroat trout, the dominant fish at the third trophic level (Figure 9). Indeed, it is apparent that the Yellowstone cutthroat trout assemblage has been substantially reduced by the presence of lake trout.

Trophic cascade theory predicts the structure of plankton communities under an even or odd number of trophic levels (Carpenter et al. 1987, 2001). Zooplankton assemblages in lakes with an odd number of trophic levels are expected to have lower biomass and a dominance of smaller zooplankton species, such as copepods. The phytoplankton assemblages in such lakes should have higher biomass, and water clarity would be lower. Conversely, lakes with an even number of trophic levels generally have higher zooplankton biomass dominated by larger species (i.e., cladocerans) and phytoplankton assemblages with lower biomass and, thus, increased water clarity.

When the zooplankton assemblages before (1977 to 1980) and after (2004) the invasion of lake trout were compared, the 2004 assemblage exhibited reduced grazing pressure (Tronstad et al. 2010). During both periods, the zooplankton assemblage was composed of three copepods (*Diacyclops bicuspidatus thomasi, Leptodiaptomus ashlandi,* and *Hesperodiaptomus*

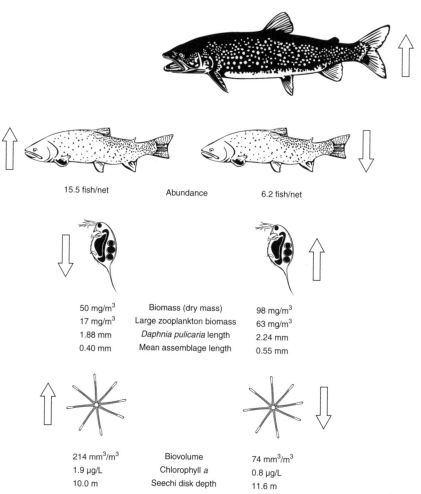

Pre-lake trout   Post-lake trout

15.5 fish/net   Abundance   6.2 fish/net

50 mg/m³	Biomass (dry mass)	98 mg/m³
17 mg/m³	Large zooplankton biomass	63 mg/m³
1.88 mm	*Daphnia pulicaria* length	2.24 mm
0.40 mm	Mean assemblage length	0.55 mm

214 mm³/m³	Biovolume	74 mm³/m³
1.9 µg/L	Chlorophyll *a*	0.8 µg/L
10.0 m	Seechi disk depth	11.6 m

Figure 9. The Yellowstone Lake food web before (left) and after (right) the invasion of lake trout. Before lake trout, Yellowstone Lake was dominated by Yellowstone cutthroat trout. With three trophic levels, zooplankton had lower biomass and were dominated by smaller species, such as copepods. Phytoplankton biomass (biovolume and chlorophyll *a*) was greater when cutthroat trout dominated. After lake trout invaded, the Yellowstone Lake food web had four trophic levels. Cutthroat trout abundance decreased. Zooplankton biomass increased, and the assemblage was dominated by larger species, such as cladocerans, under four trophic levels. Phytoplankton biomass decreased and Secchi disk depths (measure of water clarity) became shallower because of a sparser assemblage. Arrows indicate biomass predictions according to the trophic cascade theory (Carpenter et al. 1987, 2001). Line drawing is by Christine Fisher and modified from Tronstad et al. (2010).

shoshone) and two cladoceran species *(Daphnia schødleri* and *Daphnia pulicaria).* Following the expansion of the lake trout, however, the zooplankton assemblage shifted from small copepods *(Diacyclops bicuspidatus thomasi, Leptodiaptomus ashlandi,* and nauplii) prior to lake trout to an assemblage dominated by large species *(Hesperodiaptomus shoshone, Daphnia schødleri,* and *Daphnia pulicaria).* The biomass of zooplankton was two times greater following the invasion of lake trout, and the average length of individual zooplankton species was 17 percent greater. The mean length of an individual in the zooplankton assemblage increased from 0.40 millimeters prior to lake trout to 0.55 millimeters in 2004 (Tronstad et al. 2010).

Whole-lake experiments have repeatedly demonstrated that food webs dominated by fish predators (four tropic levels) are associated with zooplankton assemblages that are characterized by large species such as *Daphnia,* larger individuals, and higher total biomass (Carpenter et al. 1987, 2001; Schindler et al. 1993). Conversely, zooplankton assemblages in lakes dominated by plankton-eating fishes (three trophic levels) consist of small species, such as copepods, with smaller individual sizes and lower biomass, because plankton-eating fish consume large-bodied zooplankton (Brooks and Dodson 1965).

To evaluate changes in the phytoplankton assemblage after the introduction of lake trout, Tronstad et al. (2010) compared historical estimates of phytoplankton density and biovolume (cell volume), chlorophyll *a* (an indicator of phytoplankton biomass), and Secchi disk depths (measure of water clarity) in Yellowstone Lake with current data. Phytoplankton abundance and biovolume in Yellowstone Lake were indeed lower following the decline in cutthroat trout abundance, as predicted by trophic cascade theory. Chlorophyll *a* in West Thumb of Yellowstone Lake was twice as concentrated in 1972 (J. C. Knight 1975) as it was in recent samples (Tronstad et al. 2010). Total biovolume of phytoplankton in Yellowstone Lake was three times higher in 1972 (J. C. Knight 1975) and 1996 (Interlandi 2002) than in 2004 (Tronstad et al. 2010), and biovolume was 6.5 times higher in 1997 (Interlandi 2002) compared with 2004 (Tronstad et al. 2010). Water clarity increased 0.05 meters per year between 1976 and 2006 during summer stratification, and overall, Secchi disk depths were more than 1.5 meters deeper in 2006 compared with 30 years earlier (Tronstad et al. 2010).

A decrease in phytoplankton is common when fish eaters, such as lake trout, are the top predator. In experimental lakes with three trophic levels, phytoplankton biomass and biovolume were about three times greater than lakes with four trophic levels, and differences were attributed to increased grazing by zooplankton (Carpenter et al. 2001). For example, after introduction of northern pike *(Esox lucius)*, phytoplankton biomass and primary production decreased in a manipulated lake (Findlay et al. 2005). Water clarity increased by 3 meters when fish were removed from a Norwegian lake, changing the food web from three to two (odd to even number) trophic levels (Reinertsen et al. 1990).

The phytoplankton assemblage of Yellowstone Lake shifted to higher biovolumes of *Asterionella formosa* and lower biovolumes of *Anabaena* and *Stephanodiscus* after lake trout invaded (Tronstad et al. 2010). Increases in the biomass of *Asterionella* and a lower biovolume of large phytoplankton species have been reported for other systems with four tropic levels (Bergquist et al. 1985; Carpenter et al. 1987). In Yellowstone Lake, blooms of *Anabaena,* a nitrogen-fixing blue-green alga, were common prior to 1988 (Benson 1961; J. C. Knight 1975; Gresswell 2004), but few colonies were observed in 2004 (Tronstad et al. 2010). Anecdotal observations suggest that *Anabaena* blooms have not been common in the last two decades (Gresswell 2004). Furthermore, biomass of blue-green algae is often high in systems with three trophic levels and low in those with four trophic levels (Scavia et al. 1988). The decrease of *Stephanodiscus yellowstonsis,* a large diatom, may have been caused by the shift to a zooplankton assemblage dominated by cladocerans that are able to eat larger phytoplankton (Bergquist et al. 1985).

Evidence suggests that as lake trout became more abundant, there was a new trophic cascade in Yellowstone Lake. As predicted by trophic cascade theory (Carpenter et al. 1987), the biotic assemblage in Yellowstone Lake shifted to resemble that found in a lake dominated by a fish-eating predator. Zooplankton biomass and size have increased, and the assemblage has shifted to larger species (Tronstad et al. 2010). Phytoplankton biomass and biovolume have decreased, and water clarity has increased. Although other factors such as whirling disease and drought have probably contributed to the decrease of cutthroat trout and strengthened the trophic cascade, it appears that the current status of cutthroat trout and alterations in the food web are most closely linked to the invasion of lake trout.

There are other factors besides changes in the composition of the fish assemblage that may have influenced Yellowstone Lake in recent decades. For example, climate change has decreased the period of ice cover on lakes in the Northern Hemisphere (Magnuson et al. 2000). Water temperatures in Lake Washington from March to June have increased 0.35 °C per decade, and the spring diatom bloom has occurred earlier (Winder and Schindler 2004). At the same time, timing of maximum densities of *Daphnia pulicaria* (a grazer) have not changed, and densities have decreased over time because of the mismatch in timing with the phytoplankton bloom. Yellowstone Lake water temperatures increased 0.29 °C per decade (Tronstad et al., unpublished data), but *Daphnia* have become more abundant (Tronstad et al. 2010). Furthermore, there is some evidence that the timing of the melting of the ice on the lake surface has been occurring somewhat earlier in the past 50 years, but the average number of ice-free days on the lake does not suggest a significant trend. Although climate change may affect Yellowstone Lake, there is no strong evidence that observed changes in the food web can be attributed to a warming climate at this time.

There has been no formal analysis of the food web of the broader Yellowstone Lake ecosystem; however, decreases in Yellowstone cutthroat trout abundance appear to have had negative effects on predators throughout the Yellowstone Lake ecosystem (Varley and Schullery 1995b; Stapp and Hayward 2002a; Crait and Ben-David 2006). For example, following the introduction of lake trout, numbers of spawning Yellowstone cutthroat trout and indices of bear use decreased on streams near the developments of Grant Village and Lake (Reinhart et al. 2001). More recently, Haroldson et al. (2005) documented lakewide decreases in the number of spawning Yellowstone cutthroat trout and the number of bears fishing. Indices of grizzly bear use on monitored spawning streams have decreased (Haroldson et al. 2005), and estimates of Yellowstone cutthroat trout consumption by bears (2,226 trout annually; Felicetti et al. 2004) are less than 2 percent of estimates of cutthroat trout consumed by lake trout in the 1990s (Ruzycki et al. 2003; Felicetti et al. 2004). Furthermore, American white pelicans have maintained the breeding colony in the southeast arm of Yellowstone Lake (Baril et al. 2010), but foraging flights now appear to extend far beyond the Yellowstone Lake watershed.

A study conducted by Wengeler et al. (2010) provides insights into the diet of river otters in years following the expansion of the lake trout population

The introduction of the non-native predatory lake trout into Yellowstone Lake has resulted in a catastrophic decline in the abundance of native Yellowstone cutthroat trout that were once an important food source for many mammals, such as this river otter family, when the trout moved out of the lake and into tributary streams to spawn each year. *Photo by Cindy Goeddel.*

in Yellowstone Lake. Evidence based on scat analysis (a method that emphasizes recently consumed foods) suggests that Yellowstone cutthroat trout were the most common food in the diet of otters in the Yellowstone Lake watershed in 1999 and 2000, but longnose suckers were also very common (Wengeler et al. 2010). In contrast, isotope analysis (a method that integrates diet over longer periods) actually suggests that suckers dominated the otters' diet during that period (Wengeler et al. 2010). Results of this study support earlier predictions that lake trout have not replaced Yellowstone cutthroat trout as an accessible food source for otters (Crait 2005; Crait and Ben-David 2006), but at the same time, it appears that otters are successful in obtaining nontrout food items (Wengeler et al. 2010). Furthermore, it was apparent on the basis of carbon isotope ratios that consumption of Yellowstone cutthroat trout by lake trout was extensive.

CONSERVATION IMPLICATIONS

Shifts in the food web of Yellowstone Lake and the broader Yellowstone ecosystem underscore the consequences of the establishment of non-native lake trout in the latter decades of the twentieth century. Effects are apparent at multiple levels in the food web, both aquatic and terrestrial. Evaluating the results of the introduction from a process perspective provides important insights into the extent of the changes and the potential effects of the demise of Yellowstone cutthroat trout, a keystone species in the ecosystem.

The National Park Service implemented a suppression program the year following the discovery of lake trout in Yellowstone Lake, and substantial resources have been expended in an effort to reduce lake trout abundance. The lack of information concerning lake trout population size has made it difficult to assess the effectiveness of lake trout suppression. However, Syslo et al. (2011) recently analyzed temporal variation in individual lake trout growth, body condition, length, age at maturity, fecundity, and mortality in an effort to develop a statistical catch-at-age analysis for evaluating the program. Population metrics suggested that despite more than a decade of active lake trout removal, there was no evidence that lake trout were being overharvested. Though the lake trout population growth rate remained positive, it appeared to be lower than it would have been without suppression activities (Syslo et al. 2011). A major conclusion of the study was that suppression (gillnetting) efforts would have to be doubled (above the maximum recorded in 2007) to reduce the lake trout population growth rate below the level of replacement (Syslo et al. 2011).

In 2011, the National Park Service began implementing a Native Fish Conservation Plan that included a detailed strategy with specific benchmarks for lake trout suppression (reduced lake trout population growth rates and overall catch-per-unit-effort of lake trout) and the restoration of cutthroat trout in Yellowstone Lake based on the number of spawning Yellowstone cutthroat trout entering Clear Creek and the average annual landing rates for cutthroat trout for the period prior to the discovery of lake trout (1987 to 1991; U.S. Department of the Interior, National Park Service 2010). The goals of the plan are clearly articulated, but metrics such as a particular number of fish in a spawning run or a mean landing rate are confounded by interannual variability. Furthermore, the source of this variability may be related to factors other than lake trout predation. It is also difficult to determine how goals would differ if a different reference period were chosen and if the number of years used to assess a particular metric was changed. Indeed, other goals focused on previous Yellowstone cutthroat trout population metrics would have similar shortcomings.

One alternative would be to identify a process-based metric, such as food web structure, as a management goal for the lake trout suppression program. For example, a management goal for Yellowstone Lake could be restoring the three-trophic-level food web dominated by plankton-eating fishes. Achievement of this goal would require that lake trout be suppressed to a point where predation of Yellowstone cutthroat trout was low enough to allow reexpression of the historic food web structure in Yellowstone Lake.

This approach yields a much more robust management target that would reflect real change in the assemblage of aquatic organisms in Yellowstone Lake but would be less likely to respond to interannual population variability of either lake trout or Yellowstone cutthroat trout. Currently it appears doubtful that lake trout can be eliminated from Yellowstone Lake in the near future, but it may be possible to reduce their abundance to a point where the trophic structure of the food web is returned to historic proportions, and more important, it provides a measurable goal for assessing the lake trout suppression program. Furthermore, it is a metric that integrates the role of the Yellowstone cutthroat trout in the food web of the entire Yellowstone Lake ecosystem.

CHAPTER FOURTEEN

Balancing Bison Conservation and Risk Management of the Non-Native Disease Brucellosis

JOHN J. TREANOR

P. J. WHITE

RICK L. WALLEN

DISEASE HAS LONG been recognized as an important process influencing the distribution and abundance of animal populations (Andrewartha and Birch 1954; Anderson and May 1978). Disease organisms, much like predators, have the potential to significantly reduce the size of animal populations when they cause high mortality. However, large mammals have well-developed immune systems and are frequently able to recover from infections, thereby reducing the limiting effects of many parasites. Today the greatest disease threat to wildlife may come not directly from infectious organisms but indirectly from humans attempting to reduce health risks to people and their domestic animals. In recent years, we have witnessed a global rise in emerging and reemerging infectious diseases, which are increasingly being shared among humans, livestock, and wildlife. The rapid development of land by humans has reduced the amount of habitat available for wildlife. Much of the wildlife habitat that does remain is often fragmented or found within wildlife reserves, such as national parks. Human development along the boundaries of these preserves has also increased the proximity of humans, domestic animals, and wildlife. Consequently, the risk of infectious disease spread from wildlife to livestock and humans is a legitimate concern that is challenging wildlife conservation along the boundaries of protected areas. Because wildlife and their diseases do not recognize management or political boundaries, conservation efforts become complicated when risks to domestic animals or human health arise. This has long been the case with brucellosis management in the greater Yellowstone ecosystem.

Brucellosis is a contagious disease caused by the bacteria *Brucella abortus* that can induce abortions or the birth of nonviable calves in livestock and wildlife (Rhyan et al. 2009). The bacterium was likely introduced by European livestock to Yellowstone bison and elk before 1930 (Meagher and Meyer 1994). In wildlife and cattle, infection typically occurs through contact with infectious reproductive tissues shed during an abortion or live birth (Rhyan et al. 2009). Though rare in the United States, human brucellosis can occur if bacteria are ingested or enter through the eyes or open wounds. *Brucella abortus* infection is rarely fatal in humans and cannot be passed from humans to other humans or animals. However, if not treated early, human brucellosis can cause recurring, severe, fever-like symptoms. Early treatment is often complicated by difficulties diagnosing infection because of the low infectious dose needed to establish infection and symptoms common with other illnesses.

Infected bison and elk can pose a health risk for people, especially those that handle infected animals without proper protective equipment. There were five confirmed brucellosis cases reported to the Wyoming Department of Health from 1995 to 2005 and 17 confirmed cases reported to the Idaho Department of Health and Welfare from 1980 to 2003—though none of these cases were attributed to wildlife (Snow 2005). However, there have been two confirmed cases of hunters contracting brucellosis from elk in Montana (Zanto 2005). To minimize effects to humans, a nationwide program to eradicate brucellosis from cattle has been in place since 1934. The program has successfully eliminated *Brucella abortus* from most of the United States, with the exception of free-ranging wildlife within the greater Yellowstone ecosystem and feral swine in Texas. Over the past decade, all three states bordering Yellowstone National Park (Idaho, Montana, and Wyoming) have experienced multiple brucellosis outbreaks in cattle herds as a result of contact with infected wildlife, which has caused additional economic expenses for the state livestock industries. Thus, concerns over the risk of brucellosis transmission to cattle have led to decades of conflict regarding management of bison and elk in the greater Yellowstone ecosystem.

Traditionally, brucellosis management has focused on elk in the southern greater Yellowstone ecosystem and bison in the northern portion. This management strategy has been supported by most livestock and natural resources personnel who view supplementally fed Wyoming elk and migrating Yellowstone bison as the primary sources for brucellosis transmission to cattle (Bienen and Tabor 2006). Elk, which are widely tolerated on federal and state lands, often mingle with cattle. As a result, the recent brucellosis transmissions to cattle have been attributed to elk (Beja-Pereira et al. 2009). Resource agencies prevent bison from mingling with cattle through active management, but bison management practices such as hazing and culling are unpopular with the concerned public. Here we review the state of knowledge regarding brucellosis transmission risk and its control in bison and elk in the greater Yellowstone ecosystem.

BRUCELLOSIS TRANSMISSION RISK

The risk of brucellosis transmission to livestock or wildlife is influenced by the amount of infectious material shed onto the landscape and its ability to persist long enough to establish infection when contacted. Environmental factors that increase stress and concentrate animals, such as deep snow and the presence of predators, may increase the likelihood of exposure to shed infectious tissues. However, expelled tissues are generally removed quite quickly by scavengers (W. E. Cook et al. 2002; Aune et al. 2007). Additionally, the behavior of elk and bison during calving may limit the transmission risk to cattle. Yellowstone bison exhibit synchronous calving, with 80 percent of births occurring from late April to late May (J. D. Jones et al. 2010). Birthing females often consume shed birth tissues within two hours after calving. This behavior reduces the risk of brucellosis transmission to cattle and bison that later encounter the birth site. However, the potential for exposure is higher for bison that are more closely associated with a pregnant animal that has shed infectious tissues. During 2004 to 2007, at least one bison was observed making contact with potentially infectious birth tissues in 30 percent of observed bison births (J. D. Jones et al. 2009). Thus, brucellosis transmission requires a source of infection and relies on the behaviors of potential hosts (Cheville et al. 1998).

The role of elk in the maintenance of brucellosis in the northern portion of the greater Yellowstone ecosystem has traditionally been viewed as less important than that of bison. Elk also exhibit a high degree of birth synchrony and consume shed birth tissues. Unlike most bison, however, female elk segregate themselves from other herd members while giving birth (Johnson 1951). Thus, birth sites are dispersed, and the likelihood of other elk encountering infected birth tissues is low. However, transmission risk may be higher during late winter and early spring when elk form large aggregations on low-elevation winter ranges (Hamlin and Cunningham 2008). *Brucella*-induced abortions under these conditions could expose many susceptible elk to infectious material. During the past decade, elk abundance has increased substantially on winter ranges in many state-managed herds surrounding Yellowstone National Park (Cross et al. 2009). Concurrent with increasing abundance on managed lands, brucellosis seroprevalence (identified by the presence of antibodies to *Brucella* circulating in blood) has increased to 7 to 20 percent on some nonfeed ground areas (Cody and Buffalo Valley, Wyoming; Ruby Valley, Montana). Thus, elk populations, far from both bison and feed grounds, may be perpetuating higher levels of brucellosis as a result of increased densities and group sizes on their winter ranges (Cross et al. 2009).

The high seroprevalence (40–60 percent) of brucellosis in Yellowstone bison implies that they are a likely infection source for Yellowstone elk. However, recent data suggest that transmission between bison and elk is infrequent (Proffitt, White, and Garrott 2010). The peak bison calving period, when most of the bacteria are expected to be shed, occurs approximately one month earlier for bison than for elk, with little overlap in distribution during this time period. On wintering ranges where elk do mingle with bison, such as the Madison headwaters area in Yellowstone, elk have much lower seroprevalence rates for brucellosis (3 percent) than Yellowstone bison or elk associated with feeding programs in Wyoming. Proffitt, White, and Garrott (2010) found that brucellosis transmission risk from bison to elk was quite low in Yellowstone's Madison headwaters area, despite a high degree of spatial overlap when *Brucella abortus* is typically shed. Predation risk associated with wolves increased elk and bison spatial overlap temporarily, but these behavioral responses by elk did not have important

A female elk cleans her calf, less than 1 hour old, near Phantom Lake. Both elk and bison in the Yellowstone region carry an exotic bacterium likely transmitted to these wild populations from livestock at least a century ago. The organism causes the disease brucellosis, which may result in abortions in both wildlife and livestock and is transmitted when animals come into contact with contaminated tissues and fluids associated with the birth process. *Photo by Cindy Goeddel.*

disease implications. Thus, it appears that brucellosis in the greater Yellowstone ecosystem is a disease sustained by multiple hosts, and managing the risk of transmission to cattle will require control measures addressing bison, elk, and the factors sustaining infection.

BRUCELLOSIS CONTROL IN BISON

Management of Yellowstone bison and the brucellosis transmission risk they pose to cattle outside the park has a long, contentious history involving wildlife managers, livestock producers, and the concerned public. After intensively managing bison numbers for 60 years through husbandry and consistent culling, the National Park Service instituted a moratorium on culling in the park in 1969, which allowed bison numbers to fluctuate in response to environmental and ecological factors (G. F. Cole 1971). Bison abundance increased rapidly, with large winter migrations out of the park beginning in the late 1980s (Meagher 1989a, 1989b). These migrations led to a series of conflicts with stock growers and the state of Montana, largely because of the possibility of brucellosis transmission to cattle. As a result, in 2000, the federal government and the state of Montana agreed to a court-mediated Interagency Bison Management Plan that established guidelines for (1) cooperatively managing the risk of brucellosis transmission from bison to cattle and (2) preserving the bison population and allowing some bison to occupy winter ranges on Montana's public lands. The plan uses intensive management, such as hazing and hunting of bison migrating outside the park, to maintain separation between bison and cattle. In general, the agencies have successfully maintained spatial and temporal separation between bison and cattle with no transmission of brucellosis (P. J. White, Wallen, et al. 2011). However, this intensive management is expensive, logistically difficult, and controversial.

The enduring debate over Yellowstone bison management has largely concentrated on the culling of animals that roam outside park boundaries during the winter. In recent decades, large numbers of bison migrating into Montana have been captured by federal and state agencies, when separation between bison and cattle could not be maintained. Many of these bison were tested for brucellosis, and those animals testing positive for exposure were shipped to domestic slaughter facilities. Approximately 3,200 bison were shipped to slaughter from 2001 through 2011, to mitigate the risk of brucellosis transmission to cattle (P. J. White, Wallen, et al. 2011). Despite these actions, brucellosis prevalence in Yellowstone bison has not decreased. Intensifying bison removals to a level that may be effective at reducing brucellosis infection within the herd would be extremely expensive, unacceptable to the public, and counter to the National Park Service policy to maintain ecosystem integrity (Bienen and Tabor 2006).

A more effective and acceptable approach for brucellosis reduction might involve management actions on selected individuals in combination with brucellosis vaccination. Vaccination of female bison has the potential to reduce brucellosis prevalence by increasing herd immunity (Treanor et al. 2007). However, vaccination would require a sustained effort to consistently and reliably deliver vaccine at the appropriate time of year to a large portion

A bison calf and female forage in the sagebrush grasslands of Little America. The successful restoration of bison in Yellowstone National Park has allowed bison numbers and distribution to expand throughout suitable habitat in the park. Bison have also recently reestablished migratory movements to low-elevation areas outside the boundaries of the park, where they pose a risk of transmitting brucellosis to livestock during the calving season. *Photo by Cindy Goeddel.*

of eligible bison each year over decades (Treanor et al. 2010). Also, a vaccine with moderate efficacy is unlikely to succeed in controlling brucellosis in the long-term without the eventual inclusion of selective culling or fertility control (Treanor et al. 2010; Ebinger et al. 2011).

Contraception has also been suggested as a method to reduce brucellosis transmission because the disease is known to be transmitted only by pregnant bison or elk. National Park Service (2006) policy allows for the use of reproductive intervention in wildlife if these techniques are appropriate for achieving management goals. Thus, the use of an effective, reliable, and safe contraceptive in seropositive bison or elk has the potential to decrease brucellosis transmission, especially when combined with vaccination of seronegative animals (Ebinger et al. 2011). However, fertility control products may also cause negative long-term effects, such as permanent sterility, altered reproductive or social behaviors, and changes in the age and sex structure of a population (M. E. Gray and Cameron 2010). Though fertility control products may be promising tools for reducing brucellosis infection in free-ranging bison and elk, the risk of unacceptable side effects and their effectiveness will need to be addressed.

Until effective methods to reduce brucellosis are developed, the best approach to control the risk of brucellosis transmission from bison to cattle is to maintain spatial and temporal separation. Management agencies should continue to allow bison migration to essential winter range areas in and adjacent to Yellowstone National Park but actively prevent dispersal and range expansion to outlying private lands until there is tolerance for bison in these areas (Plumb et al. 2009). Additionally, Kilpatrick et al. (2009) recommended the cessation of cattle grazing in areas where bison leave the park in winter and compensating ranchers for lost earnings and wages. Conservation groups and government agencies have successfully used, and are still pursuing, this strategy with willing landowners (U.S. Department of the Interior et al. 2008). However, further efforts are needed from wildlife managers, livestock producers, and the public to balance long-term bison conservation with brucellosis risk management. Identifying and creating additional protected habitat for bison in Montana will help promote long-term bison conservation, while keeping bison separated from livestock and developing effective vaccination programs will help private landowners protect their cattle.

BRUCELLOSIS CONTROL IN ELK

It is generally agreed that supplemental feeding at 22 state feed grounds and one federal feed ground (National Elk Refuge, U.S. Fish and Wildlife Service) contributes to the high level of brucellosis infection in elk in the southern portion of the greater Yellowstone ecosystem (Bienen and Tabor 2006). These feed grounds sustain higher numbers of elk than remaining winter habitat could otherwise support and create large elk aggregations that facilitate brucellosis transmission (Cross et al. 2007). However, wildlife and livestock managers believe feed grounds reduce the number of elk that would otherwise mingle with cattle on local ranches.

Another factor that may influence the maintenance of brucellosis in elk throughout the greater Yellowstone ecosystem is the increasing human population and land-use practices that sustain it. Ranching and the development of rural homes in the southern portion of the greater Yellowstone ecosystem have fragmented valley bottom and flood plain habitats crucial for elk migration and use during winter. This conversion of elk habitat for human use has contributed to an increase of elk aggregations on cattle feed lines and refuges (Haggerty and Travis 2006; Cross et al. 2009). Studies of migratory elk near Yellowstone National Park suggest that they tend to select wintering areas with lower probability of wolf occupancy, lower road density, and higher forage abundance (Proffitt, Grigg et al. 2009; P. J. White et al. 2010). Many of these areas occur on private ranch lands and portions of public grazing allotments. Elk aggregating on feed grounds, natural winter ranges, or refuges influence brucellosis transmission, especially near the calving period in late winter and early spring. Thus, wildlife agencies need to explore strategies for dispersing large aggregations of elk at these times. Effective brucellosis management will require gaining enhanced cooperation from landowners to increase access for hunters; providing increased tolerance and protection of large predators, such as wolves, that disperse elk; and assisting landowners with infrastructure to isolate cattle and their feed.

CONSERVATION IMPLICATIONS

An essential first step for the conservation of bison and elk is to develop acceptable brucellosis control methods that balance risk management with wildlife conservation. Over time, the strategies discussed herein could reduce the costs and need for brucellosis risk management activities while maintaining low risk for the cattle industry. However, the potential for elk to maintain brucellosis complicates management of this disease. Tens of thousands of elk live in the greater Yellowstone ecosystem, with brucellosis in elk populations throughout the area. At this time, eradication of brucellosis from these populations is neither probable nor practical without resorting to ethically and politically unacceptable techniques, such as mass test and slaughter or depopulation (U.S. Animal Health Association 2006). Thus, management of brucellosis in elk populations might best be achieved by curtailing practices that unnaturally increase elk densities and group sizes during winter and spring.

The Yellowstone bison population will likely continue to grow and attempt to expand their range unless hunting, culling, and relocations are used to remove several hundred bison per year from the population (Hobbs et al. 2009). Thus far, Yellowstone bison have been transported to domestic slaughter or research facilities only because of the potential for *Brucella abortus* infection. These removals have differentially affected Yellowstone's central and northern breeding herds, altered gender structure, created reduced groups of females in some years, and temporarily dampened bison productivity (P. J. White, Wallen, et al. 2011). Thus, there is a need to increase tolerance for bison on key winter ranges in Montana and reduce the frequency of large culls of the population. The shipment of surplus Yellowstone bison to quarantine sites or terminal pastures operated by American Indian tribes or conservation organizations rather than to domestic slaughter facilities would be a transformational moment in the conservation of plains bison.

Brucellosis risk management in the greater Yellowstone ecosystem is one of the great challenges facing large mammal conservation in North America. Effective management practices will require a diverse range of integrated methods, which include maintaining separation of livestock and wildlife, managing habitat to reduce brucellosis transmission, and reducing disease prevalence in wildlife. The long-term success of these management practices will depend on sound science and the support of stakeholders involved. Otherwise, efforts to balance brucellosis management with wildlife conservation are unlikely to be successful.

CHAPTER FIFTEEN

Exotic Fungus Acts with Natural Disturbance Agents to Alter Whitebark Pine Communities

S. THOMAS OLLIFF

ROY A. RENKIN

DANIEL P. REINHART

KRISTIN L. LEGG

EMILY M. WELLINGTON

THE LANDSCAPE IN Yellowstone National Park is shaped, maintained, and famous for periodic landscape-scale disturbances, both ancient and ongoing. These disturbances occur at a scale larger than a single watershed, with two extreme examples being the fires of 1988, which burned over 404,685 hectares in the greater Yellowstone ecosystem, and ancient Yellowstone volcanism, which spewed ash over half of the continental United States (R. B. Smith and Siegel 2000). Over the geologic timescale (hundreds of thousands of years), changing climate and major geologic processes have dramatically restructured Yellowstone's landscape and plant and animal communities (National Research Council 2002). Thus, the landscape of the greater Yellowstone ecosystem does not remain at rest but changes continuously (Despain 1990; Pritchard 1999; National Research Council 2002).

Two native disturbance agents that evolved in the greater Yellowstone ecosystem, natural fire and the mountain pine beetle, create, change, and maintain vegetation on large scales. Fire has been a part of the ecosystem for as long as there has been vegetation to burn, and most of the ecosystem is characterized by large, infrequent, stand-replacing fires. Less than

5 percent of the fires account for 95 percent of the area burned (Westerling et al. 2011), and the time between successive fires on a specific area—the fire return interval—is between 200 and 400 years (Romme 1982; Barrett 1994). Evidence of fires that burned before the park was established in 1872 can be found in soil profiles, lake sediments, landslides, and old-growth trees. The 1988 fire season, though not unusual compared with the long history of fire in the greater Yellowstone ecosystem, has become legendary in Yellowstone folklore. Fifty fires burned almost 323,748 hectares, or almost one-third of the 890,308 hectares within the park (Despain 1990), and a total of about 566,560 hectares in the greater Yellowstone ecosystem. Romme and Despain (1989) evaluated Yellowstone's fire history and noted that fire behavior, in terms of heat release, flame height, and rate of spread, were probably similar to the fires that burned a significant percentage of the greater Yellowstone ecosystem in the early to mid-1700s. They concluded that the 1988 fires represented a nearly natural event.

While spruce budworm was the first forest insect to draw attention to the landscape-scale nature of insect disturbance in 1923 (Furniss and Renkin 2003), the mountain pine beetle is a native insect that has caused widespread mortality in Yellowstone's pine species (Despain 1990). Population levels of pine beetle are often low and inconspicuous but at times may be high and cause substantial tree mortality over large areas of the landscape in only a few years, often killing the largest trees. The first recognized outbreak of mountain pine beetle in Yellowstone began in 1925 in the central mountains of the park (Furniss and Renkin 2003) and increased over subsequent years, leading to major control efforts beginning in 1931. By 1935, the spread and extent of the infestation was so large, particularly among the park's whitebark pine forests, that control efforts were abandoned except for high-profile sites, such as campgrounds. By 1937, practically all whitebark pine trees in the park were infested, but the infestation waned so quickly, beetles were no longer considered a problem in 1942 (Despain 1990). A second serious beetle outbreak spread into the park in 1969 and lasted until around 1985. It appears that two overlapping outbreaks occurred during this period, spreading from the Bechler region in southwest Yellowstone and, by 1974, covering almost half the park in an epidemic bisecting the park diagonally northwest to southeast. By 1978, the epidemic seemed to be waning, only to reemerge and cover the entire west side of the park by 1982. The last mortality was reported in small pockets of the northern portion of

Yellowstone in 1985 (Despain 1990). The area affected by mountain pine beetle in Yellowstone at the peak of the outbreak in 1981–1982 totaled 964,000 acres, including 34,500 acres of whitebark pine (U.S. Department of Agriculture 1984).

Exotic invasive species—weeds, aquatic nuisance species, wildlife disease, and forest pathogens—are a recent (within the twentieth century) addition to the Yellowstone ecosystem that alter ecosystem processes and create landscape disturbance (D. N. Cole and Yung 2010a). Perhaps the greatest influence of modern humans on the vegetation of Yellowstone has been the introduction of over 200 new plants, some of which change whole plant communities (Whipple 2001). Timothy *(Phleum pratense)* and Kentucky bluegrass *(Poa pratensis)* are widespread as a result of the use of horses, and Canada thistle *(Cirsium arvense)* is present wherever it can grow (Despain 1990; Olliff et al. 2001). Future climate-informed projections show yellow star thistle *(Centaurea solstitialis),* cheatgrass *(Bromus tectorum),* and spotted knapweed *(Centaurea maculosa)* increasing their range in the greater Yellowstone ecosystem (B. A. Bradley et al. 2010). Aquatic nuisance species, such as New Zealand mudsnails *(Potamopyrgus antipodarum),* are spreading into greater Yellowstone ecosystem waters (McMahon et al. 2009) along with introduced pathogens, such as whirling disease (Koel et al. 2006). Bivalves, such as zebra and quagga mussels *(Dreissena polymorpha; Dreissena rostriformis bugensis),* may also spread into greater Yellowstone ecosystem waters (Independent Economic Analysis Board 2010). Exotic lake trout have taken over Yellowstone Lake and caused dramatic declines in Yellowstone cutthroat trout (Varley and Schullery 1995b; Gresswell 2009). Wildlife diseases that are not indigenous to the area have spread into Yellowstone, causing population declines and requiring management interventions. Many bison and elk in the greater Yellowstone ecosystem have been exposed to the bacterium that causes brucellosis, which originated in domestic livestock (Cheville et al. 1998). Parvovirus, distemper, mange, and hepatitis are believed to have been a major factor in wolf population declines in Yellowstone in 1999, 2005, and 2008; these diseases also appear to have affected coyotes, foxes, and possibly cougars and other smaller carnivores (D. W. Smith and Almberg 2007). Ranavirus and chytrid fungus *(Batrachochytrium dendrobatidis)* commonly cause amphibian die-offs and have been identified in Yellowstone National Park (Corn et al. 2005).

An example of an invasive species altering ecological processes across the greater Yellowstone landscape is *Cronartium ribicola,* the non-native fungal pathogen introduced to North America in 1910 that results in white pine blister rust (Spaulding 1922; Maloy 1997; Hatala et al. 2011). White pine blister rust damages and kills trees by girdling branches and trunks (Hoff 1992). Blister rust enters the stomata (pores) of whitebark pine needles and then erupts into cankers on the branches, leading to the loss of cone production and, in some cases, the death of the tree (Tomback et al. 2001). Depending on the level of infection, a tree with white pine blister rust can live for decades. However, infected saplings generally die within three years (Koteen 2002). Infection by blister rust also weakens the tree and can lead to death by cumulative factors, including mountain pine beetle, other pathogens, root diseases, and unfavorable climatic conditions (Koteen 2002). Blister rust has devastated whitebark pine populations in areas with maritime climates in the Pacific Northwest, with infection rates of 82 percent in the north Cascades and 90 percent in Glacier National Park (Kendall and Keane 2001; Koteen 2002). However, the drier climate of the greater Yellowstone ecosystem may be relatively inhospitable to the spread of blister rust (Koteen 2002). Blister rust was discovered in the Bear Creek drainage of the Gallatin National Forest, 30 kilometers north of Yellowstone National Park, in 1937 (Kendall and Asebrook 1998). It was first recognized in Yellowstone on the leaves of host plants in 1944 and discovered on pine trees in 1950 (Yellowstone National Park 1944, 1950).

Climate has fluctuated in the greater Yellowstone ecosystem for the entire period for which we have records (at least 9,000 years). Several important climate shifts strongly influenced the ecosystem during the Holocene (7,000 to 9,000 years before present); the Medieval Warm Period (A.D. 1000 to 1300), which manifested droughts in the northern plains of the United States; and the Little Ice Age (A.D. 1450 to 1890), which brought cooler temperatures (1 to 1.5 °C) and a moister climate. The past 200 years have included the warmest and coldest periods during the past 4,000 years (National Research Council 2002).

This chapter explores two natural or native landscape-scale disturbance agents—natural fire and mountain pine beetles—and one invasive disturbance organism—*Cronartium ribicola*—and how they act in tandem to affect whitebark pine abundance, distribution, and survival. Whitebark pine forests are decreasing in the greater Yellowstone ecosystem because of the landscape-scale effects of the native mountain pine beetle, white pine blister rust, and possibly fire exclusion (Keane and Arno 1993; Murray et al. 1995; Kendall and Keane 2001; U.S. Fish and Wildlife Service 2011). Recent climate warming is also affecting whitebark pine and could play a major role in further decreases in coming decades (U.S. Fish and Wildlife Service 2011).

WHITEBARK PINE IN THE YELLOWSTONE SYSTEM

Whitebark pine is one of only six conifer species that occur commonly in the greater Yellowstone ecosystem. It has a limited range and occurs near timberline in Yellowstone National Park from approximately 2,620 to 3,200 meters; in pure stands on harsh sites with poor soils, steep slopes, high winds, and extreme cold temperatures; and in mixed conifer stands just below timberline. Whitebark pine is also a common understory component in the lodgepole pine forests that dominate the Yellowstone Plateau (Despain 1990). Whitebark pine is typically 5 to 20 meters tall, with a rounded or irregularly spreading crown shape referred to as a candelabra form. On higher-density sites, whitebark pine tends to grow as tall, single-stemmed trees. On open, more exposed sites, it tends to have multiple stems (McCaughey and Tomback 2001). Above tree line, it grows in a krummholz form with stunted, shrublike growth (Arno and Hoff 1990). Whitebark pine is a slow-growing tree with a life span of up to 1,000 years (Arno and Hoff 1990). Whitebark pine stands occupy about 17 percent of Yellowstone National Park and about 10 percent or 1 million hectares of the 9.7 million hectares in the greater Yellowstone ecosystem (Greater Yellowstone Coordinating Committee Whitebark Pine Subcommittee 2011). Whitebark pine has low commercial value because it is inaccessible, with gnarled growth forms. However, it is high in ecological value and has been called a keystone species in the subalpine zone (Tomback et al. 2001). Whitebark pine can exist under conditions tolerated by few other trees and may alter the microclimate, enabling other species, such as subalpine fir, to follow (Tomback et al. 1993). Perhaps its best-known role in the greater Yellowstone ecosystem is as a food source for several wildlife species, most notably the grizzly bear.

The study of whitebark pine in relation to its environment is complex. Whitebark pine seeds are dispersed and planted by a bird called the Clark's nutcracker *(Nucifraga columbiana)*. Whitebark pine has morphological characteristics—closed cones, lack of seed wings for dispersal, and the candelabra growth form that displays cones to birds—that are adaptations for this mutually beneficial relationship (Lanner 1982, 1996; Tomback 1983). Nutcrackers tend to bury seeds in small caches across variable terrain, and unrecovered caches lead to whitebark pine regeneration (McKinney et al. 2009). The red squirrel *(Tamiasciurus hudsonicus)* is the primary seed predator of whitebark pine, taking more than 80 percent of the cone crop (McKinney and Tomback 2007). Squirrels climb whitebark pine trees, bite or tear the cones from branches, let them drop to the ground, then collect and store them in depressions on the ground or hollow logs called middens (Reinhart and Mattson 1990b). In the greater Yellowstone ecosystem, grizzly bears acquire more than 90 percent of the seeds they eat by excavating squirrel middens (Mattson and Reinhart 1997).

The high-fat, energy-rich nuts of the whitebark pine are one of the most important plant foods eaten by grizzly bears in the greater Yellowstone ecosystem. Large-cone crops are most commonly produced every three years and can be synchronous across large areas. In 2009, for example, more than 70 percent of the greater Yellowstone ecosystem experienced a large-cone crop. During years when whitebark pine seeds are abundant, bears feed almost exclusively on them from late August through early November. In years with poor whitebark pine seed production, bears must eat other foods during autumn to avoid starvation. Felicetti et al. (2003) found that only 28 percent of bears used pine nuts during years of poor cone availability, while over 90 percent of bears used pine nuts when cones were abundant.

Fire is another important component of the ecology of whitebark pine communities and its distribution across the landscape. Whitebark pine stands experience nonlethal, mixed-severity, and stand-replacing fire regimes (Arno and Hoff 1990; Barrett 1994; Walsh 2005). Large trees survive low- to moderate-severity fires, and Clark's nutcrackers cache whitebark pine seeds in newly burned areas that are created by fires. Nonlethal fires tend to maintain whitebark pine dominance by stalling succession since whitebark pine is more fire resistant than competitive conifers, such as subalpine fir (Ryan and Reinhardt 1988). However, larger, stand-replacing fires can kill mature, seed-producing whitebark pine trees. These types of fires

A Clark's nutcracker atop a tree in the upper Lamar Valley. Clark's nutcrackers provide an important ecological service for trees by dispersing and caching seeds that later germinate, helping to perpetuate the species. *Photo by Cindy Goeddel.*

are likely to increase in frequency with a warmer and drier climate (Koteen 2002; Westerling et al. 2011). Likewise, a lack of fire, in conjunction with an increase in temperature, may favor later successional species, such as subalpine fir and Engelmann spruce, and allow them to outcompete white-bark pine (Tomback et al. 2001).

A substantial reduction in whitebark pine, which inhabits the upper tree line, would seriously affect upper subalpine systems by reducing biological diversity and diminishing important ecological processes (Tomback et al. 2001). Whitebark pine stands provide food and habitat for many wildlife species, and rapid seed dispersal by Clark's nutcrackers and hardy seedling growth create early whitebark pine stands after fires (Mattson et al. 2001; Tomback and Kendall 2001). Whitebark pine seeds survive on harsh, high-elevation sites; act as nursery trees to less hardy vegetation; regulate snowmelt; and reduce soil erosion (Farnes 1990; Callaway 1998). Thus, white-bark pine is considered a foundation species for building community diversity and stability (Ellison et al. 2005; Tomback and Achuff 2010).

MONITORING AND CURRENT STATUS

There are numerous efforts to understand whitebark pine and the stressors affecting it in the greater Yellowstone ecosystem. The Whitebark Pine Sub-committee of the Greater Yellowstone Coordinating Committee provides a comprehensive list of bibliographies and publications on whitebark pine (www.fedgycc.org). Key studies to understand whitebark pine mortality began in the 1990s. Kendall, Schirokauer, et al. (1996) led a study by the U.S. Geological Survey to determine the status of whitebark pine in national parks of the Rocky Mountains, including Yellowstone and Grand Teton. A similar effort was initiated by the U.S. Forest Service, primarily in the Gallatin and Shoshone national forests (Kendall, Tyers, and Schiro-kauer 1996). Newcomb (2003) detected and described the spatial pattern of white pine blister rust, particularly in relation to its *Ribes* host species, and Bockino (2008) studied the interactions of white pine blister rust and mountain pine beetle in whitebark pine communities.

Long-term monitoring has occurred since 1980, when the Interagency Grizzly Bear Study Team began monitoring cone production on 19 transects within the grizzly bear recovery zone. This effort is used primarily as an indicator of activity and demography of bears, rather than as an indicator of whitebark pine health or production (Schwartz, Haroldson, and West 2010). The Greater Yellowstone Network, part of the National Park Service's inventory and monitoring program, collaborated in 2004 with the U.S. Forest Service, U.S. Geological Survey, and Montana State University to develop a long-term, ground-based monitoring program to understand the health of whitebark pine across the ecosystem. The objectives of this program include (1) estimating the proportion and rate of change of white-bark pines greater than 1.4 meters tall that are infected with white pine blister rust; (2) estimating the relative severity of infection and change of severity over time; (3) estimating the survival of individual whitebark pine trees more than 1.4 meters tall; and (4) estimating recruitment into the cone-producing stage (Greater Yellowstone Whitebark Pine Monitoring Working Group 2011a). From 2004 through 2007, 4,774 whitebark pine trees were tagged along 174 transects that were randomly distributed across the greater Yellowstone ecosystem. These transects are divided into panels that are revisited every four years.

Logan et al. (2010) combined U.S. Forest Health Protection aerial detection surveys with remote sensing and ground evaluation surveys to estimate the loss of larger canopy trees from mountain pine beetle. Other research is under way to understand the level of genetic resistance of whitebark pine to blister rust, what management actions can be implemented to protect trees, and effective restoration actions (Mahalovich and Dickerson 2004; Mahalovich et al. 2006; Burns et al. 2008). The incidence of blister rust on white-bark pine in the greater Yellowstone ecosystem ranges from zero to 84 percent, depending on location (Bockino 2008; Bockino and McCloskey 2010; Schwartz, Haroldson, and West 2010). For comparison, results from surveys in the 1990s showed average infection rates of less than 5 percent in Yellowstone National Park and less than 15 percent in Grand Teton National Park, with a high single-site incidence of 40 to 44 percent in Grand Teton (Kendall and Keane 2001). This was an increase from the 1 percent average infection rate found in 1967, with a high single-site incidence of 2 percent (Koteen 2002). The Greater Yellowstone Network estimated a 20 percent infection rate for trees tagged from 2004 through 2007, with blister rust being widespread through the greater Yellowstone ecosystem.

Though whitebark pine is highly vulnerable to infection by blister rust, approximately 26 percent of the greater Yellowstone population shows genetic resistance (Hoff et al. 2001; Kinloch et al. 2003; Mahalovich et al. 2006). The Greater Yellowstone Coordinating Committee Whitebark Pine Subcommittee (2011a) is exploring how to capitalize on this resistance.

The current mountain pine beetle epidemic in the greater Yellowstone ecosystem began in 1999, peaked by 2008–2009, and began to decrease in 2010. The peak of the outbreak affected 40,225 hectares in Yellowstone, including 16,592 hectares of whitebark pine (U.S. Department of Agriculture 2009). Interpretation of 2007 satellite imagery by the U.S. Department of Agriculture, Remote Sensing Applications Center, indicated that over 40 percent of whitebark pine stands contained some level of canopy mortality (Goetz et al. 2009). Data from the 2008 Forest Health Protection aerial survey found beetle activity in more than 50 percent of whitebark pine stands. Data from a 2009 aerial evaluation at a subwatershed level indicated that over 50 percent of whitebark pine stands had already suffered high to complete mortality of overstory trees, and 95 percent of forest stands containing whitebark pine had measurable mountain pine beetle activity (MacFarlane et al. 2009). By 2010, the Greater Yellowstone Network recorded 16 percent mortality of the 4,774 trees tagged between 2004 and 2007. Initial data suggest that about 52 percent of mature (diameter more than 30 centimeters at breast height) cone-producing trees were dead from all causes (beetles, fire, lightning, rust, wind). Tagged trees in smaller size classes had less mortality, with only 5, 6, and 22 percent of trees recorded as dead in size classes less than 2.5 centimeters, 2.5 to 10 centimeters, and greater than 11–30 centimeters, respectively. These data suggest that though mortality rates have been high for large cone-producing trees, a high proportion of the whitebark pine population, particularly the younger trees, remains alive (Greater Yellowstone Whitebark Pine Monitoring Working Group 2011b).

The Interagency Grizzly Bear Study Team recorded similar decreases in cone production from large trees along 19 transects. Five transects have been retired since establishment because all cone-producing trees died, most likely because of mountain pine beetles (Haroldson and Podruzny 2011). By 2010, 73 percent of transect trees surveyed since 2002 had died, and 95 percent of transects contained beetle-killed trees (Haroldson and Podruzny 2011).

Heavy infection by white pine blister rust may lead to whitebark park pine trees being more susceptible to infestation by mountain pine beetle

(Keane and Arno 1993; Burns 2006). Bockino (2008) found an increase in mountain pine beetle selection intensity for whitebark pine with heavy blister rust in the greater Yellowstone ecosystem. Six and Adams (2007) found a similar relationship in their study of whitebark pine on five study sites in Idaho and Montana.

During the summer of 1988, about 28 percent of Yellowstone's whitebark pine forest burned by way of high-intensity, stand-replacing fire (Renkin and Despain 1992). However, the influence of over a century of fire suppression in the greater Yellowstone ecosystem is largely absent in whitebark pine stands due to (1) limited capacity to suppress fires in the inaccessible areas inhabited by whitebark pine, (2) the geographic and soil properties (chemical, drainage, texture) and life history traits of whitebark pine, and (3) the long fire-return interval and infrequent nature of fire in whitebark pine stands. Both Romme and Despain (1989) and Walsh (2005) concluded that high-elevation forests were not altered fundamentally as a result of fire suppression in the twentieth century. Likewise, Bockino (2008) concluded that 80 years of effective fire exclusion in the northern Rocky Mountains did not have an influence on the decrease of whitebark pine. Although fire exclusion may be a limiting factor throughout the range of whitebark pine, mountain pine beetle and blister rust seem to be more important factors in the Yellowstone ecosystem.

FUTURE OF WHITEBARK PINE

Environmental changes resulting from climate warming are exacerbating other threats to whitebark pine, particularly in combination with disease, fire suppression, and predation, that appear to be beyond the natural adaptive capabilities and tolerances of the species (U.S. Fish and Wildlife Service 2011). The warming climate in the northern Rockies (Bartlein et al. 1997; Mote et al. 2005; Westerling et al. 2006, 2011; Intergovernmental Panel on Climate Change 2007a, 2007b) is expected to affect whitebark pine communities by (1) causing a shift in pathogen ranges, which may lead to new regions of hospitable climate for white pine blister rust and increase the potential for infection (Koteen 2002); (2) increasing habitat loss as soil moisture availability decreases and the available range for whitebark pine diminishes as a result of competitive exclusion by more heat-tolerant species, such as lodgepole pine (Mattson et al. 2001); (3) changing the frequency of severe stand-replacing fires, which could lead to overall decreases in whitebark pine numbers (Koteen 2002; Westerling et al. 2011); and (4) increasing the intensity of mountain pine beetle epidemics (Bentz and Mullins 1999; Logan and Powell 2001; Logan et al. 2010).

Climate warming has provided the favorable conditions necessary for the current mountain pine beetle epidemic in high-elevation communities across the western United States and Canada (Logan and Powell 2001, 2003; Raffa et al. 2008; Logan et al. 2010). Warming trends have resulted in intensified mountain pine beetle activity in high-elevation forests and range expansion into more northern latitudes (Logan and Powell 2003; Carroll et al. 2006; Gibson et al. 2008; Raffa et al. 2008). Winter temperatures are now warm enough for mountain pine beetles to survive and maintain the one-year life cycle that promotes epidemic population levels (Bentz and Schen-Langenheim 2007; Logan et al. 2010). Also, summers have been drier, with droughts occurring through much of the range for whitebark pine (Bentz et al. 2010). Mountain pine beetles frequently target drought-stressed trees, which are more vulnerable to attack because they are less able to mount an effective defense (Bentz et al. 2010). Despain (1990) noted that drought likely played a role in the two previous mountain pine beetle epidemics in the greater Yellowstone ecosystem during the 1930s and again during the 1970s.

The role of climate warming in driving blister rust incidence and severity is less understood. High humidity and warm weather provide better conditions for blister rust, but cold and dry conditions have not limited its spread to higher elevations in the greater Yellowstone ecosystem. In general, changes in climate can affect the resiliency of tree populations because seed production, germination, and establishment are particularly sensitive to variations in the environment. Although recruitment may decrease significantly because of climate change, persistence of adult trees can lead to a deceptively healthy looking forest (Brubaker 1986).

Climate warming is expected to significantly decrease the probability of rangewide persistence of whitebark pine. Projections from an empirically based model showed a rangewide distribution decrease of 70 percent and an average elevation loss of 333 meters for the decade beginning in 2030 (Warwell et al. 2007). By 2100, less than 3 percent of currently suitable habitat is expected to remain (Warwell et al. 2007). The area occupied by

Shrublike whitebark pine trees atop a windswept ridge. Over the last several decades, large numbers of mature whitebark pine trees have succumbed to a combination of infection by an exotic organism that causes a disease known as blister rust and infestation by mountain pine beetles. *Photo by Cindy Goeddel.*

whitebark pine in the greater Yellowstone ecosystem is predicted to be significantly reduced with increasing temperature, with near extirpation under a scenario of continued warming and an increase in precipitation (Schrag et al. 2007).

EFFECTS OF DECREASED WHITEBARK PINE ON GRIZZLY BEARS

It is impossible to predict with certainty how future changes in the abundance and distribution of whitebark pine will affect grizzly bears. These resourceful, opportunistic omnivores are capable of adapting to changes in food resources. Grizzly bears have evolved a life history strategy that includes large home ranges, highly flexible diet-switching behavior, and a generalist diet designed to buffer them against annual and seasonal variations in the availability of various foods (Mattson et al. 2001).

In years with poor whitebark pine seed production, grizzly bears in the greater Yellowstone ecosystem switch to alternate foods, such as ungulates, army cutworm moths, ants (Formicidae), and other vegetation. However, the abundance of elk and bison meat is not consistent from year to year, and some of these alternate food sources, such as elk and Yellowstone cutthroat trout, have decreased substantially in the last 15 years. Still, bears are able to attain adequate body fat levels for denning and reproduction in both good and poor seed years. In the northern continental divide ecosystem, whitebark pine seeds were a significant autumn food for grizzly and black bears during the 1950s and 1960s (Tisch 1961; Jonkel and Cowan 1971). Beginning in the mid-1970s, whitebark pine use was negligible except on the eastern front of the Rocky Mountains because of a 90 percent loss of whitebark pines to blister rust (Mace and Jonkel 1986). Grizzly bears adapted to the loss of whitebark pine seeds and switched to other autumn foods, mostly berries. The greatest influences on grizzly bear survival are human factors, such as densities of roads and developed sites, the amount of secure habitat available to bears, and human-caused mortality (Schwartz, Haroldson, and White 2010)—factors that resource managers can somewhat mitigate. Thus, the grizzly bear population will likely persist in the greater Yellowstone ecosystem, even with continued loss of whitebark pine (Servheen and Cross 2010).

MANAGEMENT RESPONSE

Blister rust control began in Yellowstone National Park in 1945 and continued until 1977. In Grand Teton National Park, control began in 1957 and ended in 1966. Treatment focused on *Ribes* species, the pathogen's alternate host, using both mechanical and chemical treatment methods. Treatment stopped when Carlson (1978) concluded that *Ribes* treatment was ineffective as a means of blister rust control. By the time the program ended in Yellowstone, however, nearly 8 million *Ribes* plants had been removed and over 1,616,370 liters of herbicide had been sprayed (Kendall and Asebrook 1998).

Large-scale control of mountain pine beetles began in Yellowstone in 1931 with a method termed *standing-burning*, in which fuel oil was sprayed upward on the trunks and branches of infested trees and then ignited. By 1933, the Civilian Conservation Corps was attempting to control mountain pine beetles by felling and burning infested trees to destroy beetle broods before they matured and flew to infest other trees. These efforts were discontinued after 1934 (Furniss and Renkin 2003). According to Gibson et al. (2008), management efforts to reduce beetle-caused mortality in the greater Yellowstone ecosystem are largely ineffective or not feasible. Individual trees can be protected from attack by insecticides or bark beetle pheromones, but longer-term efforts to reduce beetle-caused impacts are largely undeveloped and unknown.

In response to the current situation in whitebark pine communities, managers throughout the range have initiated management and restoration efforts. The Whitebark Pine Ecosystem Foundation, a science-based nonprofit group dedicated to counteracting the decrease of whitebark pine and enhancing knowledge of its communities, engages in education, promotes research, and provides technical assistance and grants for restoration projects. The Whitebark Pine Subcommittee of the Greater Yellowstone Coordinating Committee recently prepared the 2011 Whitebark Pine Strategy for the Greater Yellowstone Area to direct the protection and restoration of whitebark pine. The strategy is intended to promote the long-term persistence of whitebark pine in the greater Yellowstone ecosystem by (1) promoting effective conservation across administrative boundaries through agency cooperation; (2) protecting remaining cone-bearing trees; (3) maintaining and restoring the role of whitebark pine in the ecosystem; (4) ensuring whitebark pine regeneration and genetic variability through natural and assisted regeneration; and (5) promoting fire planning and use that protects high-value resources and provides for long-term restoration. Since 2006, cones have been harvested from rust-resistant trees treated for mountain pine beetle protection in six national forests and Grand Teton National Park. These seeds will be grown and replanted to promote rust and beetle resistance and improve whitebark pine regeneration. Other tools included in the strategy are preventing mountain pine beetle mortality of whitebark pine with carbaryl and verbenone (Bentz et al. 2005), pruning to remove blister rust infection, preventing the loss of high-value cone-producing trees to natural or prescribed fire, collecting whitebark pine seeds to establish orchards and then out-planting seedlings, and creating openings and thinning whitebark pine stands to promote regeneration.

CONSERVATION IMPLICATIONS

Prior to the recent mountain pine beetle outbreak and its association with climate warming (Logan et al. 2010), the greater Yellowstone ecosystem was considered a refugium for whitebark pine. Fire suppression, an issue throughout the range of whitebark pine, has had little impact on whitebark pine communities in the greater Yellowstone ecosystem (Bockino 2008). Also, the incidence of white pine blister rust is low compared with other areas in the northern Rocky Mountains, where mortality is as high as 90 percent (Gibson et al. 2008), and in the interior Columbia basin, where whitebark populations have decreased by at least 45 percent (Kendall and Keane 2001). Although the current mountain pine beetle epidemic has caused high mortality in mature trees, ongoing monitoring has shown much higher survival in the smaller, younger whitebark pines. In addition, the occurrence of whitebark pine in the understory of lodgepole pine stands on the Yellowstone Plateau could provide a buffer against a warming climate if those trees reach cone-bearing age. Given the long time frames associated with the most decimating effects of climate warming on the mountain pine beetle, the greater Yellowstone ecosystem may continue to be the last best place for whitebark pine for the foreseeable future.

conclusion

CHAPTER SIXTEEN

The Future of Ecological Process Management

P. J. WHITE
ROBERT A. GARROTT
GLENN E. PLUMB

THE NATIONAL PARK Service Organic Act of 1916 states that parks are "to conserve the scenery and the natural and historic objects and the wild life therein . . . by such means as will leave them unimpaired for the enjoyment of future generations." However, no guidance was provided to define "unimpaired" or indicate how scenery, nature, and history were to be preserved. Management policies later clarified that parks will preserve "components and processes in their natural condition," which was defined as "the condition of resources that would occur in the absence of human dominance over the landscape" (National Park Service 2006, 36). Guidance is reflected in the general principles adopted by the National Park Service for managing biological resources, which include (1) preserving the dynamics, distributions, habitats, and behaviors of native populations, communities, and ecosystems; (2) restoring native plant and animal populations; and (3) minimizing human intervention to native populations and the processes that sustain them (National Park Service 2006).

The Yellowstone National Park Act of 1872 (16 U.S.C. §21 et seq.) set apart a vast expanse of land in Wyoming, Montana, and Idaho "as a public park or pleasuring ground for the benefit and enjoyment of the people." The act requires the Secretary of the Interior to preserve "from injury or spoilation" the "wonders" of Yellowstone and to ensure "their retention in their natural condition." The Secretary is also required to "provide against the wanton destruction of the fish and game found within the park, and against their capture or destruction for the purposes of merchandise or profit."

The goals of preserving parks free from human effect and control, and similar to the past, are daunting and perhaps unattainable because ecosystems are continually changing in response to a multitude of factors; human impacts are widespread and accelerating; and the goal of preserving naturalness often conflicts with goals to preserve particularly valued species, places, or conditions (D. N. Cole and Yung 2010a). Also, we have entered an era in which key drivers of ecosystem change, including climate, pollution, exotic species invasions, alteration of disturbance regimes such as fire suppression, and habitat fragmentation, are exceeding the range of historic variation; as a result, we have no precedent (Stephenson et al. 2010). In some cases, ecosystems already have been altered to such an extent that "eliminating the original cause of problems" will not solve them (Cole and Yung 2010b, 3). In most cases, the future is highly uncertain, and managers cannot identify actions that will achieve desired conditions with certainty (Stephenson et al. 2010). Thus, attempts to use past conditions as targets for preservation or restoration are likely to fail or be unstable and subject to sudden shifts in conditions (Stephenson et al. 2010). Furthermore, boundaries of parks and wilderness do not encompass all the needs of various species and the ecological processes needed to sustain them (R. J. Hobbs, Cole, et al. 2010). Moreover, a warming climate is causing unprecedented ecological change that may eventually make debates about restoring ecosystems to earlier historic states moot (Stephenson et al. 2010).

To ensure that unacceptable changes do not occur to the natural resources of Yellowstone requires timely scientific information about populations, communities, and the ecological processes on which sustainable ecosystems depend. In 1993, Yellowstone established a substantial in-park research center dedicated to advancing our knowledge of the ecosystem, supporting sound natural resource management, and communicating knowledge and discoveries to the visiting public to enhance their experience and enjoyment of the park. This center is staffed by scientists with the expertise and experience to make scientifically informed decisions about increasingly difficult issues related to the management of natural systems. Also, park leadership solicits nongovernment scientists for research on basic scientific questions and to support management. The center provides facilities, funding, and logistical support for many scientists working in the park and has established extensive collaborations and partnerships with academic, federal, private, and state organizations, as well as university-based cooperative units.

In addition, staff from Yellowstone National Park collaborated with the Greater Yellowstone Inventory and Monitoring Network in 2005 to assemble more than 400 scientists and managers that identified and prioritized key indicators for assessing the health of the park's natural resources. These indicators, or vital signs, were combined into five categories according to the role they play in the greater Yellowstone ecosystem (Yellowstone National Park 2009, 2; 2011):

1) Ecosystem drivers, such as climate, fire, and the Yellowstone volcano, which are forces that create and modify the system at a regional scale and have cascading effects on virtually all park resources.

2) Landscape-scale indicators, such as air quality, amphibians, and water quality, which tell us something about the system or landscape beyond their individual status or trends.

3) Rare and sensitive species, such as bald eagles, bighorn sheep, grayling, grizzly bears, peregrine falcons, pronghorn, trumpeter swans, westslope cutthroat trout, wolves, and Yellowstone cutthroat trout, because preserving native flora and fauna is core to the park's mission and these species are of high concern to management and the public.

4) Stressors, such as aquatic nuisance species, invasive plants, land use, non-native mountain goats, visitation, warming climate, and wildlife diseases, which are agents of change that could reduce ecological diversity and the resilience of the ecosystem to respond to disturbances.

5) Focal resources, such as bison, elk, geothermal systems, and whitebark pine, which are of particular interest to management because of concerns for that resource or because of how they influence other resources.

Natural resource management in Yellowstone National Park has undergone many changes since the park was established in 1872 and as society has accrued an increasingly sophisticated understanding of the myriad ecological processes that are the fabric of the Yellowstone ecosystem. As a result of this successful linking of science, management, and policy, the public can continue to observe and enjoy scenes such as this coyote stopping to listen for prey as he travels through a wintry landscape along the Lamar River in northern Yellowstone. *Photo by Cindy Goeddel.*

Each vital sign is monitored by park and network staff, cooperators from other federal and state agencies, or university scientists for comparison to reference conditions that inform park managers about whether a resource has changed since previous years or is approaching a point or limit where more management time, energy, and effort should be directed toward that resource (Yellowstone National Park 2011).

STATE OF THE PARK

Conservation of wildlife resources in Yellowstone National Park has been relatively successful since ecological process management was adopted, with no loss of native species; the successful restoration of native species such as wolves; the recovery of other species, including bald eagles, bison, grizzly bears, and peregrine falcons; the continuation of disturbance processes, such as fire and floods; and the preservation of selection pressures that led to the diversity of species and interactions that we see today (Yellowstone National Park 2011). The majority of landscape-scale indicators (amphibians, water quality) and rare and sensitive species (bald eagles, bighorn sheep, grizzly bears, peregrine falcons, wolves) are within reference conditions, and food webs of plants, herbivores, and carnivores are generally intact (Yellowstone National Park 2011). Repeated disturbance events (fire, flooding) have created a mixture of vegetation communities with varying species compositions, structure, and trends in these attributes over time. Nutrient cycling and plant production are operating within reference conditions, and the diverse predator and scavenger associations have increased the speed of decomposition of dead and decaying material. Thus, the park provides a relatively healthy benchmark, or point of reference, and preserves options for the future (Landres 2010).

However, human intervention through culling, hunting outside the park, and restoration has been prominent during the ecological process management period for such species as bison, elk, grizzly bears, and wolves. Thus, their population dynamics are not truly naturally regulated. Also, there continue to be difficulties reaching consensus with other agencies on the conservation of species (bison, wolves) and ecological processes (migration, dispersal) across political and jurisdictional boundaries (U.S. Government Accountability Office 2008; U.S. District Court for the District of Montana Missoula Division 2009). In addition, there are indications that the system is being stressed. For example, nitrogen in precipitation has increased in recent years at many monitoring sites in western Yellowstone as a result of ammonium ion concentrations associated with fertilizer use and feedlots west of the park (Yellowstone National Park 2011). Analyses of sediment cores from lakes in and near the park found that nitrogen loading exceeded a critical threshold (1.4 kilograms per hectare per year) around 1980, after which the composition of algae began changing (Saros 2009).

Since 1969, pronghorn numbers have exhibited periods of relative stability interspersed with rapid, dramatic fluctuations (P. J. White, Bruggeman, and Garrott 2007). A decrease in counts from 536 to 235 pronghorn from 1992 to 1995 raised serious concerns about the population's long-term persistence. Likewise, the future of the resident, nonmigratory population of trumpeter swans is of grave concern. This population has decreased from almost 60 individuals in 1968 to four in 2010 as a result of changes in land use and trumpeter swan management outside Yellowstone National Park, marginal environmental conditions in the park for nesting, and the long-term desiccation of ponds that provide nesting habitat (Proffitt, McEneaney, et al. 2009). In addition, there is concern about the persistence of whitebark pine forests in the greater Yellowstone ecosystem following widespread mortality associated with outbreaks of native mountain pine beetles and non-native white pine blister rust (Logan et al. 2010). Whitebark pine seeds provide a high-energy food source for wildlife, including grizzly bears, red (or pine) squirrels, Clark's nutcrackers, and other seed-eating birds.

Furthermore, several native fish species are in serious trouble. Predation by invasive lake trout has caused a collapse of the Yellowstone cutthroat trout population in Yellowstone Lake (Gresswell 2009). Outside of the lake system, streams that historically supported Yellowstone cutthroat trout are being invaded by non-native rainbow trout and other non-natives (U.S. Department of the Interior, National Park Service 2010). Likewise, Arctic grayling were historically common in several river systems of Yellowstone, but now the only known grayling populations that remain are lake dwelling (Varley and Schullery 1998). Similarly, westslope cutthroat trout have been extirpated from 36 percent of the 1,030 kilometers of streams within the park that originally supported genetically pure fish (U.S. Department of the Interior, National Park Service 2010). The species exists in a hybridized form in most of the remaining habitat.

Stressors to the greater Yellowstone ecosystem continue to increase despite attempts to limit effects. From 1970 to 1999, the human population in the greater Yellowstone ecosystem grew by nearly 60 percent to over 370,000 residents, and the area of rural lands occupied by subdivisions increased by 350 percent (P. H. Gude et al. 2007; Hansen and DeFries 2007; Hansen 2009). Annual visitation at Yellowstone has approximated 3 million people since 1992, which has contributed to an increase in wildlife–human interactions, vehicle strikes, and the habituation of wildlife (Gunther et al. 1999; Gunther and Wyman 2008). Also, there is evidence of increases in temperature and changes in precipitation in Yellowstone over the past century (Wilmers and Getz 2005; Newman and Watson 2009). Future climate projections suggest warmer, drier summers with more frequent drought, warmer winters and earlier snowmelt, and earlier peaks in the growing season followed by earlier curing of vegetation—all of which imply considerable change in the composition of animal and plant communities, disturbance regimes (fire, flooding), and the distribution, movement, and quality of water (Ashton 2010; McWethy et al. 2010).

In Yellowstone, three aquatic nuisance species are having significant detrimental effects, including lake trout, whirling disease, and New Zealand mud snails. Lake trout were illegally introduced in Yellowstone Lake, where they feed on the native Yellowstone cutthroat trout. Biologists and contract fisherman have killed more than one million lake trout since they were first documented in Yellowstone Lake in 1994. The parasite that causes whirling disease in cutthroat trout and other species was confirmed in the park in 1998 and appears most concentrated in the Yellowstone Lake watershed, where it has reduced cutthroat trout numbers in Pelican Creek (Koel et al. 2006). Whirling disease has also been found in the Firehole and Yellowstone rivers. Also, New Zealand mud snails, which were first detected in the park in 1994 and form dense colonies that compete with native species, are now in all of the major watersheds in the park (Kerans et al. 2005). Similarly, there are concerns about the descendants of mountain goats introduced into Montana during the 1940s and 1950s that established a breeding population in northern Yellowstone during the 1990s and have reached a relatively high abundance of 200 to 300 animals (Lemke 2004). This colonization has raised concerns about adverse effects to alpine habitats and competition with bighorn sheep. Furthermore, the number of species of non-native plants discovered in Yellowstone has increased from 105 to 218 since 1986 (Yellowstone National Park 2011).

Significant diseases present in Yellowstone wildlife include brucellosis, canine distemper, and chytridiomycosis. Many bison and elk in the greater Yellowstone ecosystem have been exposed to the bacterium that causes brucellosis, which likely originated in domestic livestock (Meagher and Meyer 1994). It does not appear to have had substantial population-level impacts on wildlife, but infected females may abort their first calf and the disease can be transmitted to livestock through contact with infected birth materials (Cheville et al. 1998). Canine distemper was a major factor in wolf population decreases in Yellowstone in 1999, 2005, and 2008 (Almberg et al. 2009, 2010). Chytridiomycosis is an amphibian disease caused by a fungus of uncertain origin that has contributed to the worldwide decline in

Mountain goats are not native to Yellowstone National Park but were introduced into the mountain ranges adjacent to the park over a half century ago. Since that time, mountain goat populations have grown and expanded their range into the park. The impacts of mountain goats on native resources, such as alpine plant communities and bighorn sheep, are uncertain and the focus of ongoing research initiatives. *Photo by Cindy Goeddel.*

frogs (Yellowstone National Park 2011). Wildlife diseases that could potentially appear in Yellowstone include chronic wasting disease (affecting deer, elk, and moose), West Nile virus (affecting birds), and carotid artery worms (affecting moose).

CONSERVATION IMPLICATIONS

A healthy Yellowstone is important for achieving the park's mandate of conserving resources and values in a manner that provides for their enjoyment by people. Equally important, a healthy Yellowstone enhances the lives of people living near the park by providing sustainable resources, such as clean water, ungulate populations, and vegetation communities. Thus, park managers should continue to maintain the ecological processes that sustain the system, while monitoring ecosystem attributes that might indicate potentially serious changes (National Research Council 2002). Effective management will require strong partnerships that include the cooperative collection and sharing of information, discussion of alternate management approaches, and the implementation of measures to avoid unacceptable changes. Public education will also be important to explain that changes over time characterize natural systems and that there is no balance of nature (National Research Council 2002).

Long-term monitoring indicates that the Yellowstone system has retained its ecological integrity with a full complement of native species and the processes that facilitate the survival of these species (Woodley 2010). "Species that received the time, effort, and funding of the National Park Service and many other federal, state, and private partners as part of recovery plans under the Endangered Species Act—bald eagles, grizzly bears, peregrine falcons, and wolves—have recovered to sustainable population levels. Some issues, notably bison and wolf management and winter recreation on snowmobiles, remain controversial as the park's partners, the public, and the federal courts debate conflicting priorities. Also, the effects of three overriding ecological stressors—warming climate, invasive species, and land-use changes—that act at larger scales than the boundary of Yellowstone National Park are beginning to be seen on the landscape. Partnerships between park staff and other federal, state, and private partners have been a successful model for addressing several regional issues. Private and corporate donors, such as Canon USA, the Yellowstone Park Foundation, and the Yellowstone Association, continue to support critical park research, monitoring, and public outreach. These partnerships need to be strengthened, and new partnerships formed, to address emerging issues and continue the conservation of resources" (Yellowstone National Park 2009, 17).

Given that the future is unpredictable but will almost certainly include overriding ecological stressors (warming climate, invasive species, land-use changes) that have no historical precedent, the challenge for park managers is to embrace ecological change, maintain a diversity of native species, and preserve ecological processes across park boundaries. Managers need to address threats and promote ecosystem integrity (the retention of native ecosystem components) and resilience (the ability of the system to recover from stresses; Gunderson 2000; Stephenson et al. 2010; Woodley 2010).

Managers also need to think realistically and articulate goals that conserve the key resources (or core desired attributes or functions) that we value in the park, while avoiding undesired future conditions (Aplet and Cole 2010; Zavaleta and Chapin 2010).

Yellowstone was established to preserve the many natural curiosities, cultures, resources, and wonders of the area, including the animal, aquatic, geothermal, and vegetation communities. With regards to wildlife, we suggest that the key value or goal of Yellowstone National Park and the greater Yellowstone ecosystem is the conservation of large assemblages of large wildlife on a large landscape. Yellowstone supports intact native predator–prey–scavenger communities that move, migrate, and disperse across a vast, heterogeneous landscape and are subject to a full suite of natural selection factors, including competition, disease, predation, and substantial environmental variability. The park is not large enough to sustain these associations and these processes by itself. Thus, advocacy, monitoring, and planning to sustain this important component of the park's ecosystem must extend beyond the boundary of the park, with consideration of "regional cultural, ecological, political, and social systems" (Keiter and Boyce 1991; R. J. Hobbs, Cole, et al. 2010, 488; R. J. Hobbs, Zavaleta, et al. 2010).

Strategies to preserve key natural resources should include monitoring to detect problems leading to ecosystem degradation; research to evaluate and inform alternate management actions; strategic planning to explore possible consequences of decisions, consider plausible futures, and avoid undesired conditions; and an adaptive management framework to decide if, when, and how to intervene (G. D. Peterson et al. 2003; P. J. White, Garrott, and Olliff 2009; R. J. Hobbs, Cole, et al. 2010; R. J. Hobbs, Zavaleta, et al. 2010). Perhaps the most difficult decision managers will face is whether to "intervene with active management, such as assisting migration, culling animals, lighting fires, restoring animals and vegetation, spraying weeds, and thinning forests." R. J. Hobbs and colleagues (R. J. Hobbs, Cole, et al. 2010, 484; R. J. Hobbs, Zavaleta, et al. 2010) recommended considering historical fidelity (past states of the system), ecological integrity, and resilience when deciding on the type and extent of management intervention required. A diversity of approaches that range from nonintervention to active transformation such as restoration could be taken in Yellowstone by focusing different principles on different areas of the park and assessing how they fit into the greater Yellowstone area.

For example, approximately 95 percent of the park is currently managed as wilderness that should not be intentionally controlled or manipulated for any purpose. Perhaps the highest value that a park such as Yellowstone has is in the "authenticity of its wildness—the opportunity for us to be awed and learn from nature making its own decisions" (Schullery 2010, 12). This may be the most important lesson of ecological process management in Yellowstone, because when managers restrained themselves, they always learned more than when they intervened (Schullery 2010). Thus, management of wilderness areas in the park could continue to reflect the value of minimal human intervention with nature, while maintaining ecological integrity and resiliency. The decision not to intervene would be a deliberate management decision whereby managers would attempt to sustain the native species pool, maximize future options, allow sufficient disturbances, and sustain ecological and cultural legacies (Higgs and Hobbs 2010; R. J. Hobbs, Cole, et al. 2010; R. J. Hobbs, Zavaleta, et al. 2010; Landres 2010; Yung et al. 2010). "A hardy or resilient system may vary widely over time while retaining its core functions and the characteristic relationships that sustain them" (Zavaleta and Chapin 2010, 143). However, managers must realize that local extinctions of some species could occur under this strategy, and plant and animal communities may change substantially over time while still retaining ecological integrity. Examples include the possible switch from an elk- to a bison-dominated system in northern Yellowstone after wolf restoration and grizzly bear recovery, as well as changes in stream and river morphologies and riparian communities resulting from shifts in large mammal communities and climate (P. J. White and Garrott 2005b).

Conversely, developed areas could be managed to reflect historical conditions by using frequent human intervention to reduce exotic invasions, limit human–wildlife interactions, and restore habitats. Areas transformed by exotic species invasions, such as Yellowstone Lake, with non-native lake trout decimating the native Yellowstone cutthroat trout, will require persistent human intervention and a departure from historical conditions. Also, boundary areas could be managed to sustain connectivity between the park and the surrounding land-use matrix, while accommodating cooperative planning, management, and human intervention to achieve the diverse objectives of agencies and landowners outside the park (R. J. Hobbs, Cole, et al. 2010; R. J. Hobbs, Zavaleta, et al. 2010). The "desired outcomes of interventions or cooperative management efforts should be specified in the form of measurable operational objectives or targets that identify what elements and processes should be preserved and in what state they should be maintained" (R. J. Hobbs, Cole, et al. 2010, 484; R. J. Hobbs, Zavaleta, et al. 2010). Furthermore, park managers should continue to disclose the state of the park every five years or so, complete with detailed indicators, measures, thresholds, and references or targets for management (R. J. Hobbs, Cole, et al. 2010; Yellowstone National Park 2011).

The transition to ecosystem management in Yellowstone National Park during 1969 was a dramatic departure from the management actions that were common prior to its adoption, including the intensive culling of ungulates and predators, husbandry of bison and pronghorn, introduction of non-native fish, suppression of wildfires, and other actions. This transition has done much to restore nature and wildness in the park, and this was accomplished while hosting more than 3 million visitors a year who came to experience the historical, cultural, scenic, and wildlife resources and values that are the legacy of Yellowstone and the national park ideal. There are certainly ongoing and emerging challenges that need attention; as a result, the management of Yellowstone will continue to depend on science to lead us toward what works (Schullery 2010).

Reference List

Abu-Asab, M. S., P. M. Peterson, S. G. Shetler, and S. S. Orli. 2001. Earlier plant flowering in spring as a response to global warming in the Washington, DC, area. *Biological Conservation* 10:597–612.

Aguirre, A. A., D. E. Hansen, E. E. Starkey, and R. G. McLean. 1995. Serologic survey of wild cervids for potential disease agents in selected national parks in the United States. *Preventative Veterinary Medicine* 21:313–322.

Aho, K., and T. Weaver. 2003. *Classification of alpine plant communities of the northern Rocky Mountain volcanics.* Yellowstone National Park, Mammoth, Wyoming.

Alexander, R. R. 1964. Minimizing windfall around clear cuttings in spruce-fir forests. *Forest Science* 10:130–142.

Allen, T. F. H., and T. W. Hoekstra. 1992. *Toward a unified ecology.* Columbia University Press, New York, New York.

Almberg, E., P. Cross, and D. Smith. 2010. Persistence of canine distemper virus in the greater Yellowstone ecosystem's carnivore community. *Ecological Applications* 20:2058–2074.

Almberg, E. S., L. D. Mech, D. W. Smith, J. W. Sheldon, and R. L. Crabtree. 2009. A serological survey of infectious disease in Yellowstone National Park's canid community. *PLoS ONE* 4:e7042.

Alonzo, S. H., P. V. Switzer, and M. Mangel. 2003. Ecological games in space and time: The distribution and abundance of Antarctic krill and penguins. *Ecology* 84:1598–1607.

Ammon, E. M., and P. B. Stacey. 1997. Avian nest success in relation to past grazing regimes in a montane riparian system. *Condor* 99:7–13.

Anderson, R. M., and R. M. May. 1978. Regulation and stability of host-parasite interactions: I. Regulatory processes. *Journal of Animal Ecology* 47:219–247.

Andrewartha, H. G., and L. C. Birch. 1954. *The distribution and abundance of animals.* University of Chicago Press, Chicago, Illinois.

Aplet, G. H., and D. N. Cole. 2010. The trouble with naturalness: Rethinking park and wilderness goals. Pages 12–29 in D. N. Cole and L. Yung, editors. *Beyond naturalness: Rethinking park and wilderness stewardship in an era of rapid change.* Island Press, Washington, D.C.

Apps, C. D., B. N. McLellan, and J. G. Woods. 2006. Landscape partitioning and spatial inferences of competition between black and grizzly bears. *Ecography* 29:561–572.

Arno, S. F., and R. J. Hoff. 1990. Whitebark pine (*Pinus albicaulis* Engelm.). Pages 268–279 in R. M. Burns and B. H. Honkala, editors. *Silvics of North America: 1. Conifers; 2. Hardwoods.* Agriculture Handbook 654, U.S. Department of Agriculture, National Forest Service, Washington, D.C.

Ashton, I. W. 2010. *Observed and projected trends in climate and associated ecosystem effects in the Rocky Mountains and upper Columbia basin: A synthesis of current scientific literature.* National Park Service, Rocky Mountain Inventory and Monitoring Network, Fort Collins, Colorado.

Atwood, T. C., and E. M. Gese. 2008. Coyotes and recolonizing wolves: Social rank mediates risk-conditional behaviour at ungulate carcasses. *Animal Behaviour* 75:753–762.

Augustine, D. J., and D. A. Frank. 2001. Effects of migratory grazers on spatial heterogeneity of soil nitrogen properties in a grassland ecosystem. *Ecology* 82:3149–3162.

Augustine, D. J., and S. J. McNaughton. 1998. Ungulate effects on the functional species composition of plant communities: Herbivore selectivity and plant tolerance. *Journal of Wildlife Management* 62:1165–1183.

Aune, K. E. 1994. Comparative ecology of black and grizzly bears on the Rocky Mountain Front, Montana. *International Conference on Bear Research and Management* 9:365–374.

Aune, K., J. Rhyan, B. Corso, and T. Roffe. 2007. Environmental persistence of *Brucella* organisms in natural environments of the greater Yellowstone area—A preliminary analysis. *U.S. Animal Health Association* 110:205–212.

Bailey, V. 1930. *Animal life of Yellowstone National Park.* Charles C. Thomas Publisher, Springfield, Illinois.

Baker, B. W., H. C. Ducharme, D. C. S. Mitchell, T. R. Stanley, and H. R. Peinetti. 2005. Interaction of beaver and elk herbivory reduces standing crop of willow. *Ecological Applications* 15:110–111.

Ball, O. P., and O. B. Cope. 1961. *Mortality studies on cutthroat trout in Yellowstone Lake.* Research Report 55, U.S. Fish and Wildlife Service, Washington, D.C.

Ballard, W. B. 1991. Management of predators and their prey: The Alaskan experience. *Transactions of the North American Wildlife and Natural Resources Conference* 56:527–538.

Ballard, W. B., L. N. Carbyn, and D. W. Smith. 2003. Wolf interactions with non-prey. Pages 259–271 in L. D. Mech and L. Boitani, editors. *Wolves: Ecology, behavior, and conservation.* University of Chicago Press, Chicago, Illinois.

Ballard, W. B., D. Lutz, T. W. Keegan, L. H. Carpenter, and J. C. deVos, Jr. 2001. Deer-predator relationships: A review of recent North American studies with emphasis on mule and black-tailed deer. *Wildlife Society Bulletin* 29:99–115.

Ballard, W. B., and V. Van Ballenberghe. 1997. Predator-prey relationships. Pages 247–273 in A. W. Franzmann and C. C. Schwartz, editors. *Ecology and management of the North American moose.* Smithsonian Institution Press, Washington, D.C.

Ballard, W. B., and V. Van Ballenberghe. 1998. Moose-predator relationships: Research and management needs. *Alces* 34:91–105.

Ballard, W. B., J. S. Whitman, and C. L. Gardner. 1987. Ecology of an exploited wolf population in south-central Alaska. *Wildlife Monographs* 98.

Ballard, W. B., J. S. Whitman, and D. J. Reed. 1991. Population dynamics of moose in south-central Alaska. *Wildlife Monographs* 114.

Balling, R. C., G. A. Meyer, and S. G. Wells. 1992. Relation of surface climate and burned area in Yellowstone-National-Park. *Agricultural and Forest Meteorology* 60:285–293.

Banko, W. E. 1960. The trumpeter swan, its history, habits, and population in the United States. U.S. Department of Interior, Bureau of Sport Fish and Wildlife. *North American Fauna* 63.

Barbee, R. D., P. Schullery, and J. D. Varley. 1991. The Yellowstone vision: An experiment that failed or a vote for posterity? *Proceedings of Partnerships in Parks and Preservation,* Albany, New York, September 9–12, 1991.

Barber-Meyer, S. M., L. D. Mech, and P. J. White. 2008. Elk calf survival and mortality following wolf restoration to Yellowstone National Park. *Wildlife Monographs* 169.

Barber-Meyer, S. M., P. J. White, and L. D. Mech. 2007. Survey of selected pathogens and blood parameters of northern Yellowstone elk: Wolf sanitation effect implications. *American Midland Naturalist* 158:369–381.

Bardgett, R. D., and D. A. Wardle. 2003. Herbivore-mediated linkages between aboveground and belowground communities. *Ecology* 84:2258–2268.

Baril, L. M., L. Henry, and D. W. Smith. 2010. *Yellowstone bird program 2009 annual report.* National Park Service, Yellowstone National Park, Mammoth, Wyoming.

Barker, R. 2005. *Scorched earth: How the fires of Yellowstone changed America.* Island Press, Washington, D.C.

Barmore, W. J. 1968. *Bison and brucellosis in Yellowstone National Park: A problem analysis.* Yellowstone National Park, Mammoth, Wyoming.

Barmore, W. J. 1987. *The distribution and abundance of ungulates in the northern Yellowstone ecosystem: In pristine times and today.* Greater Yellowstone Coalition's Scientific Conference on the Northern Yellowstone: Issues and Alternatives for Ecosystem Management. Lake Hotel and Lodge, Yellowstone National Park, Wyoming, May 29, 1987.

Barmore, W. J. Jr. 2003. *Ecology of ungulates and their winter range in northern Yellowstone National Park: Research and synthesis, 1962–1970.* National Park Service, Yellowstone National Park, Mammoth, Wyoming.

Barnosky, E. H. 1994. Ecosystem dynamics through the past 2000 years as revealed by fossil mammals from Lamar Cave in Yellowstone National Park, USA. *Historical Biology* 8:71–90.

Barnosky, E. H. 1996. Late Holocene mammalian fauna of Lamar cave and its implications for ecosystem dynamics in Yellowstone National Park, Wyoming. Pages 153–164 in F. J. Singer, editor. *Effects of grazing by wild ungulates in Yellowstone National Park—A report to the director.* National Park Service, Natural Resource Information Division, Denver, Colorado.

Barnowe-Meyer, K. K. 2009. *The behavioral ecology and population genetics of pronghorn* (Antilocapra americana) *in Yellowstone National Park, Montana/Wyoming, USA.* University of Idaho, Moscow, Idaho.

Barnowe-Meyer, K. K., P. J. White, T. L. Davis, and J. A. Byers. 2009. Predator-specific mortality of pronghorn on Yellowstone's northern range. *Western North American Naturalist* 69:186–194.

Barnowe-Meyer, K. K., P. J. White, T. L. Davis, D. W. Smith, R. L. Crabtree, and J. A. Byers. 2010. Influences of wolves and high-elevation dispersion on reproductive success of pronghorn *(Antilocapra americana). Journal of Mammalogy* 91:712–721.

Barrett, S. W. 1994. Fire regimes on andesitic mountain terrain in northeastern Yellowstone National Park, Wyoming. *International Journal of Wildland Fire* 4:65–76.

Bartlein, P. J., C. Whitlock, and S. L. Shafer. 1997. Future climate in the Yellowstone National Park region and its potential impact on vegetation. *Conservation Biology* 11:782–792.

Bartlett, R. 1985. *Yellowstone: A wilderness besieged.* University of Arizona Press, Tucson, Arizona.

Bascompte, J., and C. J. Melián. 2005. Simple trophic modules for complex food webs. *Ecology* 86:2868–2873.

Becker, M. S., R. A. Garrott, P. J. White, C. N. Gower, E. J. Bergman, and R. Jaffe. 2009. Wolf prey selection in an elk-bison system: Choice or circumstance? Pages 305–337 in R. A. Garrott, P. J. White, and F. G. R. Watson, editors. *The ecology of large mammals in central Yellowstone: Sixteen years of integrated field studies.* Elsevier, San Diego, California.

Becker, M. S., R. A. Garrott, P. J. White, R. Jaffe, J. J. Borkowski, C. N. Gower, and E. J. Bergman. 2009. Wolf kill rates: Predictably variable? Pages 339–369 in R. A. Garrott, P. J. White, and F. G. R. Watson, editors. *The ecology of large mammals in central Yellowstone: Sixteen years of integrated field studies.* Elsevier, San Diego, California.

Begon, M., C. R. Townsend, and J. L. Harper. 2005. *Ecology: From individuals to ecosystems.* Wiley-Blackwell, Hoboken, New Jersey.

Behnke, R. J. 1992. *Native trout of western North America.* American Fisheries Society Monograph 6, Bethesda, Maryland.

Beja-Pereira, A., B. Bricker, S. Chen, C. Almendra, P. J. White, and G. Luikart. 2009. DNA genotyping suggests recent brucellosis outbreaks in the greater Yellowstone area originated from elk. *Journal of Wildlife Diseases* 45:1174–1177.

Belant, J. L., K. Kielland, E. H. Follmann, and L. G. Adams. 2006. Interspecific resource partitioning in sympatric ursids. *Ecological Applications* 16:2333–2343.

Bender, E. A., T. J. Case, and M. E. Gilpin. 1984. Perturbation experiments in community ecology: Theory and practice. *Ecology* 65:1–13.

Benson, N. G. 1961. *Limnology of Yellowstone Lake in relation to the cutthroat trout.* Research Report 56, U.S. Fish and Wildlife Service, Washington, D.C.

Bentz, B., C. D. Allen, M. Ayres, E. Berg, A. Carroll, M. Hansen, J. Hicke, L. Joyce, J. Logan, W. MacFarlane, S. M. MacMahon, J. Negrón, T. Paine, J. Powell, K. Raffa, J. Régnière, M. Reid, W. Romme, S. Seybold, D. T. D. Six, J. Vandygriff, T. Veblen, M. White, J. Witcosky, and D. Wood. 2009. *Bark beetle outbreaks in western North America: Causes and consequences.* University of Utah Press, Salt Lake City, Utah.

Bentz, B. J., S. Kegley, K. Gibson, and R. Thier. 2005. A test of high-dose verbenone for stand-level protection of lodgepole and whitebark pine from mountain pine beetle (Coleoptera: Curculionidae; Scolytinae) attacks. *Journal of Economic Entomology* 98:1614–1621.

Bentz, B. J., and D. E. Mullins. 1999. Ecology of mountain pine beetle (Coleopera: Scolytidae) cold hardening in the Intermountain West. *Environmental Entomology* 28:577–587.

Bentz, B., J. Regniere, C. Fettig, M. Hansen, J. Hayes, J. Hicke, R. Kelsey, J. Negron, and S. Seybold. 2010. Climate change and bark beetles of the western United States and Canada: Direct and indirect effects. *BioScience* 60:602–613.

Bentz, B., and G. Schen-Langenheim. 2007. The mountain pine beetle and whitebark pine waltz: Has the music changed? *Proceedings of the Conference on Whitebark Pine: A Pacific Coast Perspective.* USDA Forest Service, Rocky Mountain Research Station, Logan, Utah.

Berger, A. M., and R. E. Gresswell. 2009. Factors influencing coastal cutthroat trout seasonal survival rates: A spatially continuous approach among stream network habitats. *Canadian Journal of Fisheries and Aquatic Sciences* 66:613–632.

Berger, J. 2004. The last mile: How to sustain long-distance migration in mammals. *Conservation Biology* 18:320–331.

Berger, J., P. B. Stacy, L. Bellis, and M. P. Johnson. 2001. A mammalian predator-prey imbalance: Grizzly bear and wolf extinction affect avian neotropical migrants. *Ecological Applications* 11:947–960.

Berger, K. M., and E. M. Gese. 2007. Does interference competition with wolves limit the distribution and abundance of coyotes? *Journal of Animal Ecology* 76:1075–1085.

Bergerud, A. T., H. E. Butler, and D. R. Miller. 1984. Anti-predator tactics of calving caribou—Dispersion in mountains. *Canadian Journal of Zoology* 62:1566–1575.

Bergerud, A. T., and J. P. Elliot. 1986. Dynamics of caribou and wolves in northern British Columbia. *Canadian Journal of Zoology* 64:1515–1529.

Bergerud, A. T., W. Wyett, and J. B. Snider. 1983. The role of wolf predation in limiting a moose population. *Journal of Wildlife Management* 47:977–988.

Bergman, E. J., R. A. Garrott, S. Creel, J. J. Borkowski, F. G. R. Watson, and R. M. Jaffe. 2006. Assessment of prey vulnerability through analysis of wolf movements and kill sites. *Ecological Applications* 16:273–284.

Bergquist, A. M., S. R. Carpenter, and J. C. Latino. 1985. Shifts in phytoplankton size structure and community composition during grazing by contrasting zooplankton assemblages. *Limnology and Oceanography* 30:1037–1045.

Beschta, R. L. 2003. Cottonwoods, elk, and wolves in the Lamar Valley of Yellowstone National Park. *Ecological Applications* 13:1295–1309.

Beschta, R. L. 2005. Reduced cottonwood recruitment following extirpation of wolves in Yellowstone's northern range. *Ecology* 86:391–403.

Beschta, R. L., and W. J. Ripple. 2007. Increased willow heights along northern Yellowstone's Blacktail Deer Creek following wolf reintroduction. *Western North American Naturalist* 67:613–617.

Beschta, R. L., and W. J. Ripple. 2010. Recovering riparian plant communities with wolves in northern Yellowstone, USA. *Restoration Ecology* 18:380–389.

Beschta, R. L., and W. J. Ripple. 2011. Are wolves saving Yellowstone's aspen? A landscape-level test of a behaviorally mediated trophic cascade—Comment. *Ecology.* doi:10.1890/11-0063.1.

Beyer, H. L., E. H. Merrill, N. Varley, and M. S. Boyce. 2007. Willow on Yellowstone's northern range: Evidence for a trophic cascade? *Ecological Applications* 17:1563–1571.

Bienen, L., and G. Tabor. 2006. Applying an ecosystem approach to brucellosis control: Can an old conflict between wildlife and agriculture be successfully managed? *Frontiers in Ecology and the Environment* 4:319–327.

Biesinger, K. E. 1961. *Studies on the relationship of the redside shiner* (Richardsonius balteatus) *and the longnose sucker* (Catostomus catostomus) *to the cutthroat trout* (Salmo clarki) *population in Yellowstone Lake.* Thesis, Utah State University, Logan, Utah.

Bigler, W., D. R. Butler, and R. W. Dixon. 2001. Beaver-pond sequence morphology and sedimentation in northwestern Montana. *Physical Geography* 22:531–540.

Bilyeu, D. M., D. J. Cooper, and N. T. Hobbs. 2008. Water tables constrain height recovery of willow on Yellowstone's northern range. *Ecological Applications* 18:80–92.

Bjornlie, D. D., and R. A. Garrott. 2001. Effects of winter road grooming on bison in Yellowstone National Park. *Journal of Wildlife Management* 65:423–435.

Boccadori, S. J., P. J. White, R. A. Garrott, J. J. Borkowski, and T. Davis. 2008. Yellowstone pronghorn alter resource selection after sagebrush decline. *Journal of Mammalogy* 89:1031–1040.

Bockino, N. K. 2008. *Interactions of white pine blister rust, host species, and mountain pine beetle in whitebark pine ecosystems in the greater Yellowstone.* Thesis, University of Wyoming, Laramie, Wyoming.

Bockino, N., and K. McCloskey. 2010. *Whitebark pine monitoring in Grand Teton National Park, 2007–2010: White pine blister rust and mountain pine beetle.* Grand Teton National Park, Division of Science and Resource Management, Wyoming.

Boertje, R. D., M. A. Keech, D. D. Young, K. A. Kellie, and C. T. Seaton. 2009. Managing for elevated yield of moose in interior Alaska. *Journal of Wildlife Management* 73:314–327.

Boone, R. B., S. J. Thirgood, and J. G. C. Hopcraft. 2006. Serengeti wildebeest migratory patterns modeled from rainfall and new vegetation growth. *Ecology* 87:1987–1994.

Bormann, F. H., and G. E. Likens. 1979. *Pattern and process in a forested ecosystem.* Springer-Verlag, New York, New York.

Botkin, D. B. 1990. *Discordant harmonies.* Oxford University Press, New York, New York.

Boyce, M. S. 1991. Natural regulation or control of nature. Pages 183–208 in R. B. Keiter and M. S. Boyce, editors. *The greater Yellowstone ecosystem: Redefining America's wilderness heritage.* Yale University Press, New Haven, Connecticut.

Boyce, M. S. 1998. Ecological-process management and ungulates: Yellowstone's conservation paradigm. *Wildlife Society Bulletin* 26:391–398.

Boyd, D. P. 2003. *Conservation of North American bison: Status and recommendations.* Thesis, University of Calgary, Calgary, Alberta, Canada.

Bradley, B. A., D. M. Blumenthal, D. S. Wilcove, and L. H. Ziska. 2010. Predicting plant invasions in an era of global change. *Trends in Ecology and Evolution* 25:310–318.

Bradley, N. L. 1999. Phenological changes reflect climate change in Wisconsin. *Proceedings of the National Academy of Sciences of the United States of America* 96:9701–9704.

Broadbent, S. 1997. *Sportsmen and the evolution of the conservation idea in Yellowstone: 1882–1894.* Thesis, Montana State University, Bozeman, Montana.

Brockman, H. J., and C. J. Barnard. 1979. Kleptoparasitism in birds. *Animal Behavior* 27:487–514.

Brooks, J. L., and S. I. Dodson. 1965. Predation, body size, and composition of plankton. *Science* 150:28–35.

Brower, L. P., and S. B. Malcolm. 1991. Animal migrations: Endangered phenomena. *American Zoologist* 31:265–276.

Brubaker, L. B. 1986. Responses of tree populations to climate change. *Vegetation* 67:119–130.

Bruggeman, J. E. 2006. *Spatio-temporal dynamics of the central bison herd in Yellowstone National Park.* Dissertation, Montana State University, Bozeman, Montana.

Bruggeman, J. E., R. A. Garrott, D. D. Bjornlie, P. J. White, F. G. R. Watson, and J. J. Borkowski. 2006. Temporal variability in winter travel patterns of Yellowstone bison: The effects of road grooming. *Ecological Applications* 16:1539–1554.

Bruggeman, J. E., P. J. White, R. A. Garrott, and F. G. R. Watson. 2009. Partial migration in central Yellowstone bison. Pages 217–235 in R. A. Garrott, P. J. White, and F. G. R. Watson, editors. *The ecology of large mammals in central Yellowstone: Sixteen years of integrated field studies.* Elsevier, San Diego, California.

Buechner, K. 1960. The bighorn sheep of the United States: Its past, present and future. *Wildlife Monographs* 4.

Burns, K. S. 2006. *White pine blister rust surveys in the Sangre de Cristo and Wet Mountains of southern Colorado.* Biological Evaluation R2–06–05, U.S. Department of Agriculture, National Forest Service, Rocky Mountain Region.

Burns, K. S., A. W. Schoettle, W. R. Jacobi, and M. F. Mahalovich. 2008. *Options for the management of white pine blister rust in the Rocky Mountain region.* Report RMRS-GTR-206, U.S. Department of Agriculture, National Forest Service, Rocky Mountain Research Station, Fort Collins, Colorado.

Burroughs, R. D., editor. 1995. *The natural history of the Lewis and Clark expedition*. Michigan State University Press, Lansing, Michigan.

Cahalane, V. H. 1944. Restoration of wild bison. *Transactions of the North American Wildlife Conference* 9:135–143.

Callaway, R. M. 1998. Competition and facilitation on elevation gradients in subalpine forests of the northern Rocky Mountains, USA. *Oikos* 82:561–573.

Cannon, K. P. 1992. A review of archeological and paleontological evidence for the prehistoric presence of wolf and related prey species in the northern and central Rockies physiographic provinces. Pages 1–175 to 1–265 in J. D. Varley and W. G. Brewster, editors. *Wolves for Yellowstone? A report to the United States Congress*, Volume 4, *Research and analysis*. National Park Service, Yellowstone National Park, Mammoth, Wyoming.

Carbyn, L. N., and T. Trottier. 1987. Responses of bison on their calving grounds to predation by wolves in Wood Buffalo National Park. *Canadian Journal of Zoology* 65:2072–2078.

Carlisle, J. D., S. K. Skagen, B. E. Kus, C. van Riper III, K. L. Paxton, and J. F. Kelly. 2009. Landbird migration in the American west: Recent progress and future research directions. *Condor* 111:211–225.

Carlson, C. E. 1978. *Noneffectiveness of Ribes eradication as a control of white pine blister rust in Yellowstone National Park*. Report 78–18, U.S. Department of Agriculture, National Forest Service, Missoula, Montana.

Carpenter, S. R., J. J. Cole, J. R. Hodgson, J. F. Kitchell, M. L. Pace, D. Bade, K. L. Cottingham, T. E. Essington, J. N. Houser, and D. E. Schindler. 2001. Trophic cascades, nutrients, and lake productivity: Whole-lake experiments. *Ecological Monographs* 71:163–186.

Carpenter, S. R., J. F. Kitchell, J. R. Hodgson, P. A. Cochran, J. J. Elser, M. M. Elser, D. M. Lodge, D. Kretchmer, X. He, and C. N. von Ende. 1987. Regulation of lake primary productivity by food web structure. *Ecology* 68:1863–1876.

Carroll, A. L., T. L. Shore, and L. Safranyik. 2006. Direct control: Theory and practice. Pages 155–172 in L. Safranyik and B. Wilson, editors. *The mountain pine beetle: A synthesis of biology, management and impacts on lodgepole pine*. Natural Resources Canada, Canadian Forest Service, Pacific Forestry Centre, Victoria, British Columbia, Canada.

Case, T. J., and M. E. Gilpin. 1974. Interference competition and niche theory. *Proceedings of the National Academy of Sciences of the United States of America* 71:3073–3077.

Caslick, J. 1998. Yellowstone pronghorns: Relict herd in a shrinking habitat. *Yellowstone Science* 6:20–24.

Caughley, G. 1976. Wildlife management and the dynamics of ungulate populations. Pages 183–246 in T. H. Coaker, editor. *Applied Biology 1*. Academic Press, London, UK.

Caughley, G. 1979. What is this thing called carrying capacity? Pages 2–8 in M. S. Boyce and L. D. Hayden-Wing, editors. *North American elk: Ecology, behavior and management.* University of Wyoming Press, Laramie, Wyoming.

Caughley, G., and A. R. E. Sinclair. 1994. *Wildlife ecology and management.* Blackwell Scientific Publications, Boston, Massachusetts.

Chalcraft, D. R., and W. J. Resetarits, Jr. 2003. Predator identity and ecological impacts: Functional redundancy or functional diversity? *Ecology* 84:2407–2418.

Chase, A. 1987. *Playing God in Yellowstone: The destruction of America's first national park.* Harcourt Brace, San Diego, California.

Chase, J. M. 2005. Towards a really unified theory for metacommunities. *Functional Ecology* 19:182–186.

Cheville, N. F., D. R. McCullough, and L. R. Paulson. 1998. *Brucellosis in the greater Yellowstone area.* National Academy Press, Washington, D.C.

Childress, M. J., and M. A. Lung. 2003. Predation risk, gender and the group size effect: Does elk vigilance depend upon the behavior of conspecifics? *Animal Behavior* 66:389–398.

Chittenden, H. M. 1895. *The Yellowstone National Park, historical and descriptive.* Robert Clarke Company, Cincinnati, Ohio.

Christiansen, R. L. 2001. *The Quaternary and Pliocene Yellowstone Plateau volcanic field of Wyoming, Idaho, and Montana.* U.S. Geological Survey Professional Paper 729-G, Reston, Virginia.

Christianson, D. A., and S. Creel. 2007. A review of environmental factors affecting elk winter diet. *Journal of Wildlife Management* 71:164–176.

Christianson, D., and S. Creel. 2009. Fecal chlorophyll describes the link between primary production and consumption in a terrestrial herbivore. *Ecological Applications* 19:1323–1335.

Christianson, D., and S. Creel. 2010. A nutritionally mediated risk effect of wolves on elk. *Ecology* 91:1184–1191.

Cleland, E. E., I. Chuine, A. Menzel, H. A. Mooney, and M. D. Schwartz. 2007. Shifting plant phenology in response to global change. *Trends in Ecology and Evolution* 22:357–365.

Clifford, F. 2009. Wolves and the balance of nature in the Rockies. *Smithsonian* (February).

Clutton-Brock, T. H., and S. D. Albon. 1989. *Red deer in the Scottish Highlands.* Blackwell Scientific, Oxford, UK.

Clutton-Brock, T. H., M. Major, and F. E. Guinness. 1985. Population regulation in male and female red deer. *Journal of Animal Ecology* 54:831–846.

Cole, D. N., and L. Yung, editors. 2010a. *Beyond naturalness: Rethinking park and wilderness stewardship in an era of rapid change.* Island Press, Washington, D.C.

Cole, D. N., and L. Yung. 2010b. Park and wilderness stewardship: The dilemma of management intervention. Pages 1–11 in D. N. Cole and L. Yung, editors. *Beyond naturalness: Rethinking park and wilderness stewardship in an era of rapid change.* Island Press, Washington, D.C.

Cole, D. N., L. Yung, E. S. Zavaleta, G. H. Aplet, F. S. Chapin III, D. M. Graber, E. S. Higgs, R. J. Hobbs, P. B. Landres, C. I. Miller, D. J. Parsons, J. M. Randall, N. L. Stephenson, K. A. Tonnessen, P. S. White, and S. Woodley. 2008. Naturalness and beyond: Protected area stewardship in an era of global environmental change. *George Wright Forum* 25:36–56.

Cole, G. 1969. *Elk and the Yellowstone ecosystem.* National Park Service, Office of Natural Science Studies, Research Note No. 1, Yellowstone National Park, Wyoming.

Cole, G. F. 1971. An ecological rationale for the natural or artificial regulation of ungulates in parks. *Transactions of the North American Wildlife Conference* 36:417–425.

Cole, G. F. 1976. *Management involving grizzly and black bears in Yellowstone National Park, 1970–1975.* U.S. Department of the Interior, National Park Service, Yellowstone National Park, Mammoth, Wyoming.

Cole, W. E., and G. D. Amman. 1980. *Mountain pine beetle dynamics in lodgepole pine forests. Part I: Course of an infestation.* General Technical Report INT-89, U.S. Department of Agriculture, National Forest Service, Washington, D.C.

Congressional Research Service. 1987. *Greater Yellowstone ecosystem, an analysis of data submitted by federal and state agencies.* U.S. Government Printing Office, Washington, D.C.

Cook, J. G. 2002. Nutrition and food. Pages 259–349 in D. E. Toweill and J. W. Thomas, editors. *North American elk: Ecology and management.* Smithsonian Institution Press, Washington, D.C.

Cook, J. G., B. K. Johnson, R. C. Cook, R. A. Riggs, T. Delcurto, L. D. Bryant, and L. L. Irwin. 2004. Effects of summer-autumn nutrition and parturition date on reproduction and survival of elk. *Wildlife Monographs* 155.

Cook, R. C., J. G. Cook, and L. D. Mech. 2004. Nutritional condition of northern Yellowstone elk. *Journal of Mammalogy* 85:714–722.

Cook, W. E., K. W. Mills, E. S. Williams, K. D. Bardsley, and A. Boerger-Fields. 2002. *Survival of* Brucella abortus *strain RB51 on fetuses in the Wyoming environment.* Progress Report UW#5–34204, University of Wyoming, Laramie, Wyoming.

Cooper, S. M. 1991. Optimal group size: The need for lions to defend their kills against loss to spotted hyaenas. *African Journal of Ecology* 29:130–136.

Corn, P. S., B. R. Hossack, E. Muths, D. A. Patla, C. R. Peterson, and A. L. Gallant. 2005. Status of amphibians on the Continental Divide: Surveys on a transect from Montana to Colorado, USA. *Alytes* 22:85–94.

Coté, S. D., T. P. Rooney, J. P. Tremblay, C. Dussault, and D. K. Waller. 2004. Ecological impacts of deer overabundance. *Annual Review of Ecology Evolution and Systematics* 35:113–147.

Coughenour, M. B. 2005. *Spatial-dynamic modeling of bison carrying capacity in the greater Yellowstone ecosystem: A synthesis of bison movements, population dynamics, and interactions with vegetation.* Natural Resource Ecology Laboratory, Colorado State University, Fort Collins, Colorado.

Coughenour, M. B. 2008. Causes and consequences of herbivore movement in landscape ecosystems. Pages 45–91 in K. A. Galvin, R. S. Reid, R. H. Behnke, Jr., and N. T. Hobbs, editors. *Fragmentation in semi-arid and arid landscapes: Consequences for human and natural systems.* Springer, The Netherlands.

Coughenour, M. B., and F. J. Singer. 1996. Elk population processes in Yellowstone National Park under the policy of natural regulation. *Ecological Applications* 6:573–593.

Cowan, I. M., D. G. Chapman, R. S. Hoffmann, D. R. McCullough, G. A. Swanson, and R. B. Weeden. 1974. *Report of the Committee on the Yellowstone grizzlies.* National Academy of Sciences, Washington, D.C.

Crabtree, R. L., and J. W. Sheldon. 1999. Coyotes and canid coexistence in Yellowstone. Pages 127–163 in T. W. Clark, A. P. Curlee, S. C. Minta, and P. M. Kareiva, editors. *Carnivores in ecosystems: The Yellowstone experience.* Yale University Press, New Haven, Connecticut.

Craighead, J. J., G. Atwell, and B. W. O'Gara. 1972. Elk migrations in and near Yellowstone National Park. *Wildlife Monographs* 29.

Craighead, J. J., and F. C. Craighead. 1967. *Management of bears in Yellowstone National Park.* Montana Cooperative Wildlife Research Unit, University of Montana, Missoula, Montana.

Craighead, J. J., K. R. Greer, R. R. Knight, and H. I. Pac. 1988. *Grizzly bear mortalities in the Yellowstone ecosystem, 1959–1987.* Montana Department of Fish, Wildlife, and Parks, Bozeman, Montana.

Craighead, J. J., J. S. Sumner, and J. H. Mitchell. 1995. *The grizzly bears of Yellowstone: Their ecology in the Yellowstone ecosystem, 1959–1992.* Island Press, Washington, D.C.

Craighead, J. J., J. S. Sumner, and G. B. Scaggs. 1982. *A definitive system for analysis of grizzly bear habitat and other wilderness resources.* Wildlife-Wildlands Institute Monograph No. 1, University of Montana Foundation, University of Montana, Missoula, Montana.

Craighead, J. J., J. R. Varney, and F. C. Craighead, Jr. 1974. *A population analysis of the Yellowstone grizzly bears.* Montana Forest and Conservation Station Bulletin 40, University of Montana, Missoula, Montana.

Crait, J. 2005. River otters, cutthroat trout, and their future in Yellowstone National Park. *River Otter Journal* 11:1–3.

Crait, J. R., and M. Ben-David. 2006. River otters in Yellowstone Lake depend on a declining cutthroat population. *Journal of Mammalogy* 87:485–494.

Cramton, L. C. 1932. *Early history of Yellowstone National Park and its relation to national park policies.* U.S. Department of the Interior, National Park Service, Washington, D.C.

Creel, S. 2001. Four factors modifying the effect of competition on carnivore population dynamics as illustrated by African wild dogs. *Conservation Biology* 15:271–274.

Creel, S., and D. Christianson. 2009. Wolf presence and increased willow consumption by Yellowstone elk: Implications for trophic cascades. *Ecology* 90:2454–2466.

Creel, S., D. Christianson, S. Lily, and J. A. Winnie, Jr. 2007. Predation risk affects reproductive physiology and demography of elk. *Science* 315:960.

Creel, S., D. Christianson, and J. A. Winnie, Jr. 2011. A survey of the effects of wolf predation risk on pregnancy rates and calf recruitment in elk. *Ecological Applications* 21:2847–2853.

Creel, S., G. Spong, and N. Creel. 2001. Interspecific competition and the population biology of extinction-prone carnivores. Pages 35–60 in J. L. Gittleman, S. M. Funk, D. Macdonald, and R. K. Wayne, editors. *Carnivore conservation.* Cambridge University Press, Cambridge, UK.

Creel, S., J. A. Winnie, Jr., B. Maxwell, K. Hamlin, and M. Creel. 2005. Elk alter habitat selection as an antipredator response to wolves. *Ecology* 86:3387–3397.

Crête, M. 1998. Ecological correlates of regional variation in life history of the moose *Alces alces:* A comment. *Ecology* 79:1836–1838.

Crête, M. 1999. The distribution of deer biomass in North America supports the hypothesis of exploitation ecosystems. *Ecology Letters* 2:223–227.

Crête, M., and S. Larivière. 2003. Estimating the costs of locomotion in snow for coyotes. *Canadian Journal of Zoology* 81:1808–1814.

Crête, M., and M. Manseau. 1996. Natural regulation of Cervidae along a 1000 km latitudinal gradient: Change in trophic dominance. *Evolutionary Ecology* 10:51–62.

Cross, P. C., E. K. Cole, A. P. Dobson, W. H. Edwards, K. L. Hamlin, G. Luikart, A. D. Middleton, B. M. Scurlock, and P. J. White. 2009. Probable causes of increasing brucellosis in free-ranging elk of the greater Yellowstone ecosystem. *Ecological Applications* 20:278–288.

Cross, P. C., W. H. Edwards, B. M. Scurlock, E. J. Maichak, and J. D. Rogerson. 2007. Effects of management and climate on elk brucellosis in the greater Yellowstone ecosystem. *Ecological Applications* 17:957–964.

Cudmore, T. J., N. Björklund, A. L. Carroll, and B. Staffan Lindgren. 2010. Climate change and range expansion of an aggressive bark beetle: Evidence of higher beetle reproduction in naïve host tree populations. *Journal of Applied Ecology* 47:1036–1043.

Cunningham, J. A., K. L. Hamlin, and T. O. Lemke. 2009. *Northern Yellowstone elk (HD 313) 2008 annual report.* Montana Fish, Wildlife, and Parks, Bozeman, Montana.

Currens, K. P. 1997. *Evolution and risk in conservation of Pacific salmon.* Dissertation, Oregon State University, Corvallis, Oregon.

Curtis, J., and K. Grimes. 2004. *Wyoming climate atlas.* Office of the Wyoming State Climatologist, Laramie, Wyoming.

Dale, B. W., L. G. Adams, and R. T. Bowyer. 1994. Functional response of wolves preying on barren-ground caribou in a multiple-prey ecosystem. *Journal of Animal Ecology* 63:644–652.

Darwin, C. 1859. *On the origin of species by means of natural selection.* Down, Bromley, and Kent, London, UK.

Davenport, M. B. 1974. *Piscivorous avifauna on Yellowstone Lake, Yellowstone National Park.* National Park Service, Yellowstone National Park, Mammoth, Wyoming.

Dean, J. L. 1972. *Fishery management investigations in Yellowstone National Park.* Technical Report for 1971, Bureau of Sport Fisheries and Wildlife, Yellowstone National Park, Mammoth, Wyoming.

Dean, J. L., and J. D. Varley. 1974. *Fishery management investigations, Yellowstone National Park.* Technical Report for 1973, Bureau of Sport Fisheries and Wildlife, Yellowstone National Park, Mammoth, Wyoming.

DeCesare, N. J., M. Hebblewhite, H. S. Robinson, and M. Musiani. 2009. Endangered, apparently: The role of apparent competition in endangered species conservation. *Animal Conservation* 13:353–362.

DelGiudice, G. D., R. A. Moen, F. J. Singer, and M. R. Riggs. 2001. Winter nutritional restriction and simulated body condition of Yellowstone elk and bison before and after the fires of 1988. *Wildlife Monographs* 147.

Despain, D. G. 1990. *Yellowstone vegetation: Consequences of environment and history in a natural setting.* Roberts Rinehart Publishers, Boulder, Colorado.

Despain, D., editor. 1994. *Plants and their environments: Proceedings of the first biennial scientific conference on the greater Yellowstone ecosystem.* National Park Service, Natural Resources Publication Office, Denver, Colorado.

Despain, D. G. 1996. The coincidence of elk migration and flowering of bluebunch wheatgrass. Pages 139–143 in F. J. Singer, editor. *Effects of grazing by wild ungulates in Yellowstone National Park.* Technical Report NPS/NRYELL/NRTR/96–01, U.S. Department of the Interior, National Park Service, National Resource Information Division, Denver, Colorado.

Despain, D., D. Houston, M. Meagher, and P. Schullery. 1986. *Wildlife in transition: Man and nature on Yellowstone's northern range.* Roberts Rinehart, Boulder, Colorado.

Díaz, S., S. Lavorel, S. McIntyre, V. Falczuk, F. Casanoves, D. G. Milchunas, C. Skarpe, G. Rusch, M. Sternberg, I. Noy-Meir, J. Landsberg, W. Zhang, H. Clark, and B. D. Campbell. 2007. Plant trait responses to grazing—A global synthesis. *Global Change Biology* 13:313–341.

Dickson, T. 2002. Will wolves wipe out Montana's elk? *Montana Outdoors* 33:8–15.

Dizard, J. E. 1996. *Going wild: Hunting, animal rights, and the contested meaning of nature.* University of Massachusetts Press, Boston, Massachusetts.

Dolliver, R. H. 1927. The discovery of the Yellowstone. Yellowstone ranger naturalist manual. National Park Service, Yellowstone National Park, Mammoth, Wyoming.

Donadio, E., and S. W. Buskirk. 2006. Diet, morphology, and interspecific killing in Carnivora. *American Naturalist* 167:524–536.

Dunkley, S. L. 2011. *Good animals in bad places: Evaluating landscape attributes associated with elk vulnerability to wolf predation.* Thesis, Montana State University, Bozeman, Montana.

Dunne, J. A., J. Harte, and K. J. Taylor. 2003. Subalpine meadow flowering phenology responses to climate change: Integrating experimental and gradient methods. *Ecological Monographs* 73:69–86.

Du Toit, J. T., B. H. Walker, and B. M. Campbell. 2004. Conserving tropical nature: Current challenges for ecologists. *Trends in Ecology and Evolution* 19:12–17.

Eberhardt, L. L. 1977. Optimal policies for the conservation of large mammals, with special reference to marine ecosystems. *Environmental Conservation* 4:205–212.

Eberhardt, L. L. 1985. Assessing the dynamics of wild populations. *Journal of Wildlife Management* 49:997–1012.

Eberhardt, L. L. 1997. Is wolf predation ratio-dependent? *Canadian Journal of Zoology* 75:1940–1944.

Eberhardt, L. L. 2002. A paradigm for population analysis of long-lived vertebrates. *Ecology* 83:2841–2854.

Eberhardt, L. L., R. A. Garrott, D. W. Smith, P. J. White, and R. O. Peterson. 2003. Assessing the impact of wolves on ungulate prey. *Ecological Applications* 13:776–783.

Eberhardt, L. L., R. A. Garrott, P. J. White, and P. J. Gogan. 1998. Alternative approaches to aerial censusing of elk. *Journal of Wildlife Management* 62:1046–1055.

Eberhardt, L. L., P. J. White, R. A. Garrott, and D. B. Houston. 2007. A seventy-year history of trends in Yellowstone's northern elk herd. *Journal of Wildlife Management* 71:594–602.

Ebinger, M. R., P. Cross, R. Wallen, P. J. White, and J. Treanor. 2011. Simulating sterilization, vaccination, and test-and-remove as brucellosis control measures in bison. *Ecological Applications* 21:2944–2959.

Ellison, A. M., M. S. Bank, B. D. Clinton, E. A. Colburn, K. Elliott, C. R. Ford, D. R. Foster, B. D. Kloeppel, J. D. Knoepp, G. M. Lovett, J. Mohan, D. A. Orwig, N. L. Rodenhouse, W. V. Sobczak, K. A. Stinson, J. K. Stone, C. M. Swan, J. Thompson, B. von Holle, and J. R. Webster. 2005. Loss of foundation species: Consequences for the structure and dynamics of forested ecosystems. *Frontiers in Ecology and the Environment* 3:479–486.

Engstrom, D. R., C. Whitlock, S. C. Fritz, and H. E. Wright. 1991. Recent environmental change inferred from the sediments of small lakes in Yellowstone's northern range. *Journal of Paleolimnology* 5:139–174.

Engstrom, D. R., C. Whitlock, S. C. Fritz, and H. E. Wright. 1994. Reinventing erosion in Yellowstone's northern range. *Journal of Paleolimnology* 10:159–161.

Errington, P. L. 1945. Some contributions of a fifteen-year local study of the northern bobwhite to a knowledge of population phenomena. *Ecological Monographs* 15:2–34.

Errington, P. L. 1946. Predation and vertebrate populations. *Quarterly Review of Biology* 21:144–177.

Estes, J. A. 1996. Predators and ecosystem management. *Wildlife Society Bulletin* 24:390–396.

Estes, J. A., E. M. Danner, D. F. Doak, B. Konar, A. M. Springer, P. D. Steinberg, M. T. Tinker, and T. M. Williams. 2004. Complex trophic interactions in kelp forest ecosystems. *Bulletin of Marine Science* 74:621–638.

Estes, J. A., J. Terborgh, J. S. Brashares, M. E. Power, J. Berger, W. J. Bond, S. R. Carpenter, T. E. Essington, R. D. Holt, J. B. C. Jackson, R. J. Marquis, L. Oksanen, T. Oksanen, R. T. Paine, E. K. Pikitch, W. J. Ripple, S. A. Sandin, M. Scheffer, T. W. Schoener, J. B. Shurin, A. R. E. Sinclair, M. E. Soulé, R. Virtanen, and D. A. Wardle. 2011. Trophic downgrading of planet earth. *Science* 333:301–306.

Evans, S. B., L. D. Mech, P. J. White, and G. A. Sargeant. 2006. Survival of adult female elk in Yellowstone following wolf recovery. *Journal of Wildlife Management* 70:1372–1378.

Farnes, P. E. 1990. Snowtel and snow course data: Describing the hydrology of whitebark pine ecosystems. Pages 302–304 in W. C. Schmidt and K. J. MacDonald, compilers. *Proceedings—Symposium on whitebark pine ecosystems: Ecology and management of a high-mountain resource.* Bozeman, Montana, March 29–31, 1989.

Felicetti, L. A., C. C. Schwartz, R. O. Rye, K. A. Gunther, J. G. Crock, M. A. Haroldson, L. Waits, and C. T. Robbins. 2004. Use of naturally occurring mercury to determine the importance of cutthroat trout to Yellowstone grizzly bears. *Canadian Journal of Zoology* 82:493–501.

Felicetti, L. A., C. C. Schwartz, R. O. Rye, M. A. Haroldson, K. A. Gunther, and C. T. Robbins. 2003. Use of sulfur and nitrogen stable isotopes to determine the importance of whitebark pine nuts to Yellowstone grizzly bears. *Canadian Journal of Zoology* 81:763–770.

Finch, D. M. 1989. Habitat use and habitat overlap of riparian birds in three elevation zones. *Ecology* 70:866–880.

Findlay, D. L., M. J. Vanni, M. Paterson, K. H. Mills, S. E. M. Kasian, W. J. Findlay, and A. G. Salki. 2005. Dynamics of a boreal lake ecosystem during a long-term manipulation of top predators. *Ecosystems* 8:603–618.

Fitter, A. H., and R. S. R. Fitter. 2002. Rapid changes in flowering time in British plants. *Science* 296:1689–1691.

Forester, J. D., D. P. Anderson, and M. G. Turner. 2007. Do high-density patches of coarse wood and regenerating saplings create browsing refugia for aspen (*Populus tremuloides* Michx.) in Yellowstone National Park (USA)? *Forest Ecology and Management* 253:211–219.

Forrest, J., and A. J. Miller-Rushing. 2010. Toward a synthetic understanding of the role of phenology in ecology and evolution. *Philosophical Transactions of the Royal Society. Series B, Biological Sciences* 365:3101–3112.

Forsyth, D. M., and P. Caley. 2006. Testing the irruptive paradigm of large-herbivore dynamics. *Ecology* 87:297–303.

Fortin, J. K. 2011. *Niche separation of grizzly* (Ursus arctos) *and American black bears* (Ursus americanus) *in Yellowstone National Park.* Washington State University, Pullman, Washington.

Frank, D. A. 2005. The interactive effects of grazing ungulates and aboveground production on grassland diversity. *Oecologia* 143:629–634.

Frank, D. A. 2007. Drought effects on above- and belowground production of a grazed temperate grassland ecosystem. *Oecologia* 152:131–139.

Frank, D. 2008. Evidence for top predator control of a grazing ecosystem. *Oikos* 117:1718–1724.

Frank, D. A., and R. D. Evans. 1997. Effects of native grazers on grassland N cycling in Yellowstone National Park. *Ecology* 78:2238–2248.

Frank, D. A., C. A. Gehring, L. Machut, and M. Phillips. 2003. Soil community composition and the regulation of a grazed temperate grassland. *Oecologia* 442:603–609.

Frank, D. A., and P. M. Groffman. 1998. Ungulate vs. landscape control of soil C and N processes in grasslands of Yellowstone National Park. *Ecology* 79:2229–2241.

Frank, D. A., R. S. Inouye, N. Huntly, G. W. Minshall, and J. E. Anderson. 1994. The biogeochemistry of a north-temperate grassland with native ungulates: Nitrogen dynamics in Yellowstone National Park. *Biogeochemistry* 26:163–188.

Frank, D. A., M. M. Kuns, and D. R. Guido. 2002. Grazer control of grassland plant production. *Ecology* 83:602–606.

Frank, D. A., and S. J. McNaughton. 1992. The ecology of plants, large mammalian herbivores, and drought in Yellowstone National Park. *Ecology*
73:2043–2058.

Frank, D. A., and S. J. McNaughton. 1993. Evidence for promotion of aboveground grassland production by native large herbivores in Yellowstone National
Park. *Oecologia* 96:157–161.

Frank, D. A., S. J. McNaughton, and B. F. Tracy. 1998. The ecology of the earth's
grazing ecosystems. *BioScience* 48:513–521.

Franke, M. A. 1996. A grand experiment, 100 years of fisheries management in
Yellowstone: Part I. *Yellowstone Science* 4(4):2–7.

Franke, M. A. 1997. A grand experiment, The tide turns in the 1950s: Part II.
Yellowstone Science 5(1):8–13.

Franke, M. A. 2000. *Yellowstone in the afterglow: Lessons from the fires.* National
Park Service, Yellowstone National Park, Mammoth, Wyoming.

Fryxell, J. M. 1991. Forage quality and aggregation by large herbivores. *American
Naturalist* 138:478–498.

Fryxell, J. M., J. Greever, and A. R. E. Sinclair. 1988. Why are migratory ungulates
so abundant? *American Naturalist* 131:781–798.

Fryxell, J. M., and A. R. E. Sinclair. 1988. Causes and consequences of migration
by large herbivores. *Trends in Ecology and Evolution* 3:237–241.

Fujita, T. T. 1989. The Teton-Yellowstone tornado of 21 July 1987. *Monthly Weather
Review* 117:1913–1940.

Fuller, J. A., R. A. Garrott, and P. J. White. 2007. Emigration and density dependence in Yellowstone bison. *Journal of Wildlife Management* 71:1924–1933.

Furniss, M. M., and R. Renkin. 2003. Forest entomology in Yellowstone National
Park, 1923–1957: A time of discovery and learning to let live. *American
Entomologist* 49:198–209.

Gaillard, J.-M., M. Festa-Bianchet, and N. G. Yoccoz. 1998. Population dynamics
of large herbivores: Variable recruitment with constant adult survival. *Trends
in Ecology and Evolution* 13:58–64.

Gaillard, J.-M., M. Festa-Bianchet, N. G. Yoccoz, A. Loison, and C. Toigo. 2000.
Temporal variation in fitness components and population dynamics of large
herbivores. *Annual Review of Ecology and Systematics* 31:367–393.

Gale, R. S., E. O. Garton, and I. J. Ball. 1987. *The history, ecology and manage-
ment of the Rocky Mountain population of trumpeter swans.* U.S. Fish and
Wildlife Service, Montana Cooperative Wildlife Research Unit, Missoula,
Montana.

Galen, C., and M. L. Stanton. 1995. Responses of snowbed plant species to changes
in growing-season length. *Ecology* 76:1546–1557.

Garrott, R. A., L. L. Eberhardt, P. J. White, and J. Rotella. 2003. Climate-induced
variation in vital rates of an unharvested large herbivore population. *Canadian
Journal of Zoology* 81:33–45.

Garrott, R. A., J. A. Gude, E. J. Bergman, C. Gower, P. J.White, and K. L. Hamlin. 2005. Generalizing wolf effects across the greater Yellowstone area: A cautionary note. *Wildlife Society Bulletin* 33:1245–1255.

Garrott, R. A., J. Rotella, M. O'Reilly, M. Sawaya, M. Zambon, and P. J. White. 2010. *The greater Yellowstone area mountain ungulate research initiative: 2010 annual report.* Montana State University, Bozeman, Montana.

Garrott, R. A., and P. J. White. 2009. Integrated science in the central Yellowstone ecosystem. Pages 3–13 in R. A. Garrott, P. J. White, and F. G. R. Watson, editors. *The ecology of large mammals in central Yellowstone: Sixteen years of integrated field studies.* Elsevier, San Diego, California.

Garrott, R. A., P. J. White, M. S. Becker, and C. N. Gower. 2009. Apparent competition and potential regulation of large herbivores in central Yellowstone. Pages 519–540 in R. A. Garrott, P. J. White, and F. G. R. Watson, editors. *The ecology of large mammals in central Yellowstone: Sixteen years of integrated field studies.* Elsevier, San Diego, California.

Garrott, R. A., P. J. White, and J. R. Rotella. 2009a. The Madison headwaters elk herd: Stability in an inherently variable environment. Pages 191–216 in R. A. Garrott, P. J. White, and F. G. R. Watson, editors. *The ecology of large mammals in central Yellowstone: Sixteen years of integrated field studies.* Elsevier, San Diego, California.

Garrott, R. A., P. J. White, and J. R. Rotella. 2009b. The Madison headwaters elk herd: Transitioning from bottom-up regulation to top-down limitation. Pages 489–517 in R. A. Garrott, P. J. White, and F. G. R. Watson, editors. *The ecology of large mammals in central Yellowstone: Sixteen years of integrated field studies.* Elsevier, San Diego, California.

Garshelis, D. L. 1994. Density-dependent population regulation of black bears. Pages 3–14 in M. Taylor, editor. *Density-dependent population regulation of black, brown, and polar bears.* International Conference on Bear Research and Management. Monograph Series 3.

Gasaway, W. C., R. D. Boertje, D. V. Grangaard, D. G. Kelleyhouse, R. O. Stephenson, and D. G. Larsen. 1992. The role of predation in limiting moose at low densities in Alaska and Yukon and implications for conservation. *Wildlife Monographs* 120.

Gasaway, W. C., R. O. Stephenson, J. L. Davis, P. E. K. Shepherd, and O. E. Burris. 1983. Interrelationships of wolves, prey, and man in interior Alaska. *Wildlife Monographs* 84.

Gates, C. C., B. Stelfox, T. Muhly, T. Chowns, and R. J. Hudson. 2005. *The ecology of bison movements and distribution in and beyond Yellowstone National Park.* University of Calgary, Calgary, Alberta, Canada.

Geist, V. 2002. Adaptive behavioral strategies. Pages 389–433 in D. E. Toweill and J. W. Thomas, editors. *North American elk: Ecology and management.* Smithsonian Institution Press, Washington, D.C.

Geremia, C., P. J. White, R. A. Garrott, R. Wallen, K. E. Aune, J. Treanor, and J. A. Fuller. 2009. Demography of central Yellowstone bison: Effects of climate, density and disease. Pages 255–279 in R. A. Garrott, P. J. White, and F. G. R. Watson, editors. *The ecology of large mammals in central Yellowstone: Sixteen years of integrated field studies.* Elsevier, San Diego, California.

Geremia, C., P. J. White, R. L. Wallen, F. G. R. Watson, J. J. Treanor, J. Borkowski, C. S. Potter, and R. L. Crabtree. 2011. Predicting bison migration out of Yellowstone National Park using Bayesian models. *PLoS One* 6:e16848.

Gese, E. M., R. L. Ruff, and R. L. Crabtree. 1996. Foraging ecology of coyotes *(Canis latrans):* The influence of extrinsic factors and a dominance hierarchy. *Canadian Journal of Zoology* 74:769–783.

Gibson, K., K. Skov, S. Kegley, C. Jorgensen, S. Smith, and J. Witcosky. 2008. *Mountain pine beetle impacts in high-elevation five-needle pines: Current trends and challenges.* Forest Health Protection R1–08–020, U.S. Department of Agriculture, National Forest Service, Missoula, Montana.

Goetz, W., P. Maus, and E. Nielson. 2009. *Mapping whitebark pine canopy mortality in the greater Yellowstone area.* RSAC-0104-RPT1, U.S. Department of Agriculture, National Forest Service, Remote Sensing Application Center, Salt Lake City, Utah.

Good, J. D., and K. L. Pierce. 2010. *Interpreting the landscapes of Grand Teton and Yellowstone National Parks: Recent and ongoing geology.* Grand Teton National History Association, Moose, Wyoming.

Gordon, I. J., A. J. Hester, and M. Festa-Bianchet. 2004. The management of wild large herbivores to meet economic, conservation and environmental objectives. *Journal of Applied Ecology* 41:1021–1031.

Gower, C. N., R. A. Garrott, and P. J. White. 2009. Elk foraging behavior: Does predation risk reduce time for food acquisition? Pages 423–450 in R. A. Garrott, P. J. White, and F. G. R. Watson, editors. *The ecology of large mammals in central Yellowstone: Sixteen years of integrated field studies.* Elsevier, San Diego, California.

Gower, C. N., R. A. Garrott, P. J. White, S. Cherry, and N. G. Yoccoz. 2009. Elk group size and wolf predation: A flexible strategy when faced with variable risk. Pages 401–422 in R. A. Garrott, P. J. White, and F. G. R. Watson, editors. *The ecology of large mammals in central Yellowstone: Sixteen years of integrated field studies.* Elsevier, San Diego, California.

Gower, C. N., R. A. Garrott, P. J. White, F. G. Watson, S. Cornish, and M. S. Becker. 2009. Spatial responses of elk to winter wolf predation risk: Using the landscape to balance multiple demands. Pages 373–399 in R. A. Garrott, P. J. White, and F. G. R. Watson, editors. *The ecology of large mammals in central Yellowstone: Sixteen years of integrated field studies.* Elsevier, San Diego, California.

Graham, D. C. 1978. *Grizzly bear distribution, use of habitats, food habits, and habitat characterization in Pelican and Hayden valleys, Yellowstone National Park.* Thesis, Montana State University, Bozeman, Montana.

Gray, M. E., and E. Z. Cameron. 2010. Does contraceptive treatment in wildlife result in side effects? A review of quantitative and anecdotal evidence. *Reproduction* 139:45–55.

Gray, S. T., C. L. Fastie, S. T. Jackson, and J. L. Betancourt. 2004. Tree-ring-based reconstruction of precipitation in the Bighorn Basin, Wyoming, since 1260 A.D. *Journal of Climate* 17:3855–3865.

Gray, S. T., L. J. Graumlich, and J. L. Betancourt. 2007. Annual precipitation in the Yellowstone National Park region since A.D. 1173. *Quaternary Research* 68:18–27.

Greater Yellowstone Coordinating Committee. 1991. *A framework for coordination of national parks and national forests in the greater Yellowstone area.* Greater Yellowstone Coordinating Committee, Billings, Montana.

Greater Yellowstone Coordinating Committee Whitebark Pine Subcommittee. 2011. *Whitebark pine strategy for the greater Yellowstone area.* Greater Yellowstone Coordinating Committee, Bozeman, Montana.

Greater Yellowstone Whitebark Pine Monitoring Working Group. 2011a. *Interagency whitebark pine monitoring protocol for the greater Yellowstone ecosystem, Version 1.01.* Greater Yellowstone Coordinating Committee, Bozeman, Montana.

Greater Yellowstone Whitebark Pine Monitoring Working Group. 2011b. Monitoring whitebark pine in the greater Yellowstone ecosystem: 2010 annual report. Pages 56–65 in C. C. Schwartz, M. A. Haroldson, and K. West, editors. *Yellowstone grizzly bear investigations: Annual report of the Interagency Grizzly Bear Study Team, 2010.* U.S. Geological Survey, Bozeman, Montana.

Green, G. I., D. J. Mattson, and J. M. Peek. 1997. Spring feeding on ungulate carcasses by grizzly bears in Yellowstone National Park. *Journal of Wildlife Management* 61:1040–1055.

Gresswell, R. E. 1991. Use of antimycin for removal of brook trout from a tributary of Yellowstone Lake. *North American Journal of Fisheries Management* 11:83–90.

Gresswell, R. E. 1995. Yellowstone cutthroat trout. Pages 36–54 in M. Young, editor. *Conservation assessment for inland cutthroat trout.* General Technical Report RM-GTR-256, U.S. Department of Agriculture, National Forest Service, Rocky Mountain Forest and Range Experiment Station, Fort Collins, Colorado.

Gresswell, R. E. 1999. Fire and aquatic ecosystems in forested biomes of North America. *Transactions of the American Fisheries Society* 128:193–221.

Gresswell, R. E. 2004. Effects of the wildfire on growth of cutthroat trout in Yellowstone Lake. Pages 143–164 in L. Wallace, editor. *After the fires: The ecology of change in Yellowstone National Park.* Yale University Press, New Haven, Connecticut.

Gresswell, R. E. 2009. *Scientific review panel evaluation of the National Park Service lake trout suppression program in Yellowstone Lake, August 25–29.* YCR-2009-05, U.S. Geological Survey, Northern Rocky Mountain Science Center, Bozeman, Montana.

Gresswell, R. E. 2011. Biology, status, and management of the Yellowstone cutthroat trout. *North American Journal of Fisheries Management* 31:782–812.

Gresswell, R. E., and W. J. Liss. 1995. Values associated with management of Yellowstone cutthroat trout in Yellowstone National Park. *Conservation Biology* 9:159–165.

Gresswell, R. E., W. J. Liss, and G. L. Larson. 1994. Life-history organization of Yellowstone cutthroat trout *(Oncorhynchus clarki bouvieri)* in Yellowstone Lake. *Canadian Journal of Fisheries and Aquatic Sciences* 51(Supplement 1): 298–309.

Gresswell, R. E., W. J. Liss, G. L. Larson, and P. J. Bartlein. 1997. Influence of basin-scale physical variables on life-history characteristics of cutthroat trout in Yellowstone Lake. *North American Journal of Fisheries Management* 17:1046–1064.

Gresswell, R. E., and J. D. Varley. 1988. Effects of a century of human influence on the cutthroat trout of Yellowstone Lake. Pages 45–52 in R. E. Gresswell, editor. *Status and management of interior stocks of cutthroat trout.* American Fisheries Society Symposium 4, Bethesda, Maryland.

Griffin, K. A., M. Hebblewhite, H. S. Robinson, P. Zager, S. M. Barber-Meyer, D. Christianson, S. Creel, N. C. Harris, M. A. Hurley, D. H. Jackson, B. K. Johnson, L. D. Mech, W. L. Myers, J. D. Raithel, M. Schlegel, B. L. Smith, C. White, and P. J. White. 2011. Neonatal mortality of elk driven by climate, predator phenology and predator community composition. *Journal of Animal Ecology* 80:1246–1257.

Grimm, R. L. 1939. Northern Yellowstone winter range studies. *Journal of Wildlife Management* 3:295–306.

Gruell, G. 1973. *An ecological evaluation of Big Game Ridge.* U.S. Department of Agriculture, U.S. Forest Service, Intermountain Region.

Gude, J. A., R. A. Garrott, J. J. Borkowski, and F. King. 2006. Prey risk allocation in a grazing environment. *Ecological Applications* 16:285–298.

Gude, P. H., A. J. Hansen, and D. A. Jones. 2007. Biodiversity consequences of alternative future land use scenarios in greater Yellowstone. *Ecological Applications* 17:1004–1018.

Gunderson, L. H. 2000. Ecological resilience—In theory and application. *Annual Review of Ecology and Systematics* 31:425–439.

Gunther, K. A. 1991. *Grizzly bear activity and human-induced modifications in Pelican Valley, Yellowstone National Park.* Thesis, Montana State University, Bozeman, Montana.

Gunther, K. A. 1994. Bear management in Yellowstone National Park, 1960–93. *International Conference on Bear Research and Management* 9:549–560.

Gunther, K. A. 2008. Delisted but not forgotten: Management, monitoring, and conservation of grizzly bears in Yellowstone National Park after delisting. *Yellowstone Science* 16:30–34.

Gunther, K. A., M. J. Biel, N. Anderson, and L. Waits. 2002. Probable grizzly bear predation on an American black bear in Yellowstone National Park. *Ursus* 28:372–374.

Gunther, K. A., M. J. Biel, R. A. Renkin, and H. N. Zachary. 1999. *Influence of season, park visitation, and mode of transportation on the frequency of road-killed wildlife in Yellowstone National Park.* National Park Service, Yellowstone National Park, Wyoming.

Gunther, K. A., M. A. Haroldson, K. Frey, S. L. Cain, J. Copeland, and C. C. Schwartz. 2004. Grizzly bear-human conflicts in the greater Yellowstone ecosystem, 1992–2000. *Ursus* 15:10–22.

Gunther, K. A., and D. W. Smith. 2004. Interactions between wolves and female grizzly bears with cubs in Yellowstone National Park. *Ursus* 15:232–238.

Gunther, K. A., and T. Wyman. 2008. Human habituated bears, the next challenge in bear management in Yellowstone National Park. *Yellowstone Science* 16:35–41.

Hagenbarth, J. F. 2007. Testimony before the U.S. House of Representatives Subcommittee on National Parks, Forests, and Public Lands oversight hearing on Yellowstone National Park bison. March 20, 2007, Washington, D.C.

Haggerty, J. H., and W. R. Travis. 2006. Out of administrative control: Absentee owners, resident elk and the shifting nature of wildlife management in southwestern Montana. *Geoforum* 37:816–830.

Haines, A. L. 1974. *Yellowstone National Park, its exploration and establishment.* U.S. Government Printing Office, Washington D.C.

Haines, A. L. 1977. *The Yellowstone story—A history of our first national park: Volume 1.* Yellowstone Library and Museum Association, Yellowstone National Park, and Colorado Associated University Press, Boulder, Colorado.

Hamilton, E. W., and D. A. Frank. 2001. Can plants stimulate soil microbes and their own nutrient supply? Evidence from a grazing tolerant grass. *Ecology* 82:2397–2402.

Hamilton, E. W. III, D. A. Frank, P. M. Hinchey, and T. R. Murray. 2008. Defoliation induces root exudation and triggers positive rhizospheric feed-backs in a temperate grassland. *Soil Biology and Biochemistry* 40:2865–2873.

Hamlin, K. L. 2006. *Monitoring and assessment of wolf-ungulate interactions and population trends within the greater Yellowstone area, southwestern Montana, and Montana statewide.* Montana Fish, Wildlife, and Parks, Bozeman, Montana.

Hamlin, K. L., and J. A. Cunningham. 2008. *Montana elk movements, distribution, and numbers relative to brucellosis transmission risk.* Montana Department of Fish, Wildlife, and Parks, Bozeman, Montana.

Hamlin, K. L., R. A. Garrott, P. J. White, and J. A. Cunningham. 2009. Contrasting wolf-ungulate interactions in the greater Yellowstone ecosystem. Pages 541–577 in R. A. Garrott, P. J. White, and F. G. R. Watson, editors. *The ecology of large mammals in central Yellowstone: Sixteen years of integrated field studies.* Elsevier, San Diego, California.

Hampton, H. D. 1971. *How the U.S. cavalry saved our national parks.* Indiana University Press, Bloomington, Indiana.

Hansen, A. J. 2009. Species and habitats most at risk in greater Yellowstone. *Yellowstone Science* 17:27–36.

Hansen, A. J., and R. DeFries. 2007. Ecological mechanisms linking protected areas to surrounding lands. *Ecological Applications* 17:974–988.

Hansen, A. J., and J. J. Rotella. 2002. Biophysical factors, land use, and species viability in and around nature reserves. *Conservation Biology* 16:1112–1122.

Hansen, A. J., J. J. Rotella, M. L. Kraska, and D. Brown. 2000. Spatial patterns of primary productivity in the greater Yellowstone ecosystem. *Landscape Ecology* 15:505–522.

Haroldson, M. 1999. Mortalities. Pages 25–28 in C. C. Schwartz and M. A. Haroldson, editors. *Yellowstone grizzly bear investigations: Annual report of the Interagency Grizzly Bear Study Team, 1998.* U.S. Geological Survey, Bozeman, Montana.

Haroldson, M. A. 2011. Assessing trend and estimating population size from counts of unduplicated females. Pages 10–15 in C. C. Schwartz, M. A. Haroldson, and K. West, editors. *Yellowstone grizzly bear investigations: Annual report of the Interagency Grizzly Bear Study Team, 2010.* U.S. Geological Survey, Bozeman, Montana.

Haroldson, M. A., and K. Frey. 2005. Grizzly bear mortalities. Pages 24–29 in C. C. Schwartz, M. A. Haroldson, and K. West, editors. *Yellowstone grizzly bear investigations: Annual report of the Interagency Grizzly Bear Study Team, 2004.* U.S. Geological Survey, Bozeman, Montana.

Haroldson, M. A., K. A. Gunther, D. P. Reinhart, S. R. Podruzny, C. Cegelski, L. Waits, T. C. Wyman, and J. Smith. 2005. Changing numbers of spawning cutthroat trout in tributary streams of Yellowstone Lake and estimates of grizzly bears visiting streams from DNA. *Ursus* 16:167–180.

Haroldson, M. A., and S. Podruzny. 2011. Whitebark pine cone production. Pages 36–37 in C. C. Schwartz, M. A. Haroldson, and K. West, editors. *Yellowstone grizzly bear investigations: Annual report of the Interagency Grizzly Bear Study Team, 2010.* U.S. Geological Survey, Bozeman, Montana.

Haroldson, M. A., C. C. Schwartz, and K. A. Gunther. 2008. Grizzly bears in the greater Yellowstone ecosystem: From garbage, controversy, and decline to recovery. *Yellowstone Science* 16:13–24.

Haroldson, M. A., C. C. Schwartz, and G. C. White. 2006. Survival of independent grizzly bears in the greater Yellowstone ecosystem, 1983–2001. *Wildlife Monographs* 161.

Haroldson, M. A., M. Ternent, K. A. Gunther, and C. C. Schwartz. 2002. Grizzly bear denning chronology and movements in the greater Yellowstone ecosystem. *Ursus* 13:29–37.

Haroldson, M. A., M. Ternent, G. Holm, R. A. Swalley, S. Podruzny, D. Moody, and C. C. Schwartz. 1998. *Yellowstone grizzly bear investigations: Annual report of the Interagency Grizzly Bear Study Team, 1997.* U.S. Geological Survey, Bozeman, Montana.

Harris, R. B., C. C. Schwartz, M. A. Haroldson, and G. C. White. 2006. Trajectory of the Yellowstone grizzly bear population under alternative survival rates. *Wildlife Monographs* 161.

Hatala, J., M. Dietze, Interagency Whitebark Pine Monitoring Working Group, K. Kendall, D. Six, R. Crabtree, and P. Moorcroft. 2011. An ecosystem model of white pine blister rust *(Cronartium ribicola)* spread in whitebark pine *(Pinus albicaulis)* of the greater Yellowstone ecosystem. *Ecological Applications* 21:1138–1153.

Hayes, R. D., R. Farnell, R. M. P. Ward, J. Carey, M. Dehn, G. W. Kuzyk, A. M. Baer, C. L. Gardner, and M. O'Donoghue. 2003. Experimental reduction of wolves in the Yukon: Ungulate responses and management implications. *Wildlife Monographs* 152.

Hayes, R. D., and A. S. Harestad. 2000. Wolf functional response and regulation of moose in the Yukon. *Canadian Journal of Zoology* 78:60–66.

Heard, D. C. 1992. The effect of wolf predation and snow cover on musk-ox group size. *American Naturalist* 139:190–204.

Hebblewhite, M., E. H. Merrill, and G. McDermid. 2008. A multi-scale test of the forage maturation hypothesis in a partially-migratory ungulate population. *Ecological Monographs* 78:141–166.

Hebblewhite, M., E. H. Merrill, L. E. Morgantini, C. A. White, J. R. Allen, E. Bruns, L. Thurston, and T. E. Hurd. 2006. Is the migratory behaviour of montane elk herds in peril? The case of Alberta's Ya Ha Tinda elk herd. *Wildlife Society Bulletin* 34:1280–1295.

Hebblewhite, M., P. C. Paquet, D. H. Pletscher, R. B. Lessard, and C. J. Callaghan. 2003. Development and application of a ratio estimator to estimate wolf kill rates and variance in a multiple-prey system. *Wildlife Society Bulletin* 31:933–946.

Hebblewhite, M., D. H. Pletcher, and P. C. Paquet. 2002. Elk population dynamics in areas with and without predation by recolonizing wolves in Banff National Park, Alberta. *Canadian Journal of Zoology* 80:789–799.

Hebblewhite, M., and D. W. Smith. 2010. Wolf community ecology: Ecosystem effects of recovering wolves in Banff and Yellowstone National Parks. Pages

69–120 in M. Musiano, P. Paquet, and L. Boitani, editors. *The world of wolves.* University of Calgary Press, Calgary, Alberta, Canada.

Hebblewhite, M., C. A. White, C. G. Nietvelt, J. A. McKenzie, T. E. Hurd, J. M. Fryxell, S. E. Bayley, and P. C. Paquet. 2005. Human activity mediates a trophic cascade caused by wolves. *Ecology* 86:2135–2144.

Hebblewhite, M., J. Whittington, M. Bradley, G. Skinner, A. Dibb, and C. A. White. 2007. Conditions for caribou persistence in the wolf-elk-caribou systems of the Canadian Rockies. *Rangifer* Special Issue No. 17:79–91.

Herrero, S. 1978. A comparison of some features of the evolution, ecology and behavior of black and grizzly/brown bears. *Carnivore* 1:7–17.

Higgs, E. S., and R. J. Hobbs. 2010. Wild design: Principles to guide interventions in protected areas. Pages 234–251 in D. N. Cole and L. Yung, editors. *Beyond naturalness: Rethinking park and wilderness stewardship in an era of rapid change.* Island Press, Washington, D.C.

Higuera, P. E., D. G. Gavin, P. J. Bartlein, and D. J. Hallett. 2010. Peak detection in sediment-charcoal records: Impacts of alternative data analysis methods on fire-history interpretations. *International Journal of Wildland Fire* 19:996–1014.

Hilbert, D. W., D. M. Swift, J. K. Detling, and M. I. Dyer. 1981. Relative growth rates and the grazing optimization hypothesis. *Oecologia* 51:14–18.

Hilderbrand, G. V., S. G. Jenkins, C. C. Schwartz, T. A. Hanley, and C. T. Robbins. 1999. Effect of seasonal differences in dietary meat intake on changes in body mass and composition of wild and captive brown bears. *Canadian Journal of Zoology* 77:1623–1630.

Hobbs, N. T. 1996. Modification of ecosystems by ungulates. *Journal of Wildlife Management* 60:695–713.

Hobbs, N. T. 2003. Challenges and opportunities in integrating ecological knowledge across scales. *Forest Ecology and Management* 181:223–238.

Hobbs, N. T. 2006. *A model analysis of effects of wolf predation on prevalence of chronic wasting disease in elk populations of Rocky Mountain National Park.* Colorado State University, Fort Collins, Colorado.

Hobbs, N. T., R. Wallen, J. Treanor, C. Geremia, and P. J. White. 2009. *A stochastic population model of the Yellowstone bison population.* Colorado State University, Fort Collins, Colorado.

Hobbs, R. J., D. N. Cole, L. Yung, E. S. Zavaleta, G. H. Aplet, F. S. Chapin III, P. B. Landres, D. J. Parsons, N. L. Stephenson, P. S. White, D. M. Graber, E. S. Higgs, C. I. Millar, J. M. Randall, K. A. Tonnessen, and S. Woodley. 2010. Guiding concepts for park and wilderness stewardship in an era of global environmental change. *Frontiers in Ecology and Environment* 8:483–490.

Hobbs, R. J., E. S. Zavaleta, D. N. Cole, and P. S. White. 2010. Evolving ecological understandings: The implications of ecosystem dynamics. Pages 34–49 in

D. N. Cole and L. Yung, editors. *Beyond naturalness: Rethinking park and wilderness stewardship in an era of rapid change.* Island Press, Washington, D.C.

Hodges, T., and M. Bruscino. 2008. Appendix E: 2007 Wapiti and Jackson Hole Bear-Wise Community Projects Update. Pages 111–114 in C. C. Schwartz, M. A. Haroldson, and K. West, editors. *Yellowstone grizzly bear investigations: Annual report of the Interagency Grizzly Bear Study Team, 2007.* U.S. Geological Survey, Bozeman, Montana.

Hoff, R. J. 1992. *How to recognize blister rust infection on whitebark pine.* Research Note INT-406, U.S. Department of Agriculture, National Forest Service, Intermountain Research Station, Ogden, Utah.

Hoff, R. J., D. E. Ferguson, G. I. McDonald, and R. E. Keane. 2001. Strategies for managing whitebark pine in the presence of white pine blister rust. Pages 346–366 in D. F. Tomback, S. F. Arno, and R. E. Keane, editors. *Whitebark pine communities: Ecology and restoration.* Island Press, Washington, D.C.

Holdo, R. M., R. D. Holt, M. B. Coughenour, and M. E. Ritchie. 2006. Plant productivity and soil nitrogen as a function of grazing, migration, and fire in an African savanna. *Journal of Ecology* 95:115–128.

Holland, E. A., W. J. Parton, J. K. Detling, and D. L. Coppock. 1992. Physiological responses of plant populations to herbivory and their consequences for ecosystem nutrient flow. *American Naturalist* 140:685–706.

Holling, C. S. 1959a. The components of predation as revealed by a study of small-mammal predation of the European pine sawfly. *Canadian Entomologist* 91:293–320.

Holling, C. S. 1959b. Some characteristics of simple types of predation and parasitism. *Canadian Entomologist* 91:385–398.

Holling, C. S. 1978. *Adaptive environmental assessment and management.* John Wiley and Sons, London, UK.

Holm, G. W. 1998. *Interactions of sympatric black and grizzly bears in northwest Wyoming.* Thesis, University of Wyoming, Laramie, Wyoming.

Holt, R. D., and J. H. Lawton. 1994. The ecological consequences of shared natural enemies. *Annual Review of Ecology and Systematics* 25:495–520.

Hornaday, W. T. 1913. *Our vanishing wild life: Its extermination and preservation.* Clark and Fritts Printers, New York, New York.

Houston, D. B. 1971. Ecosystems of national parks. *Science* 172:648–651.

Houston, D. B. 1982. *The northern Yellowstone elk herd.* Macmillan, New York, New York.

Houston, D. C. 1979. The adaptations of scavengers. Pages 263–286 in A. R. E. Sinclair and M. Norton-Griffiths, editors. *Serengeti: Dynamics of an ecosystem.* University of Chicago Press, Chicago, Illinois.

Hoye, T. T., and M. C. Forchhammer. 2008. Phenology of high-arctic arthropods: Effects of climate on spatial, seasonal, and inter-annual variation. *Advances in Ecological Research* 40:299–324.

Huerta, M. A., C. Whitlock, and J. Yale. 2009. Holocene vegetation-fire-climate linkages in northern Yellowstone National Park, USA. *Palaeogeography, Palaeoclimatology, Palaeoecology* 271:170–181.

Huff, D. E., and J. D. Varley. 1999. Natural regulation in Yellowstone National Park's northern range. *Ecological Applications* 9:17–29.

Huggard, D. J. 1993. Effect of snow depth on predation and scavenging by gray wolves. *Journal of Wildlife Management* 57:382–388.

Hughes, F. M. R. 1997. Floodplain biogeomorphology. *Progress in Physical Geography* 21:501–529.

Hülber, K., M. Winkler, and G. Grabherr. 2010. Intraseasonal climate and habitat-specific variability controls the flowering phenology of high alpine plant species. *Functional Ecology* 24:245–252.

Hultkrantz, A. 1957. The Indians in Yellowstone Park. *Annals of Wyoming* 29:124–149.

Huth, H. 1948. Yosemite: The story of an idea. *Sierra Club Bulletin* 33:72.

Independent Economic Analysis Board. 2010. *Economic risk associated with the potential establishment of zebra and quagga mussels in the Columbia River Basin.* Task Number 159, Document IE AB 2010–1. Northwest Power and Conservation Council, Portland, Oregon.

Interagency Bison Management Plan Agencies. 2011. *Adaptive management adjustments to the interagency bison management plan.* Copy on file at Yellowstone National Park, Mammoth, Wyoming, and at http://ibmp.info.

Interagency Grizzly Bear Study Team. 2009. *Yellowstone grizzly bear mortality and conflict reduction report.* Interagency Grizzly Bear Study Team, Northern Rocky Mountain Science Center, Montana State University, Bozeman, Montana.

Intergovernmental Panel on Climate Change. 2007a. *Climate change 2007: Impacts, adaptation, and vulnerability.* Contributions of Working Group II to the fourth assessment report of the Intergovernmental Panel on Climate Change. Cambridge University Press, Cambridge, UK.

Intergovernmental Panel on Climate Change. 2007b. Summary for policymakers. Pages 1–18 in S. Solomon, D. Qin, M. Manning, Z. Chen, M. Marquis, B. Averyt, M. Tignor, and H. L. Miller, editors. *Climate change 2007: The physical science basis.* Contributions of Working Group I to the fourth assessment report of the Intergovernmental Panel on Climate Change. Cambridge University Press, Cambridge, UK.

Interlandi, S. J. 2002. Nutrient-toxicant interactions in natural and constructed phytoplankton communities: Results of experiments in semi-continuous and batch culture. *Aquatic Toxicology* 61:35–51.

Irwin, L. L., and F. M. Hammond. 1985. Managing black bear habitats for food items in Wyoming. *Wildlife Society Bulletin* 13:477–483.

Jackson, S. T., and R. J. Hobbs. 2009. Ecological restoration in the light of ecological history. *Science* 325:567–569.

Jacobsen, S. K., M. C. Monroe, and S. Marynowski. 2001. Fire at the wildland interface: The influence of experience and mass media on public knowledge, attitudes, and behavioral interactions. *Wildlife Society Bulletin* 29:929–937.

Jacoby, M. E., G. V. Hilderbrand, C. Servheen, C. C. Schwartz, S. M. Arthur, T. A. Hanley, C. T. Robbins, and R. Michener. 1999. Trophic relations of brown and black bears in several western North American ecosystems. *Journal of Wildlife Management* 63:921–929.

Jaffe, R. 2001. *Winter wolf predation in an elk-bison system in Yellowstone National Park, Wyoming.* Thesis, Montana State University, Bozeman, Montana.

Jean, C., E. Shanahan, R. Daley, S. Podruzny, J. Canfield, G. DeNitto, D. Reinhart, and C. Schwartz. 2011. *Monitoring insect and disease in whitebark pine* (Pinus albicaulis) *in the greater Yellowstone ecosystem.* Yellowstone Science Conference 2010, Yellowstone National Park, Mammoth, Wyoming.

Jędrzejewski, W., B. Jędrzejewska, H. Okarma, and A. L. Ruprecht. 1992. Wolf predation and snow cover as mortality factors in the ungulate community of the Białowieza National Park, Poland. *Oecologia* 90:27–36.

Jędrzejewski, W., K. Schmidt, J. Theuerkauf, B. Jędrzejewska, N. Selva, K. Zub, and L. Szymura. 2002. Kill rates and predation by wolves on ungulate populations in Bialowieza Primeval Forest (Poland). *Ecology* 83:1341–1356.

Johnson, D. E. 1951. Biology of the elk calf, *Cervus canadensis nelsoni. Journal of Wildlife Management* 15:396–410.

Johnston, D. B., D. J. Cooper, and N. T. Hobbs. 2007. Elk browsing increases aboveground growth of water-stressed willows by modifying plant architecture. *Oecologia* 154:467–478.

Johnston, D. B., D. J. Cooper, and N. T. Hobbs. 2011. Relationships between groundwater use, water table, and recovery of willow on Yellowstone's northern range. *Ecosphere* 2:2–11.

Jolly, W. M., R. Nemani, and S. W. Running. 2005. A generalized, bioclimatic index to predict foliar phenology in response to climate. *Global Change Biology* 11:619–632.

Jonas, R. J. 1955. *A population and ecological study of beaver* (Castor canadensis) *of Yellowstone National Park.* Dissertation, University of Idaho, Moscow, Idaho.

Jones, J. D., J. J. Treanor, and R. L. Wallen. 2009. *Parturition in Yellowstone bison.* Report YCR-2009-01, National Park Service, Yellowstone National Park, Mammoth, Wyoming.

Jones, J. D., J. J. Treanor, R. L. Wallen, and P. J. White. 2010. Timing of parturition events in Yellowstone bison—Implications for bison conservation and brucellosis transmission risk to cattle. *Wildlife Biology* 16:333–339.

Jones, R. D., R. Andrascik, D. G. Carty, E. M. Colvard, R. Ewing, W. R. Gould, R. E. Gresswell, D. L. Mahony, T. Olliff, and S. E. Relyea. 1990. *Annual Project Technical Report for 1989*. Fishery and Aquatic Management Program, U.S. Fish and Wildlife Service, Yellowstone National Park, Mammoth, Wyoming.

Jonkel, C. J., and I. M. Cowan. 1971. The black bear in the spruce-fir forest. *Wildlife Monographs* 27.

Jorgenson, J. T., M. Festa-Bianchet, J.-M. Gaillard, and W. D. Wishart. 1997. Effects of age, sex, disease, and density on survival of bighorn sheep. *Ecology* 78:1019–1032.

Kaczensky, P., R. D. Hayes, and C. Promberger. 2005. Effect of raven *Corvus corax* scavenging on the kill rates of wolf *Canis lupus* packs. *Wildlife Biology* 11:101–108.

Kaeding, L. R. 2010. *Relative contributions of climate variation, lake trout predation, and other factors to the decline of Yellowstone Lake cutthroat trout during the three recent decades*. Dissertation, Montana State University, Bozeman, Montana.

Kaeding, L. R., G. D. Boltz, and D. G. Carty. 1996. Lake trout discovery in Yellowstone Lake threaten native cutthroat trout. *Fisheries* 21:16–20.

Kaplinski, M. A. 1991. *Geomorphology and geology of Yellowstone Lake, Yellowstone National Park, Wyoming*. Thesis, Northern Arizona University, Flagstaff, Arizona.

Karanth, K. U., and M. E. Sunquist. 1995. Prey selection by tiger, leopard and dhole in tropical forests. *Journal of Animal Ecology* 64:439–450.

Kauffman, M. J., J. F. Brodie, and E. S. Jules. 2010. Are wolves saving Yellowstone's aspen? A landscape test of a behaviorally mediated trophic cascade. *Ecology* 91:2742–2755.

Kauffman, M. J., N. Varley, D. W. Smith, D. R. Stahler, D. R. MacNulty, and M. S. Boyce. 2007. Landscape heterogeneity shapes predation in a newly restored predator-prey system. *Ecological Letters* 10:690–700.

Kaufmann, M. R., G. H. Aplet, M. Babler, W. L. Baker, B. Bentz, M. Harrington, B. C. Hawkes, L. Stroh Huckaby, M. J. Jenkins, D. M. Kashian, R. E. Keane, C. M. Kulakowski, J. Negron, J. Popp, W. H. Romme, T. Schoennagel, W. Shepperd, F. W. Smith, E. Kennedy Sutherland, D. Tinker, and T. T. Veblen. 2008. *The status of our scientific understanding of lodgepole pine and mountain pine beetles—A focus on forest ecology and fire behavior*. GFI Technical Report 2008–2, Nature Conservancy, Arlington, Virginia.

Kay, C. E. 1990. *Yellowstone's northern elk herd: A critical evaluation of the "natural regulation" paradigm*. Dissertation, Utah State University, Logan, Utah.

Kay, C. E. 1993. Aspen seedlings in recently burned areas of Grand Teton and Yellowstone National Parks. *Northwest Science* 67:94–104.

Kay, C. 1995. Aboriginal overkill and native burning: Implications for modern ecosystem management. *Western Journal of Applied Forestry* 10:121–126.

Kay, C. E. 1998. Are ecosystems structured from the top-down or bottom-up: A new look at an old debate. *Wildlife Society Bulletin* 26:484–498.

Kay, C. E. 2007. Testimony before the U.S. House of Representatives Subcommittee on National Parks, Forests, and Public Lands oversight hearing on Yellowstone National Park bison. March 20, 2007, Washington, D.C.

Kaya, C. 2000. Arctic grayling in Yellowstone: Status, management, and recent restoration efforts. *Yellowstone Science* 8:12–17.

Keane, R. E., and S. F. Arno. 1993. Rapid decline of whitebark pine in western Montana: Evidence from 20-year remeasurements. *Western Journal of Applied Forestry* 8:44–47.

Keating, K. A. 1982. *Population ecology of Rocky Mountain bighorn sheep in the upper Yellowstone River drainage, Montana/Wyoming.* Thesis, Montana State University, Bozeman, Montana.

Keating, K. 2002. *History of pronghorn population monitoring, research, and management in Yellowstone National Park.* U.S. Geological Survey, Northern Rocky Mountain Science Center, Bozeman, Montana.

Keiter, R. B., and M. S. Boyce. 1991. *The greater Yellowstone ecosystem: Redefining America's wilderness heritage.* Yale University Press, New Haven, Connecticut.

Keith, L. B. 1974. Some features of population dynamics in mammals. *Proceedings of the International Congress of Game Biologists* 11:17–58.

Keith, L. B. 1983. Population dynamics of wolves. Pages 66–77 in L. N. Carbyn, editor. *Wolves in Canada and Alaska.* Canadian Wildlife Service Report Series 45, Ottawa, Ontario, Canada.

Kendall, K. C., and J. Asebrook. 1998. The war against blister rust in Yellowstone National Park, 1945–1978. *George Wright Forum* 15:36–49.

Kendall, K. C., and R. E. Keane. 2001. Whitebark pine decline: Infection, mortality, and population trends. Pages 221–242 in D. F. Tomback, S. F. Arno, and R. E. Keane, editors. *Whitebark pine communities: Ecology and restoration.* Island Press, Washington, D.C.

Kendall, K. C., D. Schirokauer, E. Shanahan, R. Watt, D. Reinhart, R. Renkin, S. Cain, and G. Green. 1996. Whitebark pine health in the northern Rockies national park ecosystems: A preliminary report. *Nutcracker Notes* 7:16–18.

Kendall, K. C., D. Tyers, and D. Schirokauer. 1996. Preliminary status report on whitebark pine in Gallatin National Forest, Montana. *Nutcracker Notes* 7:19.

Kerans, B. L., M. F. Dybdahl, M. M. Gangloff, and J. E. Jannot. 2005. *Potamopyrgus antipodarum:* Distribution, density, and effects on native macroinvertebrate assemblages in the greater Yellowstone ecosystem. *Journal of the North American Benthological Society* 24:123–148.

Kie, J. G., R. T. Bowyer, and K. M. Stewart. 2003. Ungulates in western coniferous forests: Habitat relationships, population dynamics, and ecosystem processes. Pages 296–339 in C. J. Zabel and R. G. Anthony, editors. *Mammal community dynamics.* Cambridge University Press, Cambridge, UK.

Kilpatrick, A. M., C. M. Gillin, and P. Daszak. 2009. Wildlife-livestock conflict: The risk of pathogen transmission from bison to cattle outside Yellowstone National Park. *Journal of Applied Ecology* 46:476–485.

Kinloch, B. B., Jr., R. A. Sniezko, and G. E. Dupper. 2003. Origin and distribution of Cr2, a gene for resistance to white pine blister rust in natural populations of western white pine. *Phytopathology* 93:691–694.

Knapp, A. K., J. M. Blair, J. M. Briggs, S. L. Collins, D. C. Hartnett, L. C. Johnson, and E. G. Towne. 1999. The keystone role of bison in North American tallgrass prairie—Bison increase habitat heterogeneity and alter a broad array of plant, community, and ecosystem processes. *BioScience* 49:39–50.

Knapp, A. K., and M. D. Smith. 2001. Variation among biomes in temporal dynamics of aboveground primary production. *Science* 291:481–484.

Knight, D. H., S. W. Weaver, C. R. Starr, and W. H. Romme. 1979. Differential response of subalpine meadow vegetation to snow augmentation. *Journal of Range Management* 32:356–359.

Knight, J. C. 1975. *The limnology of the West Thumb of Yellowstone Lake, Yellowstone National Park, Wyoming.* Thesis, Montana State University, Bozeman, Montana.

Knight, R. R., B. M. Blanchard, and K. C. Kendall. 1982. *Yellowstone grizzly bear investigations: Annual report of the Interagency Study Team 1981.* U. S. Department of the Interior, National Park Service.

Knowles, N., M. D. Dettinger, and D. R. Cayan. 2006. Trends in snowfall versus rainfall in the western United States. *Journal of Climate* 19:4545–4559.

Koch, E. 1941. Big game in Montana from early historical records. *Journal of Wildlife Management* 5:357–370.

Koel, T. M., J. L. Arnold, P. E. Bigelow, P. D. Doepke, B. D. Ertel, and M. E. Ruhl. 2008. *Yellowstone fisheries and aquatic sciences annual report, 2007.* YCR-2008-02, National Park Service, Yellowstone National Park, Mammoth, Wyoming.

Koel, T. M., P. E. Bigelow, P. D. Doepke, B. D. Ertel, and D. L Mahony. 2005. Nonnative lake trout result in Yellowstone cutthroat trout decline and impacts to bears and anglers. *Fisheries* 30:10–19.

Koel, T. M., D. L. Mahony, K. L. Kinnan, C. Rasmussen, C. J. Hudson, S. Murcia, and B. L. Kerans. 2006. *Myxobolus cerebralis* in native cutthroat trout of the Yellowstone Lake ecosystem. *Journal of Aquatic Animal Health* 18:157–175.

Koteen, L. 2002. Climate change, whitebark pine, and grizzly bears in the greater Yellowstone ecosystem. Pages 343–414 in S. H. Schneider and T. L. Root, editors. *Wildlife responses to climate change: North American case studies.* Island Press, Washington, D.C.

Kruten, B., and E. Anderson. 1980. *Pleistocene mammals of North America.* Columbia University Press, New York, New York.

Kunkel, K. E., and D. H. Pletscher. 1999. Species-specific population dynamics of cervids in a multipredator system. *Journal of Wildlife Management* 63:1082–1093.

Lackey, R. T. 2007. Science, scientists, and policy advocacy. *Conservation Biology* 21:12–17.

Ladle, R. J., and L. Gillson. 2009. The (im)balance of nature: A public perception time-lag? *Public Understanding of Science* 18:229–242.

Lake, P. S. 2000. Disturbance, patchiness, and diversity in streams. *Journal of the North American Benthological Society* 19:573–592.

Landhäusser, S. M, D. Deshaies, and V. J. Lieffers. 2010. Disturbance facilitates rapid range expansion of aspen into higher elevations of the Rocky Mountains under a warming climate. *Journal of Biogeography* 37:68–76.

Landres, P. 2010. Let it be: A hands-off approach to preserving wildness in protected areas. Pages 88–105 in D. N. Cole and L. Yung, editors. *Beyond naturalness: Rethinking park and wilderness stewardship in an era of rapid change.* Island Press, Washington, D.C.

Langford, N. 1905. *Diary of the Washburn expedition to the Yellowstone and Firehole rivers in the year 1870.* F. J. Haynes Company, St. Paul, Minnesota.

Lanner, R. M. 1982. Adaptations of whitebark pine seed dispersal by Clark's nutcracker. *Canadian Journal of Forest Research* 12:391–402.

Lanner, R. M. 1996. *Made for each other: A symbiosis of birds and pines.* Oxford University Press, New York, New York.

Laundré, J. W. 1990. *The status, distribution, and management of mountain goats in the greater Yellowstone ecosystem.* Yellowstone National Park, Mammoth, Wyoming.

Laundré, J. W. 1992. Are wolves native to Yellowstone National Park? Pages 1–267 to 1–274 in J. D. Varley and W. G. Brewster, editors. *Wolves for Yellowstone? A report to the United States Congress,* Volume 4, *Research and analysis.* U.S. Department of the Interior, National Park Service, Yellowstone National Park, Mammoth, Wyoming.

Laundré, J. W., L. Hernández, and K. B. Altendorf. 2001. Wolves, elk, and bison: Reestablishing the "landscape of fear" in Yellowstone National Park, USA. *Canadian Journal of Zoology* 79:1401–1409.

Lee, K. N. 1993. *Compass and gyroscope: Integrating science and politics for the environment.* Island Press, Washington, D.C.

Legg, K. L. 1996. *Movements and habitat use of bighorn sheep along the upper Yellowstone river valley.* Thesis, Montana State University, Bozeman, Montana.

Leigleiter, C. J., R. L. Lawrence, M. A. Fonstad, W. A. Marcus, and R. Aspinall. 2003. Fluvial response a decade after wildfire in the northern Yellowstone ecosystem: A spatially explicit analysis. *Geomorphology* 54:119–136.

Lemke, T. O. 2004. Origin, expansion, and status of mountain goats in Yellowstone National Park. *Wildlife Society Bulletin* 32:532–541.

Lemke, T. 2009. *Spring 2009 antelope survey (Rock Creek to Big Creek).* Montana Fish, Wildlife, and Parks, Bozeman, Montana.

Lemke, T. O., J. A. Mack, and D. B. Houston. 1998. Winter range expansion by the northern Yellowstone elk herd. *Intermountain Journal of Sciences* 4:1–9.

Leonard, J. A., R. K. Wayne, and A. Cooper. 2000. Population genetics of Ice Age brown bears. *Proceedings of the National Academy of Sciences of the United States of America* 97:1651–1654.

Leopold, A. 1933. *Game management.* Charles Scribner's Sons, New York, New York.

Leopold, A. 1949. *A Sand County almanac.* Oxford University Press, New York, New York.

Leopold, A. S., S. A. Cain, D. M. Cottam, I. N. Gabrielson, and T. L. Kimball. 1963. Wildlife management in the national parks. *Transactions of the North American Wildlife and Natural Resources Conference* 28:28–45.

Lichtman, P. 1998. The politics of wildfire: Lessons from Yellowstone. *Journal of Forestry* 96:4–9.

Liley, S., and S. Creel. 2008. What best explains vigilance in elk: Characteristics of prey, predator, or the environment. *Behavioral Ecology* 19:245–254.

Linnell, J. D. C., and O. Strand. 2000. Interference iterations, co-existence and conservation of mammalian carnivores. *Diversity and Distribution* 6:169–176.

Littell, J., M. Elsner, and G. Mauger. 2011. *Regional climate and hydrologic change in the northern US Rockies and Pacific northwest: Internally consistent projections of future climate for resource management.* Climate Impacts Group, University of Washington, Seattle, Washington.

Lodge, D. M., and C. Hamlin, editors. 2006. *Religion and the new ecology: Environmental responsibility in a world in flux.* University of Notre Dame, South Bend, Indiana.

Loendorf, L. L., and N. M. Stone. 2006. *Mountain spirit: The Sheep Eater Indians of Yellowstone.* University of Utah Press, Salt Lake City, Utah.

Logan, J. A., W. W. Macfarlane, and L. Willcox. 2010. Whitebark pine vulnerability to climate-driven mountain pine beetle disturbance in the greater Yellowstone ecosystem. *Ecological Applications* 20:895–902.

Logan, J. A., and J. A. Powell. 2001. Ghost forests, global warming and the mountain pine beetle (Coleoptera: Scolytidae). *American Entomology* 47:160–172.

Logan, J. A., and J. A. Powell. 2003. Modelling mountain pine beetle phenological response to temperature. Pages 210–222 in T. L. Shore, J. E. Brooks, and J. E. Stone, editors. *Mountain pine beetle symposium: Challenges and solutions.* Report BC-X-399, Natural Resources Canada, Canadian Forest Service, Pacific Forestry Centre, Victoria, British Columbia, Canada.

Lovaas, A. L. 1970. *People and the Gallatin elk herd.* Montana Fish and Game Department, Helena, Montana.

Lubow, B. C., and B. L. Smith. 2004. Population dynamics of the Jackson elk herd. *Journal of Wildlife Management* 68:810–829.

Lung, M. A., and M. J. Childress. 2006. The influence of conspecifics and predation risk on the vigilance of elk *(Cervus elaphus)* in Yellowstone National Park. *Behavioral Ecology* 18:12–20.

Luyssaert, S., I. A. Janssens, M. Sulkava, D. Papale, A. J. Dolman, M. Reichstein, J. Hollmen, J. G. Martin, T. Suni, T. Vesala, D. Loustau, B. E. Law, and E. J. Moors. 2007. Photosynthesis drives anomalies in net carbon-exchange of pine forests at different latitudes. *Global Change Biology* 13:2110–2127.

Mace, R. D., and C. J. Jonkel. 1986. Local food habits of the grizzly bear in Montana. *International Conference on Bear Research and Management* 6:105–110.

MacFarlane, W. W., J. A. Logan, and W. R. Kern. 2009. *Using the landscape assessment system (LAS) to assess mountain pine beetle-caused mortality of whitebark pine, greater Yellowstone ecosystem, 2009.* Greater Yellowstone Coordinating Committee, Whitebark Pine Subcommittee, Jackson, Wyoming.

MacNulty, D. R., L. D. Mech, and D. W. Smith. 2007. A proposed ethogram of large-carnivore predatory behavior, exemplified by the wolf. *Journal of Mammalogy* 88:595–605.

Magnuson, J. J., D. M. Robertson, B. J. Benson, R. H. Wynne, D. M. Livingstone, T. Arai, R. A. Assel, R. G. Barry, V. Card, K. M. Stewart, and V. S. Vuglinski. 2000. Historical trends in lake and river ice cover in the northern hemisphere. *Science* 289:1743–1746.

Mahalovich, M. F., K. E. Burr, and D. L. Foushee. 2006. *Whitebark pine germination, rust resistance, and cold hardiness among seed sources in the inland northwest: Planting strategies for restoration.* Proceedings RMRS-P-43, U.S. Department of Agriculture, National Forest Service.

Mahalovich, M. F., and G. A. Dickerson. 2004. Whitebark pine genetic restoration program for the intermountain west (United States). Pages 181–187 in R. A. Sniezko, S. Samman, S. E. Schlarbaum, E. Scott, and H. B. Kriebel, editors. *Breeding and genetic resources of five-needle pines: Growth, adaptability and pest resistance.* RMRS-P-32, U.S. Department of Agriculture, National Forest Service, Rocky Mountain Research Station, Fort Collins, Colorado.

Malison, R. L., and C. V. Baxter. 2010. The "fire pulse": Wildfire stimulates flux of aquatic prey to terrestrial habitats driving increases in riparian consumers. *Canadian Journal of Fisheries and Aquatic Sciences* 67:570–579.

Maloy, O. C. 1997. White pine blister rust control in North America: A case history. *Annual Review of Phytopathology* 35:87–109.

Mao, J. S., M. S. Boyce, D. W. Smith, F. J. Singer, D. J. Vales, J. M. Vore, and E. H. Merrill. 2005. Habitat selection by elk before and after wolf reintroduction in Yellowstone National Park. *Journal of Wildlife Management* 69:1691–1707.

Marcus, W. A., G. A. Meyer, and D. R. Nimmo. 2001. Geomorphic control of persistent mine impacts in a Yellowstone Park stream and implications for the recovery of fluvial systems. *Geology* 29:355–358.

Marcus, W. A., J. Rasmussen, and M. A. Fonstad. 2011. Response of the fluvial wood system to fire and floods in northern Yellowstone. *Annals of the Association of American Geographers* 101:21–44.

Maschinski, J., and T. G. Whitham. 1989. The continuum of plant responses to herbivory: The influence of plant association, nutrient availability, and timing. *American Naturalist* 134:1–19.

Matthews, A. 1906. The word park in the United States. *Publications of the Colonial Society of Massachusetts* 7:373–399.

Mattson, D. 1997. Use of ungulates by Yellowstone grizzly bears *Ursus arctos*. *Biological Conservation* 81:161–177.

Mattson, D. J., B. M. Blanchard, and R. R. Knight. 1991. Food habits of Yellowstone grizzly bears, 1977–1987. *Canadian Journal of Zoology* 69:1619–1629.

Mattson, D. J., B. M. Blanchard, and R. R. Knight. 1992. Yellowstone grizzly bear mortality, human habituation, and whitebark pine seed crops. *Journal of Wildlife Management* 56:432–442.

Mattson, D. J., S. Herrero, and T. Merrill. 2005. Are black bears a factor in the restoration of North American grizzly bear populations? *Ursus* 16:11–30.

Mattson, D. J., K. C. Kendall, and D. P. Reinhart. 2001. Whitebark pine, grizzly bears, and red squirrels. Pages 121–136 in D. F. Tomback, S. F. Arno, and R. E. Keane, editors. *Whitebark pine communities: Ecology and restoration.* Island Press, Washington, D.C.

Mattson, D. J., and D. P. Reinhart. 1995. Influences of cutthroat trout (*Oncorhynchus clarki*) on behavior and reproduction of Yellowstone grizzly bears (*Ursus arctos*), 1975–1989. *Canadian Journal of Zoology* 73:2072–2079.

Mattson D. J., and D. P. Reinhart. 1997. Excavation of red squirrel middens by Yellowstone grizzly bears. *Journal of Applied Ecology* 34:926–940.

Mattson, D. J., R. G. Wright, K. C. Kendall, and C. J. Martinka. 1995. Grizzly bears. Pages 103–105 in E. T. LaRoe, G. S. Farris, C. E. Puckett, P. D. Doran, and M. J. Mac, editors. *Our living resources: A report to the nation on the distribution, abundance, and health of U.S. plants, animals, and ecosystems.* U.S. Department of the Interior, National Biological Service, Washington, D.C.

May, B. E., S. E. Albeke, and T. Horton. 2007. *Range-wide status of Yellowstone cutthroat trout* (Oncorhynchus clarkii bouvieri): *2006.* Montana Department of Fish, Wildlife and Parks, Helena, Montana.

McCaughey, W. W., and D. F. Tomback. 2001. The natural regeneration process. Pages 105–120 in D. F. Tomback, S. F. Arno, and R. E. Keane, editors. *Whitebark pine communities: Ecology and restoration.* Island Press, Washington, D.C.

McEneaney, T. 2002. Piscivorous birds of Yellowstone Lake: Their history, ecology, and status. Pages 121–134 in R. J. Anderson and D. Harmon, editors. *Yellowstone Lake: Hotbed of chaos or reservoir of resilience? Proceedings of the 6th Biennial Scientific Conference on the Greater Yellowstone Ecosystem.* Yellowstone Center for Resources and the George Wright Society, Yellowstone National Park, Wyoming, and Hancock, Michigan.

McEneaney, T. 2006. *Yellowstone bird report, 2005.* YCR-2006-2, National Park Service, Yellowstone National Park, Mammoth, Wyoming.

McKinney, S. T., C. E. Fiedler, and D. F. Tomback. 2009. Invasive pathogen threatens bird-pine mutualism: Implications for sustaining a high-elevation ecosystem. *Ecological Applications* 19:597–607.

McKinney, S. T., and D. F. Tomback. 2007. The influence of white pine blister rust on seed dispersal in whitebark pine. *Canadian Journal of Forest Research* 37:1044–1057.

McLellan, B. N. 1994. Density-dependent population regulation of brown bears. Pages 15–24 in M. Taylor, editor. *Density-dependent population regulation of black, brown, and polar bears.* International Conference on Bear Research and Management, Monograph Series 3.

McMahon, R., S. Kumar, M. Sytsma, R. Hall, D. Britton, S. Spaulding, E. Willimas, A. Farag, S. O'Ney, and J. Carpurso. 2009. *AIS inventory and monitoring framework for the GYA.* GYCC Aquatic Invasive Species Cooperative. http://www.cleaninspectdry.com/internal.asp.

McMenamin, S. K., E. A. Hadly, and C. K. Wright. 2008. Climatic change and wetland desiccation cause amphibian decline in Yellowstone National Park. *Proceedings of the National Academy of Sciences of the United States of America* 105:16988–16993.

McNaughton, S. J. 1976. Serengeti migratory wildebeest: Facilitation of energy flow by grazing. *Science* 278:1798–1800.

McNaughton, S. J. 1979. Grazing as an optimization process: Grass-ungulate relationships in the Serengeti. *American Naturalist* 113:691–703.

McNaughton, S. J. 1985. Ecology of a grazing ecosystem: The Serengeti. *Ecological Monographs* 55:259–295.

McNaughton, S. J. 1990. Mineral nutrition and seasonal movements of African migratory ungulates. *Nature* 345:613–615.

McNaughton, S. J., F. F. Banyikwa, and M. M. McNaughton. 1997. Promotion of the cycling of diet enhancing nutrients by African grazers. *Science* 278:1798–1800.

McNaughton, S. J., R. W. Ruess, and S. W. Seagle 1988. Large mammals and process dynamics in African ecosystems. *BioScience* 38:794–800.

McWethy, D. B., S. T. Gray, P. E. Higuera, J. S. Littell, G. T. Pederson, A. Ray, and C. Whitlock. 2010. *Climate and terrestrial ecosystem change in the U.S. Rocky Mountains and upper Columbia basin: Historic and future perspectives for natural resource management.* Natural Resource Report NPS/GRYN/NRR-2010/260. National Park Service, Fort Collins, Colorado.

Meagher, M. 1973. *The bison of Yellowstone National Park.* Science Monographs 1, National Park Service, Government Printing Office, Washington, D.C.

Meagher, M. 1989a. Evaluation of boundary control for bison of Yellowstone National Park. *Wildlife Society Bulletin* 17:15–19.

Meagher, M. 1989b. Range expansion by bison of Yellowstone National Park. *Journal of Mammalogy* 70:670–675.

Meagher, M. 1993. *Winter recreation-induced changes in bison numbers and distribution in Yellowstone National Park.* National Park Service, Yellowstone National Park, Wyoming.

Meagher, M. 2008. Bears in transition, 1959–1970s. *Yellowstone Science* 16:5–12.

Meagher, M. M., and D. B. Houston. 1998. *Yellowstone and the biology of time: Photographs across a century.* University of Oklahoma Press, Norman, Oklahoma.

Meagher, M., and M. E. Meyer. 1994. On the origin of brucellosis in bison of Yellowstone National Park: A review. *Conservation Biology* 8:645–653.

Meagher, M., and J. R. Phillips. 1983. Restoration of natural populations of grizzly and black bears in Yellowstone National Park. *International Conference on Bear Research and Management* 5:152–158.

Meagher, M., W. J. Quinn, and L. Stackhouse. 1992. Chlamydial-caused infectious keratoconjunctivitis in bighorn sheep of Yellowstone National Park. *Journal of Wildlife Diseases* 28:171–176.

Mealey, S. P. 1975. *The natural food habits of free ranging grizzly bears in Yellowstone National Park.* Thesis, Montana State University, Bozeman, Montana.

Mealey, S. P. 1980. The natural food habits of grizzly bears in Yellowstone National Park, 1973–1974. Pages 281–292 in C. J. Martinka and K. L. MacArthur, editors. *International Conference on Bear Research and Management,* Volume 4, Kalispell, Montana.

Mech, L. D. 1970. *The wolf: The ecology and behavior of an endangered species.* University of Minnesota Press, Minneapolis, Minnesota.

Mech, L. D., L. G. Adams, T. J. Meier, J. W. Burch, and B. W. Dale. 1998. *The wolves of Denali.* University of Minnesota Press, Minneapolis, Minnesota.

Mech, L. D., D. W. Smith, K. M. Murphy, and D. R. MacNulty. 2001. Winter severity and wolf predation on a formerly wolf-free elk herd. *Journal of Wildlife Management* 65:998–1003.

Meentemeyer, R. K., and D. R. Butler. 1999. Hydrogeomorphic effects of beaver dams in Glacier National Park, Montana. *Physical Geography* 20:436–446.

Merkle, J. A., D. R. Stahler, and D. W. Smith. 2009. Interference competition between gray wolves and coyotes in Yellowstone National Park. *Canadian Journal of Zoology* 87:56–63.

Merrill, E. H., and M. S. Boyce. 1991. Summer range and elk population dynamics in Yellowstone National Park. Pages 263–274 in R. B. Keiter and M. S. Boyce, editors. *The greater Yellowstone ecosystem: Redefining America's wilderness heritage.* Yale University Press, New Haven, Connecticut.

Merrill, E. H., M. K. Bramblebrodahl, R. W. Marrs, and M. S. Boyce. 1993. Estimation of green herbaceous phytomass from Landsat MSS data in Yellowstone National Park. *Journal of Range Management* 46:151–157.

Messier, F. 1994. Ungulate population models with predation: A case study with the North American moose. *Ecology* 75:478–488.

Messier, F., and M. Crête. 1985. Moose-wolf dynamics and the natural regulation of moose populations. *Oecologia* 65:503–512.

Metz, M. C. 2010. *Seasonal patterns in foraging and predation of gray wolves in Yellowstone National Park.* Michigan Technological University, Houghton, Michigan.

Metz, M. C., D. W. Smith, J. A. Vucetich, D. R. Stahler, and R. O. Peterson. 2012. Seasonal patterns of predation for gray wolves in the multi-prey system of Yellowstone National Park. *Journal of Animal Ecology.* doi:10.1111/j.1365-2656.2011.01945.x.

Meyer, G. A., S. Wells, R. Balling, and A. J. Jull. 1992. Response of alluvial systems to fire and climate change in Yellowstone National Park. *Nature* 357:147–149.

Meyer, G. A., S. G. Wells, and A. J. T. Jull. 1995. Fire and alluvial chronology in Yellowstone National Park—Climatic and intrinsic controls on Holocene geomorphic processes. *Geological Society of America Bulletin* 107:1211–1230.

Milius, S. 2010. Climate change may favor couch potato elk. *Science News* 178(2).

Miller, C. R., and L. P. Waits. 2003. The history of effective population size and genetic diversity in the Yellowstone grizzly *(Ursus arctos):* Implications for conservation. *Proceedings of the National Academy of Sciences of the United States of America* 100:4334–4339.

Miller, M. L. 2005. The wolf returns to Idaho. *Bugle* 22:79–83.

Miller-Rushing, A. J., T. Katsuki, R. B. Primack, Y. Ishii, S. D. Lee, and H. Higuchi. 2007. Impact of global warming on a group of related species and their hybrids: Cherry tree (Rosaceae) flowering at Mt. Takao, Japan. *American Journal of Botany* 94:1470–1478.

Millspaugh, S. H., and C. Whitlock. 1995. A 750-year fire history based on lake sediment records in central Yellowstone National Park, USA. *The Holocene* 5:283–292.

Millspaugh, S. H., C. Whitlock, and P. J. Bartlein. 2000. Variations in fire frequency and climate over the past 17,000 yr in central Yellowstone National Park. *Geology* 28:211–214.

Millspaugh, S. H., C. Whitlock, and P. J. Bartlein. 2004. Postglacial fire, vegetation, and climate history of the Yellowstone-Lamar and Central Plateau Provinces, Yellowstone National Park. Pages 10–28 in L. L. Wallace, editor. *After the fires: The ecology of change in Yellowstone National Park.* Yale University Press, New Haven, Connecticut.

Minshall, G. W., C. T. Robinson, and D. E. Lawrence. 1997. Postfire responses of lotic ecosystems in Yellowstone National Park, U.S.A. *Canadian Journal of Fisheries and Aquatic Sciences* 54:2509–2525.

Minshall, G. W., C. T. Robinson, and T. V. Royer. 1998. Stream ecosystem responses to the 1988 wildfires. *Yellowstone Science* 6:15–22.

Minshall, G. W., T. V. Royer, and C. T. Robinson. 2001. Response of the Cache Creek macroinvertebrates during the first 10 years following disturbance by the 1988 Yellowstone wildfires. *Canadian Journal of Fisheries and Aquatic Sciences* 58:1077–1088.

Mock, C. J. 1996. Climatic controls and spatial variations of precipitation in the western United States. *Journal of Climate* 9:1111–1125.

Molvar, E. M., and T. R. Bowyer. 1994. Costs and benefits of group living in a recently social ungulate: The Alaskan moose. *Journal of Mammalogy* 75:621–630.

Moore, H. L., O. B. Cope, and R. E. Beckwith. 1952. *Yellowstone Lake creel censuses, 1950–1951.* Special Scientific Report—Fisheries Number 81, U.S. Fish and Wildlife Service, Yellowstone National Park, Mammoth, Wyoming.

Mote, P. W., A. F. Hamlet, M. P. Clark, and D. P. Lettenmaier. 2005. Declining mountain snowpack in western North America. *Bulletin of the American Meteorological Society* 86:39–44.

Munro, A. R., T. E. McMahon, and J. R. Ruzycki. 2005. Natural chemical markers identify source and date of introduction of an exotic species: Lake trout *(Salvelinus namaycush)* in Yellowstone Lake. *Canadian Journal of Fisheries and Aquatic Sciences* 62:79–87.

Murdoch, W. W. 1994. Population regulation in theory and practice. *Ecology* 75:271–287.

Murie, A. 1940. *Ecology of the coyote in the Yellowstone.* Fauna of the National Parks No. 4, U.S. Government Printing Office, Washington, D.C.

Murie, O. J. 1951. *The elk of North America.* Stackpole Company and the Wildlife Management Institute, Harrisburg, Pennsylvania.

Murphy, K. M. 1998. *Ecology of the cougar in the cougar* (Puma concolor) *in the northern Yellowstone ecosystem: Interactions with prey, bears, and humans.* Dissertation, University of Idaho, Moscow, Idaho.

Murphy, K., and T. K. Ruth. 2010. Diet and prey selection of a perfect predator. Pages 118–137 in M. Hornocker and S. Negri, editors. *Cougar: Ecology and conservation.* University of Chicago Press, Chicago, Illinois.

Murray, M. P., S. C. Bunting, and P. Morgan. 1995. *Whitebark pine and fire suppression in small wilderness areas.* General Technical Report INT-GTR-320, U.S. Department of Agriculture, National Forest Service, Washington, D.C.

Mysterud, A., and R. Langvatn. 2001. Plant phenology, migration and geographical variation in body weight of a large herbivore: The effect of a variable topography. *Journal of Animal Ecology* 70:915–923.

Nabokov, P., and L. Loendorf. 2004. *Restoring a presence: American Indians and Yellowstone National Park.* University of Oklahoma Press, Norman, Oklahoma.

Naiman, R. J., and H. Decamps. 1997. The ecology of interfaces: Riparian zones. *Annual Review of Ecology and Systematics* 28:621–658.

National Park Service. 1997. *Yellowstone's northern range: Complexity & change in a wildland ecosystem.* National Park Service, Yellowstone National Park, Mammoth, Wyoming.

National Park Service. 2006. *Management policies 2006.* U.S. Department of the Interior, Washington, D.C.

National Park Service. 2010. *Restoration of lands on the Gardiner basin winter range.* Greater Yellowstone Science Learning Center, http://www.greater yellowstonescience.org/topics/biological/mammals/pronghorn/projects /gardiner.

National Park Service. 2011. *A call to action: Preparing for a second century of stewardship and engagement.* U.S. Department of the Interior, Washington, D.C.

National Research Council. 2002. *Ecological dynamics on Yellowstone's northern range.* National Academies Press, Washington, D.C.

Nelson, M. E., and L. D. Mech. 1986. Relationship between snow depth and gray wolf predation on white-tailed deer. *Journal of Wildlife Management* 50:471–474.

Nelson, R. A., G. E. Folk, Jr., E. W. Pfeiffer, J. J. Craighead, C. J. Jonkel, and D. L. Steiger. 1983. Behavior, biochemistry, and hibernation in black, grizzly, and polar bears. *International Conference for Bear Research and Management* 5:284–290.

Newcomb, M. 2003. *White pine blister rust, whitebark pine, and Ribes species in the greater Yellowstone area.* Thesis, University of Montana, Missoula, Montana.

Newman, W. B., and F. G. R. Watson. 2009. The central Yellowstone landscape: Terrain, geology, climate, vegetation. Pages 17–35 in R. A. Garrott, P. J. White, and F. G. R. Watson, editors. *The ecology of large mammals in central Yellowstone: Sixteen years of integrated field studies.* Elsevier, San Diego, California.

Norris, P. W. 1878. *Report upon the Yellowstone National Park to the Secretary of the Interior by P. W. Norris, Superintendent, for the year 1877.* U.S. Government Printing Office, Washington, D.C.

Oberbauer, S. F., G. Starr, and E. W. Pop. 1998. Effects of extended growing season and soil warming on carbon dioxide and methane exchange of tussock tundra in Alaska. *Journal of Geophysical Research* 103:29075–29082.

O'Gara, B. W. 1968. *A study of the reproductive cycle of the female pronghorn* (Antilocapra americana *Ord.).* Dissertation, University of Montana, Missoula, Montana.

Olff, H., and M. E. Ritchie. 1998. Effects of herbivores on grassland plant diversity. *Trends in Ecology and Evolution* 13:261–265.

Olliff, T., R. Renkin, C. McClure, P. Miller, D. Price, D. Reinhart, and J. Whipple. 2001. Managing a complex exotic vegetation program in Yellowstone National Park. *Western North American Naturalist* 61:347–358.

Ornstein, R., and P. R. Ehrlich. 1989. *New world. New mind.* Doubleday, New York, New York.

Ostovar, K. 1998. *Impacts of human activity on bighorn sheep in Yellowstone National Park.* Thesis, Montana State University, Bozeman, Montana.

Ostovar, K., and L. R. Irby. 1998. High predator densities in Yellowstone National Park may limit recovery of bighorn sheep populations from the 1981 *Chlamydial*-caused die off. *Biennial Symposium of the Wild Sheep and Goat Council* 11:96–103.

Owen-Smith, N. 1983. Dispersal and the dynamics of large herbivore populations in enclosed areas: Implications for management. Page 127–143 in R. N. Owen-Smith, editor. *Management of large mammals in African conservation areas.* Haum Educational Publishers, Pretoria, South Africa.

Oyler-McCance, S. J., F. A. Ransler, L. K. Berkman, and T. W. Quinn. 2007. A rangewide population genetic study of trumpeter swans. *Conservation Genetics* 8:1339–1353.

Paetkau, D., L. P. Waits, P. Waser, L. Clarkson, L. Craighead, E. Vyse, R. Ward, and C. Stobeck. 1998. Variation in genetic diversity across the range of North American brown bears. *Conservation Biology* 12:418–429.

Paine, R. T., and S. A. Levin. 1981. Intertidal landscapes: Disturbance and the dynamics of pattern. *Ecological Monographs* 51:145–178.

Palomares, F., and T. M. Caro. 1999. Interspecific killing among mammalian carnivores. *American Naturalist* 153:492–508.

Paquet, P. C. 1991. Scent-marking behavior of sympatric wolves *(Canis lupus)* and coyotes *(C. latrans)* in Riding Mountain National Park. *Canadian Journal of Zoology* 69:1721–1727.

Parmenter, A. W., A. Hansen, R. E. Kennedy, W. Kohen, U. Langner, R. Lawrence, B. Maxwell, A. Gallant, and R. Aspinall. 2003. Land use and land cover change in the greater Yellowstone ecosystem: 1975–1995. *Ecological Applications* 13:687–703.

Parsons, A. J., and P. D. Penning. 1988. The effect of the duration of regrowth on photosynthesis, leaf death, and the average rate of regrowth in a rotationally grazed sward. *Grass Forage Science* 43:15–27.

Parsons, D. J. 2000. The challenge of restoring natural fire to wilderness. Pages 276–282 in D. N. Cole, S. F. McCool, W. T. Borrie, and J. O'Loughlin, compilers. *Wilderness science in a time of change conference. Volume 5: Wilderness ecosystems, threats, and management, May 23–27, 1999, Missoula, Montana.* Proceedings RMRS-P-15-VOL-5, U.S. Department of Agriculture, Forest Service, Rocky Mountain Research Station, Ogden, Utah.

Pastor, J., and Y. Cohen. 1997. Herbivores, the functional diversity of plants, species, and the cycling of nutrients of ecosystems. *Theoretical Population Biology* 51:165–179.

Pastor, J., B. Dewey, R. J. Naiman, P. F. McInnes, and Y. Cohen. 1993. Moose browsing and soil fertility in the boreal forests of Isle Royale National Park. *Ecology* 74:467–480.

Pastor, J. R., R. Moen, and Y. Cohen. 1997. Spatial heterogeneities, carrying capacity, and feedbacks in animal-landscape interactions. *Journal of Mammalogy* 78:1040–1052.

Patten, D. T. 1991. Defining the greater Yellowstone ecosystem. Pages 19–25 in R. B. Keiter and M. S. Boyce, editors. *The greater Yellowstone ecosystem: Redefining America's wilderness heritage.* Yale University Press, New Haven, Connecticut.

Pederson, G. T., L. J. Graumlich, D. B. Fagre, T. Kipfer, and C. C. Muhlfeld. 2010. A century of climate and ecosystem change in western Montana: What do temperature trends portend? *Climatic Change* 98:133–154.

Pengelly, W. L. 1963. Thunder on the Yellowstone. *Naturalist* 14:18–25.

Persico, L., and G. Meyer. 2009. Holocene beaver damming, fluvial geomorphology, and climate in Yellowstone National Park, Wyoming. *Quaternary Research* 71:340–353.

Person, B. T., M. P. Herzog, R. W. Ruess, J. S. Sedinger, R. M. Anthony, and C. A. Babcock. 2003. Feedback dynamics of grazing lawns: Coupling vegetation change with animal growth. *Oecologia* 135:583–592.

Peterson, G. D., G. S. Cumming, and S. R. Carpenter. 2003. Scenario planning: A tool for conservation in an uncertain world. *Conservation Biology* 17:358–366.

Peterson, R. O. 1977. *Wolf ecology and prey relationships on Isle Royale.* National Park Service Scientific Monograph Series 11, Washington, D.C.

Peterson, R. O. 1995. *The wolves of Isle Royale.* Willow Creek Press, Minocqua, Wisconsin.

Peterson, R. O., and R. E. Page. 1983. Wolf-moose fluctuations at Isle Royale National Park, Michigan, U.S.A. *Acta Zoologica Fennica* 174:251–253.

Pettorelli, N., F. Pelletier, A. von Hardenberg, M. Festa-Biachet, and S. D. Cote. 2007. Early onset of vegetation growth vs. rapid green-up: Impacts on juvenile mountain ungulates. *Ecology* 88:381–390.

Pettorelli, N., R. B. Weladji, O. Holand, A. Mysterud, H. Breie, and N.-C. Stenseth. 2005. The relative role of winter and spring conditions: Linking climate and landscape-scale plant phenology to alpine reindeer body mass. *Biology Letters* 1:24–26.

Pianka, E. R. 1978. *Evolutionary ecology.* Harper and Row, New York, New York.

Pickett, S. T. A., and P. S. White. 1985. *The ecology of natural disturbance and patch dynamics.* Academic Press, Orlando, Florida.

Pielmeier, C., and M. Schneebeli. 2003. Stratigraphy and changes in hardness of snow measured by hand, ramsonde and snow micro penetrometer: A comparison with planar sections. *Cold Regions Science and Technology* 37:393–405.

Pierce, J. L., G. A. Meyer, and A. J. T. Jull. 2004. Fire-induced erosion and millennial scale climate change in northern ponderosa pine forests. *Nature* 432:87–90.

Pierce, K. L. 1979. *History and dynamics of glaciation in the northern Yellowstone National Park area.* U.S. Geological Survey Professional Paper 729 F:91.

Plumb, G. E. 1991. *Foraging ecology of bison and cattle on a northern mixed prairie.* University of Wyoming, Laramie, Wyoming.

Plumb, G. E., and R. Sucec. 2006. A bison conservation history in the U.S. National Parks. *Journal of the West* 45:22–28.

Plumb, G. E., P. J. White, M. B. Coughenour, and R. L. Wallen. 2009. Carrying capacity, migration, and dispersal in Yellowstone bison. *Biological Conservation* 142:2377–2387.

Polis, G. A., W. B. Anderson, and R. D. Holt. 1997. Toward an integration of landscape and food web ecology: The dynamics of spatially subsidized food webs. *Annual Review of Ecology and Systematics* 28:289–316.

Polis, G. A., and R. D. Holt. 1992. Intraguild predation: The dynamics of complex trophic interactions. *Trends in Ecology and Evolution* 7:151–154.

Polis, G. A., C. A. Myers, and R. D. Holt. 1989. The ecology and evolution of intraguild predation: Potential competitors that eat each other. *Annual Review of Ecology and Systematics* 20:297–330.

Post, D. M., M. L. Pace, and N. G. Hairston. 2000. Ecosystem size determines food-chain length in lakes. *Nature* 405:1047–1049.

Post, E., and M. C. Forchhammer. 2008. Climate change reduces reproductive success of an Arctic herbivore through trophic mismatch. *Philosophical Transactions of the Royal Society. Series B, Biological Sciences* 363:2369–2375.

Post, E. S., C. Pedersen, C. C. Wilmers, and M. C. Forchhammer. 2008a. Phenological sequences reveal aggregate life history response to climatic warming. *Ecology* 89:363–370.

Post, E., C. Pedersen, C. C. Wilmers, and M. C. Forchhammer. 2008b. Warming, plant phenology and the spatial dimension of trophic mismatch for large herbivores. *Proceedings of the Royal Society. B, Biological Sciences* 275:2005–2013.

Powers, J., and M. Wild. 2004. *A National Park Service manager's reference notebook to understanding chronic wasting disease.* National Park Service, Biological Resource Management Division, Fort Collins, Colorado.

Preisser, E. L, and D. I. Bolnick. 2008. When predators don't eat their prey: Nonconsumptive predator effects on prey dynamics. *Ecology* 89:2414–2415.

Primack, D., C. Imbres, R. B. Primack, A. J. Miller-Rushing, and P. Del Tredici. 2004. Herbarium specimens demonstrate earlier flowering times in response to warming in Boston. *American Journal of Botany* 91:1260–1264.

Pritchard, J. 1999. *Preserving Yellowstone's natural conditions, science and the perception of nature.* University of Nebraska Press, Lincoln, Nebraska.

Proffitt, K., J. Grigg, R. A. Garrott, and K. L. Hamlin. 2009. Elk behavioral responses to predation risk: Contrasting effects of wolves and human hunters. *Journal of Wildlife Management* 73:345–356.

Proffitt, K. M., T. P. McEneaney, P. J. White, and R. A. Garrott. 2009. Trumpeter swan abundance and growth rates in Yellowstone National Park. *Journal of Wildlife Management* 73:728–736.

Proffitt, K. M., T. P. McEneaney, P. J. White, and R. A. Garrott. 2010. Productivity and fledging success of trumpeter swans in Yellowstone National Park, 1987–2007. *Waterbirds* 33:341–348.

Proffitt, K. M., P. J. White, and R. A. Garrott. 2010. Spatio-temporal overlap between Yellowstone bison and elk—Implications for wolf restoration and other factors for brucellosis transmission risk. *Journal of Applied Ecology* 47:281–289.

Pulliam, H. R. 1973. On the advantage of flocking. *Journal of Theoretical Biology* 38:419–422.

Pulliam, H. R., and T. Caraco. 1984. Living in groups: Is there an optimal group size? Pages 122–147 in J. R. Krebs and N. B. Davies, editors. *Behavioural ecology: An evolutionary approach.* Blackwell Scientific, Oxford, UK.

Raffa, K. F., B. H. Aukema, B. J. Bentz, A. L. Carroll, J. A. Hicke, M. G. Turner, and W. H. Romme. 2008. Cross-scale drivers of natural disturbances prone to anthropogenic amplification: Dynamics of biome-wide bark beetle eruptions. *BioScience* 58:501–517.

Ray, A., J. Barsugli, K. Averyt, K. Wolter, M. Hoerling, N. Doesken, B. Udall, and R. S. Webb. 2008. *Climate change in Colorado: A synthesis to support water resources management and adaptation.* Report by the Western Water Assessment for the Colorado Water Conservation Board. University of Colorado, Boulder, Colorado.

Ray, A. J., J. J. Barsugli, K. E. Wolter, and J. K. Wischeid. 2010. *Rapid-response climate assessment to inform the FWS status review of the American pike.* U.S. Fish and Wildlife Service and NOAA Earth System Research Laboratory, Boulder, Colorado.

Read, M. A. 1958. *Silvical characteristics of plains cottonwood.* Rocky Mountain Forest and Range Experiment Station, Fort Collins, Colorado.

Reeves, G. H., L. E. Benda, K. M. Burnett, P. A. Bisson, and J. R. Sedell. 1995. A disturbance-based ecosystem approach to maintaining and restoring freshwater habitats of evolutionarily significant units of anadromous salmonids in the Pacific Northwest. *American Fisheries Society Symposium* 17:334–349.

Regonda, S. K., B. Rajagopalan, M. Clark, and J. Pitlick. 2005. Seasonal cycle shifts in hydroclimatology over the western United States. *Journal of Climate* 18:372–384.

Reiger, J. 2001. *American sportsmen and the origins of conservation.* Oregon State University Press, Corvallis, Oregon.

Reinertsen, H., A. Jensen, J. I. Koksvik, A. Langeland, and Y. Olsen. 1990. Effects of fish removal on the limnetic ecosystem of a eutrophic lake. *Canadian Journal of Fisheries and Aquatic Sciences* 47:166–173.

Reinhart, D. P., M. A. Haroldson, D. J. Mattson, and K. A. Gunther. 2001. Effects of exotic species on Yellowstone's grizzly bears. *Western North American Naturalist* 61:277–288.

Reinhart, D. P., and D. J. Mattson. 1990a. Bear use of cutthroat trout spawning streams in Yellowstone National Park. *International Conference on Bear Research and Management* 8:343–350.

Reinhart, D. P., and D. J. Mattson. 1990b. Red squirrels in the whitebark pine zone. Pages 256–263 in W. C. Schmidt and K. J. McDonald, compilers. *Proceedings of a symposium on whitebark pine ecosystems: Ecology and management of a high-mountain resource.* General Technical Report INT-270, U.S.

Department of Agriculture, National Forest Service, Intermountain Research Station, Ogden, Utah.

Renkin, R. A., and D. G. Despain. 1992. Fuel moisture, forest type, and lightning-caused fire in Yellowstone National Park. *Canadian Journal of Forest Research* 22:37–45.

Resh, V. H., A. V. Brown, A. P. Covich, M. E. Gurtz, H. W. Li, G. W. Minshall, S. R. Reice, A. L. Sheldon, J. B. Wallace, and R. C. Wissmar. 1988. The role of disturbance in stream ecology. *Journal of the North American Benthological Society* 7:433–455.

Rhyan, J. C., K. Aune, T. Roffe, D. Ewalt, S. Hennager, T. Gidlewski, S. Olsen, and R. Clarke. 2009. Pathogenesis and epidemiology of brucellosis in Yellowstone bison: Serologic and culture results from adult females and their progeny. *Journal of Wildlife Diseases* 45:729–739.

Richardson, A. D., D. Y. Hollinger, D. B. Dail, J. T. Lee, J. W. Munger, and J. O'Keefe. 2009. Influence of spring phenology on seasonal and annual carbon balance in two contrasting New England forests. *Tree Physiology* 29:321–331.

Richmond, G. M. 1976. *Surficial geological history of the Canyon Village quadrangle, Yellowstone National Park, Wyoming, for use with map I-642.* Bulletin 1427, U.S. Geological Survey, Reston, Virginia.

Ripple, W. J., and R. L. Beschta. 2003. Wolf reintroduction, predation risk, and cottonwood recovery in Yellowstone National Park. *Forest Ecology and Management* 184:299–313.

Ripple, W. J., and R. L. Beschta. 2004a. Wolves and the ecology of fear: Can predation risk structure ecosystems? *BioScience* 54:755–766.

Ripple, W. J., and R. L. Beschta. 2004b. Wolves, elk, willows, and trophic cascades in the upper Gallatin Range of southwestern Montana, USA. *Forest Ecology and Management* 200:161–181.

Ripple, W. J., and R. L. Beschta. 2006. Linking wolves to willows via risk-sensitive foraging by ungulates in the northern Yellowstone ecosystem. *Forest Ecology and Management* 230:96–106.

Ripple, W. J., and R. L. Beschta. 2007. Restoring Yellowstone's aspen with wolves. *Biological Conservation* 138:514–519.

Ripple, W. J., and E. J. Larsen. 2000. Historic aspen recruitment, elk, and wolves in northern Yellowstone National Park, USA. *Biological Conservation* 95:361–370.

Ripple, W. J., E. J. Larsen, R. A. Renkin, and D. W. Smith. 2001. Trophic cascades among wolves, elk and aspen on Yellowstone National Park's northern range. *Biological Conservation* 102:227–234.

Risser, P. J., and W. J. Parton. 1982. Ecosystem analysis of the tallgrass prairie: Nitrogen cycle. *Ecology* 63:1342–1351.

Ritchie, M. E., D. Tilman, and J. M. H. Knops. 1998. Herbivore effects on plant and nitrogen dynamics in oak savanna. *Ecology* 79:165–177.

Robbins, C. T., C. C. Schwartz, and L. A. Felicetti. 2004. Nutritional ecology of ursids: A review of newer methods and management implications. *Ursus* 15:161–171.

Robbins, W., E. Ackerman, M. Bates, S. Cain, F. Darling, J. Fogg, T. Gill, J. Gillson, E. Hall, C. Hubbs, and C. Durham. 1963. *A report by the advisory committee to the National Park Service on research. National Academy of Sciences—National Research Council.* National Academy of Sciences, Washington, D.C.

Rode, K. D., C. T. Robbins, and L. A. Shipley. 2001. Constraints on herbivory by grizzly bears. *Oecologia* 128:62–71.

Rodman, A., H. Shovic, and D. Thomas. 1996. *Soils of Yellowstone National Park.* Report Number YCR-NRSR-96-2. National Park Service, Yellowstone National Park, Wyoming.

Rolston, H. 1990. Biology and philosophy in Yellowstone. *Biology and Philosophy* 5:241–258.

Romme, W. 1982. Fire and landscape diversity in subalpine forests of Yellowstone National Park. *Ecological Monographs* 52:199–221.

Romme, W. H, M. S. Boyce, R. E. Gresswell, E. H. Merrill, G. W. Minshall, C. Whitlock, and M. G. Turner. 2011. Twenty years after the 1988 Yellowstone fires: Lessons about disturbance and ecosystems. *Ecosystems* 14:1196–1215.

Romme, W. H., and D. G. Despain. 1989. Historical perspective on the Yellowstone fires of 1988. *BioScience* 39:695–706.

Romme, W. H., and D. H. Knight. 1981. Fire frequency and subalpine forest succession along a topographic gradient in Wyoming. *Ecology* 62:319–326.

Romme, W. H., and M. G. Turner. 1991. Implications of global climate change for biogeographic patterns in the greater Yellowstone ecosystem. *Conservation Biology* 5:373–386.

Romme, W. H., M. G. Turner, R. H. Gardner, W. W. Hargrove, G. A. Tuskan, D. G. Despain, and R. A. Renkin. 1997. A rare episode of sexual reproduction in aspen (*Populus tremuloides* Michx.) following the 1988 Yellowstone fires. *Natural Areas Journal* 17:17–25.

Romme, W. H., M. G. Turner, G. A. Tuskan, and R. A. Reed. 2005. Establishment, persistence, and growth of aspen *(Populus tremuloides)* seedlings in Yellowstone National Park. *Ecology* 86:404–418.

Romme, W. H., M. G. Turner, L. L. Wallace, and J. S. Walker. 1995. Aspen, elk, and fire in northern Yellowstone Park. *Ecology* 76:2097–2106.

Rood, S. B., J. H. Braatne, and F. M. R. Hughes. 2003. Ecophysiology of riparian cottonwoods: Stream flow dependency, water relations and restoration. *Tree Physiology* 23:1113–1124.

Ruess, R. W. 1984. Nutrient movement and grazing: Experimental effects of clipping and nitrogen source on nutrient uptake in *Kyllinga nervosa*. *Oikos* 43:183–188.

Ruess, R. W., and S. J. McNaughton. 1987. Grazing and the dynamics of nutrient and energy related microbial processes in the Serengeti grasslands. *Oikos* 49:101–110.

Rush, W. M. 1932. *Northern Yellowstone elk study*. Montana Fish and Game Commission, Helena, Montana.

Russell, C. P. 1960. Museum prospectus for the Madison Junction visitor center dated June 3, 1960. Box D-66, Yellowstone National Park archives, Heritage and Research Center, Yellowstone National Park, Gardiner, Montana.

Rutberg, A. T. 1987. Adaptive hypotheses of birth synchrony in ruminants—An interspecific test. *American Naturalist* 130:692–710.

Ruth, T. K. 2004. Ghost of the Rockies: The Yellowstone cougar project. *Yellowstone Science* 12:13–24.

Ruth, T. K., and P. C. Buotte. 2007. *Cougar ecology and cougar-carnivore interactions in Yellowstone National Park*. Final Technical Report, Hornocker Wildlife Institute/Wildlife Conservation Society, Bozeman, Montana.

Ruth, T. K., and K. M. Murphy. 2009a. Competition with other carnivores for prey. Pages 163–172 in M. Hornocker and S. Negri, editors. *Cougar: Ecology and conservation*. University of Chicago Press, Chicago, Illinois.

Ruth, T. K., and K. M. Murphy. 2009b. Cougar-prey relationships. Pages 138–162 in M. Hornocker and S. Negri, editors. *Cougar: Ecology and conservation*. University of Chicago Press, Chicago, Illinois.

Ruzycki, J. R., D. A. Beauchamp, and D. L. Yule. 2003. Effects on introduced lake trout on native cutthroat trout in Yellowstone Lake. *Ecological Applications* 13:23–37.

Ryan, K. C., and E. D. Reinhardt. 1988. Predicting postfire mortality of seven western conifers. *Canadian Journal of Forest Research* 18:1291–1297.

Sæther, B.-E. 1997. Environmental stochasticity and population dynamics of large herbivores: A search for mechanisms. *Trends in Ecology and Evolution* 12:143–149.

Saros, J. 2009. *Inferring critical nitrogen deposition loads to alpine lakes of western national parks with diatom fossil records*. National Park Service, Air Resources Division, Washington, D.C.

Scavia, D., G. A. Lang, and J. F. Kitchell. 1988. Dynamics of Lake Michigan plankton: A model evaluation of nutrient loading, competition, and predation. *Canadian Journal of Fisheries and Aquatic Sciences* 45:165–177.

Scheffer, M. 2009. *Critical transitions in nature and society*. Princeton University Press, Princeton, New Jersey.

Schindler, D. E., J. F. Kitchell, X. He, S. R. Carpenter, J. R. Hodgson, and K. L. Cottingham. 1993. Food web structure and phosphorus cycling in lakes. *Transactions of the American Fisheries Society* 122:759–772.

Schleyer, B. 1983. *Activity patterns of grizzly bears in the Yellowstone ecosystem and their reproductive behavior, predation and the use of carrion.* Thesis, Montana State University, Bozeman, Montana.

Schoennagel, T., T. T. Veblen, and W. H. Romme. 2004. The interaction of fire, fuels, and climate across Rocky Mountain forests. *BioScience* 54:661–676.

Schrag, A. M., A. G. Bunn, and L. J. Graumlich. 2007. Influence of bioclimatic variables on tree-line conifer distribution in the greater Yellowstone ecosystem: Implications for species of conservation concern. *Journal of Biogeography* 35:698–710.

Schullery, P. 1989. The fires and fire policy. *BioScience* 39:686–694.

Schullery, P. 1992. *The bears of Yellowstone.* High Plains Publishing Company, Worland, Wyoming.

Schullery, P. 2004. *Searching for Yellowstone: Ecology and wonder in the last wilderness.* Houghton Mifflin, Boston, Massachusetts.

Schullery, P. 2010. Greater Yellowstone science: Past, present, and future. *Yellowstone Science* 18:7–13.

Schullery, P., W. Brewster, and J. Mack. 1998. Bison in Yellowstone: A historical overview. Pages 326–336 in L. Irby and J. Knight, editors. *International symposium on bison ecology and management in North America.* Montana State University, Bozeman, Montana.

Schullery, P., and J. D. Varley. 1995. Cutthroat trout and the Yellowstone Lake ecosystem. Pages 12–21 in J. D. Varley and P. Schullery, editors. *The Yellowstone Lake crisis: Confronting a lake trout invasion.* National Park Service, Yellowstone National Park, Mammoth, Wyoming.

Schullery, P., and L. Whittlesey. 1992. The documentary record of wolves and related wildlife species in the Yellowstone National Park area prior to 1882. Pages 1–4 to 1–174 in J. D. Varley and W. G. Brewster, editors. *Wolves for Yellowstone? A report to the United States Congress,* Volume 4, *Research and analysis.* National Park Service, Yellowstone National Park, Mammoth, Wyoming.

Schullery, P., and L. Whittlesey. 1999a. *Early wildlife history of the greater Yellowstone ecosystem: An interim research report presented to National Research Council, National Academy of Sciences, Committee on Ungulate Management in Yellowstone National Park.* Yellowstone Center for Resources, Yellowstone National Park, Mammoth, Wyoming.

Schullery, P., and L. Whittlesey. 1999b. Greater Yellowstone carnivores: A history of changing attitudes. Pages 10–49 in T. P. Clark, P. Curlee Griffin, S. Minta, and P. Kareiva, editors. *Carnivores in ecosystems: The Yellowstone experience.* Yale University Press, New Haven, Connecticut.

Schullery, P., and L. Whittlesey. 2001. Mountain goats in the greater Yellowstone ecosystem: A prehistoric and historical context. *Western North American Naturalist* 61:289–307.

Schullery, P., and L. Whittlesey. 2003. *Myth and history in the creation of Yellowstone National Park.* University of Nebraska Press, Lincoln, Nebraska.

Schullery, P., and L. H. Whittlesey. 2006. Greater Yellowstone bison distribution and abundance in the early historical period. Pages 135–140 in A. W. Biel, editor. *Greater Yellowstone public lands: Proceedings of the Eighth Biennial Scientific Conference on the Greater Yellowstone Ecosystem.* Yellowstone National Park, Mammoth, Wyoming.

Schwartz, C. C., and A. W. Franzmann. 1991. Interrelationship of black bears to moose and forest succession in the northern coniferous forest. *Wildlife Monographs* 113.

Schwartz, C. C., M. A. Haroldson, and S. Cherry. 2006. Reproductive performance for grizzly bears in the greater Yellowstone ecosystem, 1983–2002. *Wildlife Monographs* 161.

Schwartz, C. C., M. A. Haroldson, K. A. Gunther, and D. Moody. 2002. Distribution of grizzly bears in the greater Yellowstone ecosystem, 1990–2000. *Ursus* 13:203–212.

Schwartz, C. C., M. A. Haroldson, K. A. Gunther, and D. Moody. 2006. Distribution of grizzly bears in the greater Yellowstone ecosystem in 2004. *Ursus* 17:63–66.

Schwartz, C. C., M. A. Haroldson, and K. West, editors. 2010. *Yellowstone grizzly bear investigations: Annual report of the Interagency Grizzly Bear Study Team, 2009.* U.S. Geological Survey, Bozeman, Montana.

Schwartz, C. C., M. A. Haroldson, and G. C. White. 2006. Survival of cub and yearling grizzly bears in the greater Yellowstone ecosystem, 1983–2001. *Wildlife Monographs* 161.

Schwartz, C. C., M. A. Haroldson, and G. C. White. 2010. Hazards affecting grizzly bear survival in the greater Yellowstone ecosystem. *Journal of Wildlife Management* 74:654–667.

Schwartz, C. C., M. A. Haroldson, G. C. White, R. B. Harris, S. Cherry, K. A. Keating, D. Moody, and C. Servheen. 2006. Temporal, spatial, and environmental influences on the demographics of grizzly bears in the greater Yellowstone ecosystem. *Wildlife Monographs* 161.

Schwartz, C. C., R. B. Harris, and M. A. Haroldson. 2006. Impacts of spatial and environmental heterogeneity on grizzly bear demographics in the greater Yellowstone ecosystem: A source-sink dynamic with management consequences. *Wildlife Monographs* 161.

Schweitzer, B. 2007. Testimony before the U.S. House of Representatives Subcommittee on National Parks, Forests, and Public Lands oversight hearing on Yellowstone National Park bison. March 20, 2007, Washington, D.C.

Scott, J. M., F. W. Davis, R. G. McGhie, R. G. Wright, C. Groves, and J. Estes. 2001. Nature reserves: Do they capture the full range of America's biological diversity? *Ecological Applications* 11:999–1007.

Scott, J. M, D. D. Goble, J. A. Wiens, D. S. Wilcove, M. Bean, and T. Male. 2005. Recovery of imperiled species under the Endangered Species Act: The need for a new approach. *Frontiers in Ecology and the Environment* 3:383–389.

Scott, M. D., and H. Geisser. 1996. Pronghorn migration and habitat use following the 1988 Yellowstone fires. Pages 123–132 in J. M. Greenlee, editor. *Ecological implications of fire in the greater Yellowstone.* National Park Service, Yellowstone National Park, Wyoming.

Sellars, R. W. 1997. *Preserving nature in the national parks: A history.* Yale University Press, New Haven, Connecticut.

Senft, R. L., M. B. Coughenour, D. W. Baily, R. L. Rittenhouse, O. E. Sala, and D. M. Swift. 1987. Larger herbivore foraging and ecological hierarchies. *BioScience* 37:789–799.

Servheen, C., and M. Cross. 2010. *Climate change impacts on grizzly bears and wolverines in the northern U.S. and transboundary Rockies: Strategies for conservation.* Report on a workshop held September 13–15, 2010, in Fernie, British Columbia, Canada. www.cfc.umt.edu/GrizzlyBearRecovery/pdfs /Servheen%20and%20Cross%202010.pdf.

Servheen, C., S. Herrero, and B. Peyton, editors. 1999. *Status survey of the bears of the world and global conservation action plan.* International Union for Conservation of Nature, Gland, Switzerland.

Seton, E. T. 1927. *Lives of game animals.* Doubleday, Page, New York, New York.

Shafer, S. H., P. J. Bartlein, and C. Whitlock. 2005. Understanding the spatial heterogeneity of global environmental change in mountainous regions. Pages 21–31 in U. M. Huber, H. K. M. Bugmann, and M. A. Reasoner, editors. *Global change and mountain regions: An overview of current knowledge.* Kluwer, New York, New York.

Shariff, A. R., M. E. Biondini, and C. E. Grygiel. 1994. Grazing intensity effects on litter decomposition and soil nitrogen mineralization. *Journal of Range Management* 47:444–449.

Shaver, G. R., and J. Kummerov. 1992. Phenology, resource allocation, and growth of arctic vascular plants. Pages 191–212 in F. S. Chapin, R. L. Jefferies, and J. F. Reynolds, editors. *Ecosystems in a changing climate: An ecophysiological perspective.* Academic Press, New York, New York.

Shea, R. 1979. *The ecology of the trumpeter swan in Yellowstone National Park and vicinity.* Thesis, University of Montana, Missoula, Montana.

Sibley, D. A. 2003. *The Sibley field guide to birds of western North America*. Turtleback, New York, New York.

Sikes, D. S. 1994. *Influence of ungulate carcasses on coleopteran communities in Yellowstone National Park*. Thesis, Montana State University, Bozeman, Montana.

Simon, J. R. 1962. *Yellowstone fishes*. Yellowstone Library and Museum Association, Yellowstone Interpretive Series 3, Yellowstone National Park, Mammoth, Wyoming.

Sinclair, A. R. E. 1975. The resource limitation of trophic levels in tropical grassland ecosystems. *Journal of Animal Ecology* 44:497–520.

Sinclair, A. R. E. 1998. Natural regulation of ecosystems in protected areas as ecological baselines. *Wildlife Society Bulletin* 26:399–409.

Sinclair, A. R. E. 2003. Mammal population regulation, keystone processes and ecosystem dynamics. *Philosophical Transactions of the Royal Society of London. Series B, Biological Sciences* 358:1729–1740.

Sinclair, A. R. E., R. P. Pech, C. R. Dickman, D. Hik, P. Mahon, and A. E. Newsome. 1998. Predicting effects of predation on conservation of endangered prey. *Conservation Biology* 12:564–575.

Singer, F. J., A. Harting, K. K. Symonds, and M. B. Coughenour. 1997. Density dependence, compensation, and environmental effects on elk calf mortality in Yellowstone National Park. *Journal of Wildlife Management* 61:12–25.

Singer, F. J., L. C. Mark, and R. C. Cates. 1994. Ungulate herbivory of willows on Yellowstone northern winter range. *Journal of Range Management* 47:435–443.

Singer, F. J., and J. E. Norland. 1994. Niche relationships within a guild of ungulate species in Yellowstone National Park, Wyoming, following release from artificial controls. *Canadian Journal of Zoology* 72:1383–1394.

Singer, F. J., and R. A. Renkin. 1995. Effects of browsing by native ungulates on the shrubs in big sagebrush communities in Yellowstone National Park. *Great Basin Naturalist* 55:201–212.

Singer, F. J., and K. A. Schoenecker. 2003. Do ungulates accelerate or decelerate nitrogen cycling? *Forest Ecology and Management* 181:189–204.

Singer, F. J., W. Schreir, J. Oppenheim, and E. O. Garton. 1989. Drought, fires and large mammals. *BioScience* 39:716–722.

Six, D. L., and J. C. Adams. 2007. Relationships between white pine blister rust and the selection of individual whitebark pine by the mountain pine beetle. *Journal of Entomological Science* 42:345–353.

Skagen, S. K., R. Hazelwood, and M. L. Scott. 2005. *The importance and future condition of western riparian ecosystems as migratory bird habitat*. General Technical Report PSW-GTR-191.2005, U.S. Department of Agriculture, National Forest Service.

Skinner, M. P. 1922. The prong-horn. *Journal of Mammalogy* 3:82–105.

Skinner, M. P. 1925. Migration routes in Yellowstone Park. *Journal of Mammalogy* 6:184–192.

Skinner, M. P. 1927. The predatory and fur bearing animals of the Yellowstone National Park. *Roosevelt Wildlife Bulletin* 4:163–281.

Skinner, M. P. 1928. The elk situation. *Journal of Mammalogy* 9:309–317.

Smith, B. L., E. S. Williams, K. C. McFarland, T. L. McDonald, G. Wang, and T. D. Moore. 2006. *Neonatal mortality of elk in Wyoming: Environmental, population, and predator effects.* Biological Technical Publication BTP-R6007-2006, U.S. Department of the Interior, U.S. Fish and Wildlife Service, Washington, D.C.

Smith, C. 1996. Media coverage of fire ecology in Yellowstone after 1988. Pages 25–34 in J. Greenlee, editor. *The ecological implications of fire in Greater Yellowstone.* Proceedings of the Second Biennial Scientific Conference on the Greater Yellowstone Ecosystem. International Association of Wildland Fire, Fairfield, Washington.

Smith, D. W. 2005. Ten years of Yellowstone wolves. *Yellowstone Science* 13:7–33.

Smith, D. W., and E. Almberg. 2007. Wolf diseases in Yellowstone National Park. *Yellowstone Science* 15:17–19.

Smith, D. W., T. D. Drummer, K. M. Murphy, D. S. Guernsey, and S. B. Evans. 2004. Winter prey selection and estimation of wolf kill rates in Yellowstone National Park, 1995–2000. *Journal of Wildlife Management* 68:153–166.

Smith, D. W., L. D. Mech, M. Meagher, W. E. Clark, R. Jaffe, M. K. Phillips, and J. A. Mack 2000. Wolf-bison interactions in Yellowstone National Park. *Journal of Mammalogy* 81:1128–1135.

Smith, D. W., R. O. Peterson, and D. B. Houston. 2003. Yellowstone after wolves. *BioScience* 53:330–340.

Smith, D., D. Stahler, E. Albers, R. McIntyre, M. Metz, J. Irving, R. Raymond, C. Anton, K. Cassidy-Quimby, and N. Bowersock. 2011. *Yellowstone wolf project annual report 2010.* YCR-2011-06, National Park Service, Yellowstone National Park, Mammoth, Wyoming.

Smith, D. W., D. R. Stahler, E. Albers, M. Metz, L. Williamson, N. Ehlers, K. Cassidy, J. Irving, R. Raymond, E. Almberg, and R. McIntyre. 2008. *Yellowstone wolf project: Annual report, 2008.* National Park Service, Yellowstone National Park, Wyoming.

Smith, D. W., D. Stahler, and M. S. Becker. 2009. Wolf recolonization of the Madison headwaters area in Yellowstone. Pages 283–303 in R. A. Garrott, P. J. White, and F. G. R. Watson, editors. *The ecology of large mammals in central Yellowstone: Sixteen years of integrated field studies.* Elsevier, San Diego, California.

Smith, R. B., and L. J. Siegel. 2000. *Windows into the earth: The geologic story of Yellowstone and Grand Teton national parks.* Oxford University Press, New York, New York.

Snow, J. 2005. November 3, 2005, electronic mail message to J. MacDonald, Greystone Environmental Consultants, Greenwood Village, Colorado regarding the number of human cases of brucellosis in Wyoming. State Public Health Veterinarian, Wyoming Department of Public Health, Cheyenne, Wyoming.

Solomon, M. E. 1949. The natural control of animal populations. *Journal of Animal Ecology* 18:1–35.

Southwood, T. R. E. 1977. Habitat, the template for ecological strategies? *Journal of Animal Ecology* 46:337–365.

Spaulding, P. 1922. *Investigations of the white-pine blister rust.* Bulletin 957, U.S. Department of Agriculture, National Forest Service, Washington, D.C.

Spence, M. 1999. *Dispossessing the wilderness: Indian removal and the making of the national parks.* Oxford University Press, New York, New York.

Stahler, D. R. 2000. *Interspecific interactions between the common raven* (Corvus corax) *and the gray wolf* (Canis lupus) *in Yellowstone National Park, Wyoming: Investigations of a predator and scavenger relationship.* Thesis, University of Vermont, Burlington, Vermont.

Stahler, D., B. Heinrich, and D. Smith. 2002. Common ravens, *Corvus corax,* preferentially associate with grey wolves, *Canis lupus,* as a foraging strategy in winter. *Animal Behaviour* 64:283–290.

Stahler, D. R., D. W. Smith, and D. S. Guernsey. 2006. Foraging and feeding ecology of the gray wolf *(Canis lupus):* Lessons from Yellowstone National Park, Wyoming, USA. *Journal of Nutrition* 136:1923S–1926S.

Stanford, J. A., M. S. Lorang, and F. R. Hauer. 2005. The shifting habitat mosaic of river ecosystems. *Verhandlungen Interntionale für Theoretische und Ange-wandte Limnologie* 29:123–136.

Stapp, P., and G. D. Hayward. 2002a. Effects of an introduced piscivore on native trout: Insights from a demographic model. *Biological Invasions* 4:299–316.

Stapp, P., and G. D. Hayward. 2002b. Estimates of predator consumption of Yellowstone cutthroat trout *(Oncorhynchus clarki bouvieri)* in Yellowstone Lake. *Journal of Freshwater Ecology* 17:319–329.

Stenseth, N. C., and W. Z. Lidicker, Jr., editors. 1992. *Animal dispersal—Small mammals as a model.* Chapman and Hall, London, UK.

Stephenson, N. L., C. I. Millar, and D. N. Cole. 2010. Shifting environmental foundations: The unprecedented and unpredictable future. Pages 50–66 in D. N. Cole and L. Yung, editors. *Beyond naturalness: Rethinking park and wilderness stewardship in an era of rapid change.* Island Press, Washington, D.C.

Stevens, D. R., and N. J. Goodson. 1993. Assessing effects of removals for trans-planting on a high-elevation bighorn sheep population. *Conservation Biology* 7:908–915.

Stevens, L. R., and W. E. Dean. 2008. Geochemical evidence for hydroclimatic variability over the last 2460 years from Crevice Lake in Yellowstone National Park, USA. *Quaternary International* 188:139–148.

Stewart, I. T. 2009. Changes in snowpack and snowmelt runoff for key mountain regions. *Hydrological Processes* 23:78–94.

Stewart, I. T., D. R. Cayan, and M. D. Dettinger. 2005. Changes toward earlier streamflow timing across western North America. *Journal of Climate* 18:1136–1155.

Stewart, K. M., R. T. Bowyer, B. L. Dick, B. K. Johnson, and J. G. Kie. 2005. Density-dependent effects on physical condition and reproduction in North American elk: An experimental test. *Oecologia* 143:85–93.

Stewart, K. M., R. T. Bowyer, J. G. Kie, B. L. Dick, and R. W. Ruess. 2009. Population density of North American elk: Effects on plant diversity. *Oecologia* 161:303–312.

Stewart, K. M., R. T. Bowyer, R. W. Ruess, B. L. Dick, and J. G. Kie. 2006. Herbivore optimization by North American elk: Consequences for theory and management. *Wildlife Monographs* 167.

Storer, T. I., and L. P. Tevis. 1955. *California grizzly.* University of California Press, Berkeley, California.

Stromberg, J. C. 2001. Restoration of riparian vegetation in the south-western United States: Importance of flow regimes and fluvial dynamism. *Journal of Arid Environments* 49:17–34.

Suding, K. N., K. L. Gross, and G. R. Houseman. 2004. Alternative states and positive feedbacks in restoration ecology. *Trends in Ecology and Evolution* 19:46–53.

Swenson, J. E. 1978. Prey and foraging behavior of ospreys on Yellowstone Lake, Wyoming. *Journal of Wildlife Management* 42:87–90.

Swenson, J. E., K. L. Alt, and R. L. Eng. 1986. Ecology of bald eagles in the greater Yellowstone ecosystem. *Wildlife Monographs* 95.

Syslo, J. M., C. S. Guy, P. Bigelow, P. D. Doepke, B. D. Ertel, and T. M. Koel. 2011. Response of non-native lake trout *(Salvelinus namaycush)* to 15 years of harvest in Yellowstone Lake, Yellowstone National Park. *Canadian Journal of Fisheries and Aquatic Sciences* 68:2132–2145.

Taper, M. L., and P. J. P. Gogan. 2002. The northern Yellowstone elk: Density dependence and climatic conditions. *Journal of Wildlife Management* 66:106–122.

Taper, M. L., M. Meagher, and C. L. Jerde. 2000. *The phenology of space: Spatial aspects of bison density dependence in Yellowstone National Park.* U.S. Geological Service, Biological Resources Division, Bozeman, Montana.

Telfer, E. S., and J. P. Kelsall. 1984. Adaptation of some large North American mammals for survival in snow. *Ecology* 65:1828–1834.

Terborgh, J., L. Lopez, P. Nunez, M. Rao, G. Shahabuddin, G. Orhuela, M. Riveros, R. Ascanio, G. R. Adler, T. D. Lambert, and L. Balbas. 2001. Ecological meltdown in predator-free forest fragments. *Science* 294:1923–1926.

Tercek, M. T., R. S. Stolltemyer, and R. Renkin. 2010. Bottom-up factors influencing riparian willow recovery in Yellowstone National Park. *Western North American Naturalist* 70:387–399.

Thein, T. R., F. G. R. Watson, S. S. Cornish, T. N. Anderson, W. B. Newman, and R. E. Lockwood. 2009. Vegetation dynamics of Yellowstone's grazing system. Pages 113–133 in R. A. Garrott, P. J. White, and F. G. R. Watson, editors. *The ecology of large mammals in central Yellowstone: Sixteen years of integrated field studies.* Elsevier, San Diego, California.

Thorhallsdottir, T. E. 1998. Flowering phenology in the central highland of Iceland and implications for climatic warming in the Arctic. *Oecologia* 114:43–49.

Thorne, E. T., E. S. Williams, W. M. Samuel, and T. P. Kistner. 2002. Diseases and parasites. Pages 351–387 in D. E. Toweill and J. W. Thomas, editors. *North American elk: Ecology and management.* Smithsonian Institution Press, Washington, D.C.

Tietje, W. D., B. O. Pelchat, and R. L. Ruff. 1986. Cannibalism of denned black bears. *Journal of Mammalogy* 67:762–766.

Tisch, E. L. 1961. *Seasonal food habitats of the black bear in the Whitefish Range of northwestern Montana.* Thesis, Montana State University, Bozeman, Montana.

Tomback, D. F. 1983. Nutcrackers and pines: Coevolution or coadaption? Pages 179–223 in M. H. Nitecki, editor. *Coevolution.* University of Chicago Press, Chicago, Illinois.

Tomback, D. F., and P. Achuff. 2010. Blister rust and western forest biodiversity: Ecology, values, and outlook for white pines. *Forest Pathology* 40:186–225.

Tomback, D. F., S. F. Arno, and R. E. Keane. 2001. The compelling case for management intervention. Pages 3–25 in D. F. Tomback, S. F. Arno, and R. E. Keane, editors. *Whitebark pine communities: Ecology and restoration.* Island Press, Washington, D.C.

Tomback, D. F., and K. C. Kendall. 2001. Biodiversity losses: The downward spiral. Pages 243–262 in D. F. Tomback, S. F. Arno, and R. E. Keane, editors. *Whitebark pine communities: Ecology and restoration.* Island Press, Washington, D.C.

Tomback, D. F., S. K. Sund, and L. A. Hoffman. 1993. Post-fire regeneration of *Pinus albicaulis:* Height-age relationships, age structure, and microsite characteristics. *Canadian Journal of Forest Research* 23:113–119.

Tooke, F., and N. H. Battey. 2010. Temperate flowering phenology. *Journal of Experimental Botany* 61:2853–2862.

Treanor, J. J., C. Geremia, P. H. Crowley, J. J. Cox, P. J. White, R. L. Wallen, and D. W. Blanton. 2011. Estimating probabilities of active brucellosis infection in Yellowstone bison through quantitative serology and tissue culture. *Journal of Applied Ecology* 48:1324–1332.

Treanor, J. J., J. S. Johnson, R. L. Wallen, S. Cilles, P. H. Crowley, J. J. Cox, D. S. Maehr, P. J. White, and G. E. Plumb. 2010. Vaccination strategies for managing brucellosis in Yellowstone bison. *Vaccine* 28S:F64–F72.

Treanor, J. J., R. L. Wallen, D. S. Maehr, and P. H. Crowley. 2007. Brucellosis in Yellowstone bison: Implications for conservation management. *Yellowstone Science* 15:20–24.

Tronstad, L. M., R. O. Hall, T. M. Koel, and K. G. Gerow. 2010. Introduced lake trout produced a four-level trophic cascade in Yellowstone Lake. *Transactions of the American Fisheries Society* 139:1536–1550.

Tucker, B., S. Mahoney, B. Greene, E. Menchenton, and L. Russell. 2010. The influence of snow depth and hardness on winter habitat selection by caribou on the southwest coast of Newfoundland. *Rangifer* 11:160–163.

Turner, M. G. 2010. Disturbance and landscape dynamics in a changing world. *Ecology* 91:2833–2849.

Turner, M. G., and R. H. Gardner. 1994. Landscape disturbance models and the long-term dynamics of natural areas. *Natural Areas Journal* 14:3–11.

Turner, M. G., W. H. Romme, and D. B. Tinker. 2003. Surprises and lessons from the 1988 Yellowstone fires. *Frontiers in Ecology and the Environment* 1:351–358.

Turner, M. G., E. A. H. Smithwick, K. L. Metzger, D. B. Tinker, and W. H. Romme. 2007. Inorganic nitrogen availability after severe stand-replacing fire in the greater Yellowstone ecosystem. *Proceedings of the National Academy of Sciences of the United States of America* 104:4782–4789.

Turner, M. G., D. B. Tinker, W. H. Romme, D. M. Kashian, and C. M. Litton. 2004. Landscape patterns of sapling density, leaf area, and aboveground net primary production in postfire lodgepole pine forests, Yellowstone National Park (USA). *Ecosystems* 7:751–775.

Tyers, D. B. 1981. *The condition of the northern winter range in Yellowstone National Park—A discussion of the controversy.* Thesis professional paper, Montana State University, Bozeman, Montana.

Tyers, D. B. 2003. *Winter ecology of moose on the northern Yellowstone winter range.* Dissertation, Montana State University, Bozeman, Montana.

U.S. Animal Health Association. 2006. *Enhancing brucellosis vaccines, vaccine delivery, and surveillance diagnostics for elk and bison in the greater Yellowstone area: A technical report from a working symposium held August 16–18, 2005, at the University of Wyoming.* T. Kreeger and G. Plumb, editors. University of

Wyoming Haub School and Ruckelshaus Institute of Environment and Natural Resources, Laramie, Wyoming.

U.S. Department of Agriculture. 1984. *Montana forest pest conditions and program highlights, 1983.* Report 84–2, National Forest Service, Missoula, Montana.

U.S. Department of Agriculture. 2009. *Insect and disease aerial detection survey, Region 1.* National Forest Service, Missoula, Montana.

U.S. Department of the Interior. 2009. *Secretarial order 3289. Addressing the impacts of climate change on America's water, land, and other natural and cultural resources.* Available at http://elips.doi.gov/app_so/act_getfiles.cfm?order_number=3289A1.

U.S. Department of the Interior, National Park Service. 2010. *Native fish conservation plan environmental assessment.* Yellowstone National Park, Mammoth, Wyoming.

U.S. Department of the Interior, National Park Service, and U.S. Department of Agriculture, Forest Service, Animal and Plant Health Inspection Service. 2000. *Record of decision for final environmental impact statement and bison management plan for the State of Montana and Yellowstone National Park.* Washington, D.C.

U.S. Department of the Interior, National Park Service, and U.S. Department of Agriculture, Forest Service, Animal and Plant Health Inspection Service, and the State of Montana, Department of Fish, Wildlife, and Parks, Department of Livestock. 2008. *Adaptive adjustments to the interagency bison management plan.* National Park Service, Yellowstone National Park, Wyoming.

U.S. District Court for the District of Montana Missoula Division. 2009. Defenders of Wildlife et al., Plaintiffs, v. Salazar et al., Defendants. Case 9:09-cv-00077-DWM, Missoula, Montana.

U.S. Fish and Wildlife Service. 1975. Grizzly bear. *Federal Register* 40:31734–31736.

U.S. Fish and Wildlife Service. 1993. *Grizzly bear recovery plan.* Missoula, Montana.

U.S. Fish and Wildlife Service. 1998. *Pacific flyway management plan for the Rocky Mountain population of trumpeter swans.* Subcommittee on Rocky Mountain trumpeter swans, Pacific Flyway Study Committee, Portland, Oregon.

U.S. Fish and Wildlife Service. 2003. *Trumpeter swan survey of the Rocky Mountain population, winter 2003.* Division of Migratory Birds and State Programs, Denver, Colorado.

U.S. Fish and Wildlife Service. 2007a. Endangered and threatened wildlife and plants; designating the greater Yellowstone ecosystem population of grizzly bears as a distinct population segment; removing the Yellowstone distinct population segment of grizzly bears from the federal list of endangered and threatened wildlife; 90-day finding on a petition to list as endangered the Yellowstone distinct population segment of grizzly bears. *Federal Register* 72:14866–14938.

U.S. Fish and Wildlife Service. 2007b. Endangered and threatened wildlife and plants; removing the bald eagle in the lower 48 states from the list of endangered and threatened wildlife. *Federal Register* 72:37346–37372.

U.S. Fish and Wildlife Service. 2011. Endangered and threatened wildlife and plants; 12-month finding on a petition to list *Pinus albicaulis* as endangered or threatened with critical habitat. *Federal Register* 76:42631–42654.

U.S. Geological Survey. 1972. *Geologic map of Yellowstone National Park.* Miscellaneous Geological Investigations, 1–711, Washington, D.C.

U.S. Government Accountability Office. 2008. *Yellowstone bison—Interagency plan and agencies' management need improvement to better address bison-cattle brucellosis controversy.* Report GAO-08-291 to Congressional Requesters. Washington, D.C.

Vale, T. R. 1998. The myth of the humanized landscape: An example from Yosemite National Park. *Natural Areas Journal* 18:231–236.

Van Ballenberghe, V. 1987. Effects of predation on moose numbers: A review of recent North American studies. *Swedish Wildlife Research* (Supplement 1): 431–460.

van Mantgem, P. J., and N. L. Stephenson. 2007. Apparent climatically induced increase of tree mortality rates in a temperate forest. *Ecology Letters* 10:909–916.

van Mantgem, P. J., N. L. Stephenson, J. C. Byrne, L. D. Daniels, J. F. Franklin, P. Z. Fule, M. E. Harmon, A. J. Larson, J. M. Smith, A. H. Taylor, and T. T. Veblen. 2009. Widespread increase of tree mortality rates in the western United States. *Science* 323:521–524.

Varley, J. D., and P. Schullery. 1995a. Socioeconomic values associated with the Yellowstone Lake cutthroat trout. Pages 22–27 in J. D. Varley and P. Schullery, editors. *The Yellowstone Lake crisis: Confronting a lake trout invasion.* National Park Service, Yellowstone National Park, Mammoth, Wyoming.

Varley, J. D., and P. Schullery. 1995b. *The Yellowstone Lake crisis: Confronting a lake trout invasion.* National Park Service, Yellowstone National Park, Mammoth, Wyoming.

Varley, J. D., and P. Schullery. 1998. *Yellowstone fishes: Ecology, history, and angling in the park.* Stackpole Books, Mechanicsburg, Pennsylvania.

Varley, N. 1996. *Ecology of mountain goats in the Absaroka Range, south-central Montana.* Thesis, Montana State University, Bozeman, Montana.

Varley, N., and M. S. Boyce. 2006. Adaptive management for reintroductions. Updating a wolf recovery model for Yellowstone National Park. *Ecological Modeling* 193:315–339.

Vitousek, P. M., and R. W. Howarth. 1991. Nitrogen limitation on land and in the sea: How can it occur? *Biogeochemistry* 13:87–115.

Vucetich, J. A., R. O. Peterson, and T. A. Waite. 2004. Raven scavenging favours group foraging in wolves. *Animal Behaviour* 67:1117–1126.

Vucetich, J. A., D. W. Smith, and D. R. Stahler. 2005. Influence of harvest, climate, and wolf predation on Yellowstone elk, 1961–2004. *Oikos* 111:259–270.

Wagner, F. H. 2006. *Yellowstone's destabilized ecosystem: Elk effects, science, and policy conflict.* Oxford University Press, Oxford, UK.

Wagner, F. H., R. Foresta, R. B. Gill, D. R. McCullough, M. R. Pelton, W. F. Porter, and H. Salwasser. 1995. *Wildlife policies in the U.S. national parks.* Island Press, Washington, D.C.

Wallace, L. L. 1990. Comparative photosynthetic responses of big bluestem to clipping vs. grazing. *Journal of Range Management* 43:58–61.

Wallace, L. L. 1996. *Grazing and competition in montane grasslands.* Report YCR-NR-96-6, National Park Service, Yellowstone National Park, Mammoth, Wyoming.

Wallace, L. L., editor. 2004. *After the fires: The ecology of change in Yellowstone National Park.* Yale University Press, New Haven, Connecticut.

Wallace, L. L., M. G. Turner, W. H. Romme, R. V. O'Neill, and Y. Wu. 1995. Scale of heterogeneity of forage production and winter foraging by elk and bison. *Landscape Ecology* 10:75–83.

Walpole, K. V. 1997. Letter dated April 1 from Kyle V. Walpole to Kevin Hurley, Wyoming Game and Fish Department regarding the historical record of mountain goats in Wyoming. Cody, Wyoming.

Walsh, J. R. 2005. *Fire regimes and stand dynamics of whitebark pine* (Pinus albicaulis) *communities in the greater Yellowstone ecosystem.* Thesis, Colorado State University, Fort Collins, Colorado.

Walters, C. 1986. *Adaptive management of renewable resources.* Macmillan, New York, New York.

Walters, C. J., and C. S. Holling. 1990. Large-scale management experiments and learning by doing. *Ecology* 71:2060–2068.

Wambolt, C. L., and H. W. Sherwood. 1999. Sagebrush response to ungulate browsing in Yellowstone. *Journal of Range Management* 52:363–369.

Wang, G., N. T. Hobbs, R. B. Boone, A. W. Illius, I. J. Gordon, J. E. Gross, and K. L. Hamlin. 2006. Spatial and temporal variability modify density dependence in populations of large herbivores. *Ecology* 87:95–102.

Warren, C. E., and W. J. Liss. 1980. Adaptation to aquatic environments. Pages 15–40 in R. T. Lackey and L. Nielsen, editors. *Fisheries management.* Blackwell Scientific, Oxford, UK.

Warren, E. R. 1926. A study of the beaver in the Yancey region of Yellowstone National Park. *Roosevelt Wildlife Annals* 1, Syracuse, New York.

Warwell, M. V., G. E. Rehfeldt, and N. L. Crookston. 2007. Modeling contemporary climate profiles of whitebark pine *(Pinus albicaulis)* and predicting responses to global warming. *Proceedings of the Conference on Whitebark Pine: A Pacific Coast Perspective.* R6-NR-FHP-2007-01, U.S. Department of Agriculture, National Forest Service. http://www.fs.fed.us/r6/nr/fid/wbpine/papers /2007-wbp-climate-warwell.pdf.

Watson, F. G. R., T. N. Anderson, M. Kramer, J. Detka, T. Masek, S. S. Cornish, and S. W. Moore. 2009. Effects of wind, terrain, and vegetation on snow pack. Pages 67–84 in R. A. Garrott, P. J. White, and F. G. R. Watson, editors. *The ecology of large mammals in central Yellowstone: Sixteen years of integrated field studies.* Elsevier, San Diego, California.

Watson, F. G. R., T. N. Anderson, W. B. Newman, S. E. Alexander, and R. A. Garrott. 2006. Optimal sampling schemes for estimating mean snow water equivalents in stratified heterogeneous landscapes. *Journal of Hydrology* 328:432–452.

Watson, F. G. R., T. N. Anderson, W. B. Newman, S. S. Cornish, and T. R. Thein. 2009. Modeling spatial snow pack dynamics. Pages 85–112 in R. A. Garrott, P. J. White, and F. G. R. Watson, editors. *The ecology of large mammals in central Yellowstone: Sixteen years of integrated field studies.* Elsevier, San Diego, California.

Watson, F. G. R., and W. B. Newman. 2009. Mapping mean annual precipitation using trivariate kriging. Pages 37–52 in R. A. Garrott, P. J. White, and F. G. R. Watson, editors. *The ecology of large mammals in central Yellowstone: Sixteen years of integrated field studies.* Elsevier, San Diego, California.

Watson, F. G. R., W. B. Newman, T. N. Anderson, R. E. Lockwood, and R. A. Garrott. 2009. Effects of Yellowstone's unique geothermal landscape on snow pack. Pages 53–66 in R. A. Garrott, P. J. White, and F. G. R. Watson, editors. *The ecology of large mammals in central Yellowstone: Sixteen years of integrated field studies.* Elsevier, San Diego, California.

Watson, F. G. R., W. B. Newman, J. C. Coughlan, and R. A. Garrott. 2006. Testing a distributed snowpack simulation model against spatial observations. *Journal of Hydrology* 328:453–466.

Welch, C. A., J. Keay, K. C. Kendall, and C. T. Robbins. 1997. Constraints on frugivory by bears. *Ecology* 78:1105–1119.

Wengeler, W. R., D. A. Kelt, and M. L. Johnson. 2010. Ecological consequences of invasive lake trout on river otters in Yellowstone National Park. *Biological Conservation* 143:144–153.

Wenger, S. J., D. J. Isaak, C. H. Luce, H. M. Neville, K. D. Fausch, J. B. Dunham, D. C. Dauwalter, M. K. Young, M. M. Elsner, B. E. Rieman, A. F. Hamlet, and J. E. Williams. 2011. Flow regime, temperature, and biotic interactions drive differential declines of trout species under climate change. *Proceedings of the National Academy of Sciences of the United States of America* 108:14175–14180.

Westerling, A. L., H. G. Hidalgo, D. R. Cayan, and T. W. Swetnam. 2006. Warming and earlier spring increase western US forest wildfire activity. *Science* 313:940–943.

Westerling, A. L, M. G. Turner, E. A. H. Smithwick, W. H. Romme, and M. G. Ryan. 2011. Continued warming could transform greater Yellowstone fire regimes by mid-21st century. *Proceedings of the National Academy of Sciences of the United States of America* 108:13165–13170.

Whipple, J. 2001. Annotated checklist of exotic vascular plants in Yellowstone National Park. *Western North American Naturalist* 61:336–346.

White, P. J., J. E. Bruggeman, and R. A. Garrott. 2007. Irruptive population dynamics in Yellowstone pronghorn. *Ecological Applications* 17:1598–1606.

White, P. J., T. L. Davis, K. K. Barnowe-Meyer, R. L. Crabtree, and R. A. Garrott. 2007. Partial migration and philopatry of Yellowstone pronghorn. *Biological Conservation* 135:518–526.

White, P. J., and R. A. Garrott. 2005a. Northern Yellowstone elk after wolf restoration. *Wildlife Society Bulletin* 33:942–955.

White, P. J., and R. A. Garrott. 2005b. Yellowstone's ungulates after wolves—Expectations, realizations, and predictions. *Biological Conservation* 125:141–152.

White, P. J., R. A. Garrott, J. J. Borkowski, J. G. Berardinelli, D. R. Mertens, and A. C. Pils. 2009. Diet and nutrition of central Yellowstone elk during winter. Pages 157–176 in R. A. Garrott, P. J. White, and F. G. R. Watson, editors. *The ecology of large mammals in central Yellowstone: Sixteen years of integrated field studies.* Elsevier, San Diego, California.

White, P. J., R. A. Garrott, J. J. Borkowski, K. L. Hamlin, and J. G. Berardinelli. 2009. Elk nutrition after wolf recolonization of central Yellowstone. Pages 477–488 in R. A. Garrott, P. J. White, and F. G. R. Watson, editors. *The ecology of large mammals in central Yellowstone: Sixteen years of integrated field studies.* Elsevier, San Diego, California.

White, P. J., R. A. Garrott, S. Cherry, F. G. R. Watson, C. N. Gower, M. S. Becker, and E. Meredith. 2009. Changes in elk resource selection and distribution with the reestablishment of wolf predation risk. Pages 451–476 in R. A. Garrott, P. J. White, and F. G. R. Watson, editors. *The ecology of large mammals in central Yellowstone: Sixteen years of integrated field studies.* Elsevier, San Diego, California.

White, P. J., R. A. Garrott, K. L. Hamlin, R. C. Cook, J. G. Cook, and J. A. Cunningham. 2011. Body condition and pregnancy in northern Yellowstone elk—Evidence for predation risk effects? *Ecological Applications* 21:3–8.

White, P. J., R. A. Garrott, and S. T. Olliff. 2009. Science in Yellowstone: Contributions, limitations, and recommendations. Pages 671–688 in R. A. Garrott, P. J. White, and F. G. R. Watson, editors. *The ecology of large mammals in central Yellowstone: Sixteen years of integrated field studies.* Elsevier, San Diego, California.

White, P. J., T. O. Lemke, D. B. Tyers, and J. A. Fuller. 2008. Initial effects of reintroduced wolves *Canis lupus* on bighorn sheep *Ovis canadensis* dynamics in Yellowstone National Park. *Wildlife Biology* 14:138–146.

White, P. J., K. M. Proffitt, and T. O. Lemke. 2012. Changes in elk distribution and group sizes after wolf restoration. *American Midland Naturalist* 167:174–187.

White, P. J., K. M. Proffitt, L. D. Mech, S. B. Evans, J. A. Cunningham, and K. L. Hamlin. 2010. Migration of northern Yellowstone elk—Implications of spatial structuring. *Journal of Mammalogy* 91:827–837.

White, P. J., R. L. Wallen, C. Geremia, J. J. Treanor, and D. W. Blanton. 2011. Management of Yellowstone bison and brucellosis transmission risk—Implications for conservation and restoration. *Biological Conservation* 144:1322–1334.

White, P. S., and S. P. Bratton. 1980. After preservation: Philosophical and practical problems of change. *Biological Conservation* 18:241–255.

White, R. G. 1983. Foraging patterns and their multiplier effects on productivity of northern ungulates. *Oikos* 40:377–384.

Whitlock, C. 1993. Postglacial vegetation and climate of Grand Teton and southern Yellowstone National Parks. *Ecological Monographs* 63:173–198.

Whitlock, C., and P. J. Bartlein. 1993. Spatial variations of Holocene climatic change in the Yellowstone region. *Quaternary Research* 39:231–238.

Whitlock, C., W. Dean, J. Rosenbaum, L. Stevens, S. Fritz, B. Bracht, and M. Power. 2008. A 2650-year-long record of environmental change from northern Yellowstone National Park based on a comparison of multiple proxy data. *Quaternary International* 188:126–138.

Whitlock, C., S. Fritz, and D. R. Engstrom. 1991. A prehistoric perspective on the northern range. Pages 289–308 in R. B. Keiter and M. S. Boyce, editors. *The greater Yellowstone ecosystem: Redefining America's wilderness heritage.* Yale University Press, New Haven, Connecticut.

Whitlock, C., J. Marlon, C. Briles, A. Brunelle, C. Long, and P. J. Bartlein. 2008. Long-term relations among fire, fuel, and climate in the north-western US based on lake-sediment studies. *International Journal of Wildland Fire* 17:72–83.

Williams, B. K., R. C. Szaro, and C. D. Shapiro. 2007. *Adaptive management: The U.S. Department of Interior technical guide.* Adaptive Management Working Group, U.S. Department of the Interior, Washington, D.C.

Wilmers, C. C., R. L. Crabtree, D. W. Smith, K. M. Murphy, and W. M. Getz. 2003. Trophic facilitation by introduced top predators: Grey wolf subsidies to scavengers in Yellowstone National Park. *Journal of Animal Ecology* 72:909–916.

Wilmers, C. C., and W. M. Getz. 2005. Gray wolves as climate change buffers in Yellowstone. *PLoS Biology* 3:571–576.

Wilmers, C. C., D. R. Stahler, R. L. Crabtree, D. W. Smith, and W. M. Getz. 2003. Resource dispersion and consumer dominance: Scavenging at wolf- and hunter-killed carcasses in Greater Yellowstone, USA. *Ecology Letters* 6:996–1003.

Winder, M., and D. E. Schindler. 2004. Climate change uncouples trophic interactions in an aquatic ecosystem. *Ecology* 85:2100–2106.

Winnie, J. A., Jr., and S. Creel. 2007. Sex-specific behavioural responses of elk to spatial and temporal variation in the threat of wolf predation. *Animal Behaviour* 73:215–225.

Wipf, S. 2009. Phenology, growth, and fecundity of eight subarctic tundra species in response to snowmelt manipulations. *Plant Ecology* 207:53–66.

Wishart, W., B. MacCallum, and J. Jorgenson. 1998. Lessons learned from rates of increase in bighorn herds. *Biennial Symposium of the Wild Sheep and Goat Council* 11:126–132.

Wittmer, H. U., A. R. E. Sinclair, and B. N. Mclellan. 2005. The role of predation in the decline and extirpation of woodland caribou. *Oecologia* 144:257–267.

Wolf, E. C., D. J. Cooper, and N. T. Hobbs. 2007. Hydrologic regime and herbivory stabilize an alternative state in Yellowstone National Park. *Ecological Applications* 17:1572–1587.

Wolfe, M. L., J. F. Kimball, Jr., and G. T. M. Schildwatcher. 2002. Refuges and elk management. Pages 583–616 in D. E. Toweill and J. W. Thomas, editors. *North American elk: Ecology and management.* Smithsonian Institution Press, Washington, D.C.

Wolff, J. O., and T. Van Horn. 2003. Vigilance and foraging patterns of American elk during the rut in habitats with and without predators. *Canadian Journal of Zoology* 81:266–271.

Wondrak Biel, A. 2006. *Do (not) feed the bears: The fitful history of wildlife and tourists in Yellowstone.* University Press of Kansas, Lawrence, Kansas.

Woodley, S. 2010. Ecological integrity: A framework for ecosystem-based management. Pages 106–124 in D. N. Cole and L. Yung, editors. *Beyond naturalness: Rethinking park and wilderness stewardship in an era of rapid change.* Island Press, Washington, D.C.

Wright, G. J., R. O. Peterson, D. W. Smith, and T. O. Lemke. 2006. Selection of northern Yellowstone elk by gray wolves and hunters. *Journal of Wildlife Management* 70:1070–1078.

Wright, G. M., J. S. Dixon, and B. H. Thompson. 1933. *Fauna of the national parks of the United States.* U.S. Government Printing Office, Washington, D.C.

Wright, R. G. 1992. *Wildlife research and management in the national parks.* University of Illinois Press, Urbana, Illinois.

Yellowstone National Park. 1944. *Annual forestry report.* Yellowstone National Park archives, Heritage Research Center, Gardiner, Montana.

Yellowstone National Park. 1950. *Annual forestry report.* Yellowstone National Park archives, Heritage Research Center, Gardiner, Montana.

Yellowstone National Park. 2009. *Superintendent's 2008 Report on Natural Resource Vital Signs.* YCR-2009-04, National Park Service, Yellowstone Center for Resources, Mammoth, Wyoming.

Yellowstone National Park. 2011. *Yellowstone National Park: Natural resource vital signs.* YCR-2011-07, National Park Service, Yellowstone Center for Resources, Mammoth, Wyoming.

Yochim, M. J. 2001. Aboriginal overkill overstated: Errors in Charles Kay's hypothesis. *Human Nature* 12:141–167.

Yung, L., D. N. Cole, D. M. Graber, D. J. Parsons, and K. A. Tonnessen. 2010. Changing policies and practices: The challenge of managing for naturalness. Pages 67–84 in D. N. Cole and L. Yung, editors. *Beyond naturalness: Rethinking park and wilderness stewardship in an era of rapid change.* Island Press, Washington, D.C.

Zager, P., C. White, and G. Pauley. 2007. *Study IV. Factors influencing elk calf recruitment.* Federal Aid in Wildlife Restoration, Job Progress Report, W-160-R-33, Subproject 31, Idaho Department of Fish and Game, Boise, Idaho.

Zanto, S. 2005. Montana Public Health and Safety Division, Public Health Laboratory, personal communication with L. Bambrey, Greystone, August 3, 2005.

Zavaleta, E. S., and F. S. Chapin III. 2010. Resilience frameworks: Enhancing the capacity to adapt to change. Pages 142–158 in D. N. Cole and L. Yung, editors. *Beyond naturalness: Rethinking park and wilderness stewardship in an era of rapid change.* Island Press, Washington, D.C.

Acknowledgments

Special thanks to Cindy Goeddel for sharing her beautiful photographs and to Janine Waller for preparing figures. Robert Garrott's contributions to the book were partially supported by National Science Foundation Grant DEB-0716188. Glenn Plumb thanks his family, Sally, Neva, Kit, and Skye, for making 12 years at Yellowstone wonderful. The views and opinions in this book are those of the authors and should not be construed to represent any views, determinations, or policies of the National Park Service.

Contributors

LISA M. BARIL, National Park Service, Yellowstone National Park

COLDEN V. BAXTER, Stream Ecology Center, Department of Biological Sciences, Idaho State University

MATTHEW S. BECKER, Zambian Carnivore Programme

DAVID J. COOPER, Department of Forest, Rangeland and Watershed Stewardship, Colorado State University

WYATT F. CROSS, Department of Ecology, Montana State University–Bozeman

DOUGLAS A. FRANK, Department of Biology, Syracuse University

ROBERT A. GARROTT, Fish and Wildlife Ecology and Management Program, Department of Ecology, Montana State University–Bozeman

ROBERT E. GRESSWELL, U.S. Geological Survey, Northern Rocky Mountain Science Center

KERRY A. GUNTHER, National Park Service, Yellowstone National Park

MARK A. HAROLDSON, U.S. Geological Survey, Northern Rocky Mountain
Science Center, Interagency Grizzly Bear Study Team

N. THOMPSON HOBBS, Natural Resource Ecology Laboratory, Warner
College of Natural Resources, Colorado State University

KRISTIN L. LEGG, National Park Service, Greater Yellowstone Inventory and
Monitoring Network

TAAL LEVI, Department of Environmental Studies, University of California,
Santa Cruz

DAVID B. MCWETHY, Department of Earth Sciences, Montana State
University–Bozeman

S. THOMAS OLLIFF, National Park Service, Great Northern Landscape
Conservation Cooperative

GLENN E. PLUMB, National Park Service, Natural Resource Stewardship and
Science Directorate, Biological Resource Management Division

KARTHIK RAM, Department of Environmental Studies, University of California, Santa Cruz

DANIEL P. REINHART, National Park Service, Yellowstone National Park

ROY A. RENKIN, National Park Service, Yellowstone National Park

CHARLES T. ROBBINS, Department of Natural Resources and School of
Biological Sciences, Washington State University

PAUL SCHULLERY, National Park Service (retired)

CHARLES C. SCHWARTZ, U.S. Geological Survey, Northern Rocky Mountain Science Center, Interagency Grizzly Bear Study Team

DOUGLAS W. SMITH, National Park Service, Yellowstone National Park

DANIEL R. STAHLER, National Park Service, Yellowstone National Park

JOHN J. TREANOR, National Park Service, Yellowstone National Park

LUSHA M. TRONSTAD, Wyoming Natural Diversity Database, University of Wyoming

RICK L. WALLEN, National Park Service, Yellowstone National Park

FRED G. R. WATSON, Division of Science and Environmental Policy, California State University, Monterey Bay

EMILY M. WELLINGTON, National Park Service, Great Northern Landscape Conservation Cooperative

P. J. WHITE, National Park Service, Yellowstone National Park

CATHY WHITLOCK, Department of Earth Sciences, Montana State University–Bozeman

LEE H. WHITTLESEY, National Park Service, Yellowstone National Park

CHRISTOPHER C. WILMERS, Department of Environmental Studies, University of California, Santa Cruz

EDWARD O. WILSON, University Research Professor Emeritus, Harvard University

Index